MARINE 2011, IV International Conference on Computational Methods in Marine Engineering

Computational Methods in Applied Sciences

Volume 29

Series Editor

E. Oñate
International Center for Numerical Methods in Engineering (CIMNE)
Technical University of Catalonia (UPC)
Edificio C-1, Campus Norte UPC
Gran Capitán, s/n
08034 Barcelona, Spain
onate@cimne.upc.edu
www.cimne.com

For further volumes:
www.springer.com/series/6899

Luís Eça · Eugenio Oñate ·
Julio García-Espinosa · Trond Kvamsdal ·
Pål Bergan
Editors

MARINE 2011, IV International Conference on Computational Methods in Marine Engineering

Selected Papers

Springer

Editors
Luís Eça
Mechanical Engineering
Instituto Superior Técnico
Lisbon, Portugal

Eugenio Oñate
CIMNE
Technical University of Catalonia (UPC)
Barcelona, Spain

Julio García-Espinosa
Universitat Politècnica de Catalunya
Barcelona, Spain

Trond Kvamsdal
Norwegian Univ. of Science & Technology
Trondheim, Norway

Pål Bergan
Norwegian Univ. of Science & Technology
Trondheim, Norway

ISSN 1871-3033 Computational Methods in Applied Sciences
ISBN 978-94-007-6142-1 ISBN 978-94-007-6143-8 (eBook)
DOI 10.1007/978-94-007-6143-8
Springer Dordrecht Heidelberg New York London

Library of Congress Control Number: 2013935214

Printed on acid-free paper

Springer is part of Springer Science+Business Media (www.springer.com)

Preface

This book contains selected papers from the *Fourth International Conference on Computational Methods in Marine Engineering*, held at Instituto Superior Técnico from the Technical University of Lisbon in Portugal in September 2011. The previous editions of the conference were held in Oslo, Norway (2005), Barcelona, Spain (2007) and Trondheim, Norway (2009).

Nowadays, computational methods are an essential tool of Engineering, which includes a major field of interest in Marine applications, such as the maritime and offshore industries and engineering challenges related to the marine environment and renewable energies. The 2011 Conference included 8 invited plenary lectures and 86 presentations distributed by 10 thematic sessions that covered many of the most relevant topics of marine engineering of nowadays. This book contains 16 selected papers from the Conference that cover "CFD for Offshore Applications", "Fluid-Structure Interaction", "Isogeometric Methods for Marine Engineering", "Marine/Offshore Renewable Energy", "Maneuvering and Seakeeping", "Propulsion and Cavitation" and "Ship Hydrodynamics". The papers were selected with the help of the recognized experts that collaborated in the organization of the thematic sessions of the Conference, which guarantees the high quality of the papers included in this book.

MARINE 2011 was organized in the framework of the Thematic Conferences of the European Community on Computational Methods in Applied Sciences (ECCOMAS). MARINE 2011 is also a Special Interest Conference of the IACM.

The conference was jointly organized by Instituto Superior Técnico of the Technical University of Lisbon, the Norwegian University of Science and Technology, SINTEF, and International Center for Numerical Methods in Engineering (CIMNE) in co-operation with the Technical University of Catalonia (UPC). We acknowledge the financial support from the sponsors: Maritime Research Institute Netherlands, Office for Naval Research Global, EDP Inovação and Caixa Geral de Depósitos.

Lisbon

Luís Eça
Eugenio Õnate
Julio García
Trond Kvamsdal
Pål Bergan

Contents

Part I
CFD for Offshore Applications

Viscous-Flow Calculations for Model and Full-Scale Current Loads on Typical Offshore Structures

A.H. Koop, C.M. Klaij, and G. Vaz

Abstract In this paper, CFD calculations for current loads on an LNG carrier and a semi-submersible are presented, both for model and full-scale situations, for current angles ranging from 180 to 0 degrees. MARIN's in-house URANS code ReFRESCO is used. Numerical studies are carried out concerning iterative convergence and grid refinement. In total, more than 100 calculations have been performed. Detailed verification analysis is carried out using modern techniques, and numerical uncertainties are calculated. Afterwards, quantitative validation for model-scale Reynolds number is done taking into account numerical and experimental uncertainties. Scale effects on the current coefficients are investigated, having in mind the estimated numerical uncertainties, and unsteady effects are briefly studied. Good iterative convergence is obtained in most calculations, i.e. a decrease in residuals of more than 5 orders is achieved. The sensitivity to grid resolution has been investigated for both model and full scale using five consecutively refined grids and for 3 current headings. The differences in the solution between two consecutive refinements converge for all cases. The numerical uncertainties are larger for angles with small values of the loads. Comparison with experiments shows that ReFRESCO provides good quantitative prediction of the current loads at model scale: for angles with larger forces the CFD results are validated with 15 % of uncertainty. To determine scale effects the numerical uncertainties must be considered in order to prevent wrong conclusions drawn on basis of numerical differences rather than on physical differences. For the full-scale results larger numerical uncertainties are found than for model scale and for absolute values for scale effects this uncertainty should be improved. For the LNG carrier significant scale effects, i.e. more than 40 %, have been obtained for current angles where the friction component is dominant. For these cases the numerical uncertainty is relatively low. For the other current angles differences of 8–30 % between model and full scale can be observed, but here larger numerical uncertainties are found. For the semi-submersible the numerical uncertainties for the full-scale results are larger than for the LNG carrier. For

A.H. Koop (✉) · C.M. Klaij · G. Vaz
Maritime Research Institute Netherlands (MARIN), P.O. Box 28, 6700 AA Wageningen, The Netherlands
e-mail: a.koop@marin.nl
url: http://www.marin.nl

L. Eça et al. (eds.), *MARINE 2011, IV International Conference on Computational Methods in Marine Engineering*, Computational Methods in Applied Sciences 29,
DOI 10.1007/978-94-007-6143-8_1, © Springer Science+Business Media Dordrecht 2013

the semi-submersible the pressure component of the force is highly dominant, i.e. larger than 90 % of the total force. On average the full-scale current coefficients are 20 % lower than at model scale, but larger differences for a number of angles can be observed.

Keywords CFD · RANS · Current loads · Verification · Validation · Scale effects · LNG carrier · Semi-submersible

1 Introduction

At present, most of the design of offshore structures is done based on current loads coming from empirical methods or from model-scale experiments. These are usually conservative and therefore adequate in this phase. However, there are some counter arguments to this reasoning: (1) for example, a fixed cylinder experiences higher loads at super-critical Reynolds numbers than in the drag-crisis at lower Reynolds numbers. Or, as observed in the Current Affairs JIP [1], full-scale average forces on a schematic semi-submersible are lower than at model scale, but both frequency and amplitudes of full-scale loads are larger; (2) if one can improve the accuracy of the loads used in the design phase, the safety margins can be reduced decreasing the manufacturing costs of the structures and improving the dynamic-positioning capabilities. Therefore, there is a real need for full-scale experiments, full-scale calculations or (general) scaling rules. However, full-scale experiments are scarce, difficult to design and to carry out, and when performed kept confidential. General scaling rules such as used for ship resistance are not easily devised for these kind of complex flows. Thus, currently one is left to full-scale CFD calculations.

Nowadays, most engineers, including the authors, perform CFD at model scale and in steady mode to calculate current coefficients on offshore constructions. Again, the usual reasoning is that scale effects are small for this type of structures, that model-scale calculations are necessary for validation anyhow, and that unsteady calculations are not needed and/or too expensive. Moreover, usually there is no time/money to perform thorough numerical sensitivity variations, to achieve sufficient iterative convergence, and to perform verification studies.

The major objectives of this paper are then fourfold: (1) perform detailed verification studies for model and full-scale calculations of current loads, on two typical offshore structures, for several current angles from 180 to 0 degrees; (2) validate the model-scale numerical results with experimental data; (3) study the scale effects on the current loads; (4) perform a preliminary study on possible unsteady effects on model and full-scale loads.

Modern verification and validation techniques [2] are used in order to quantitatively asses numerical, experimental and validation uncertainties. Without those, the accuracy of the numerical results cannot be determined and conclusions on scale effects cannot be drawn. However, this requires many calculations and in total more than 100 calculations have been performed.

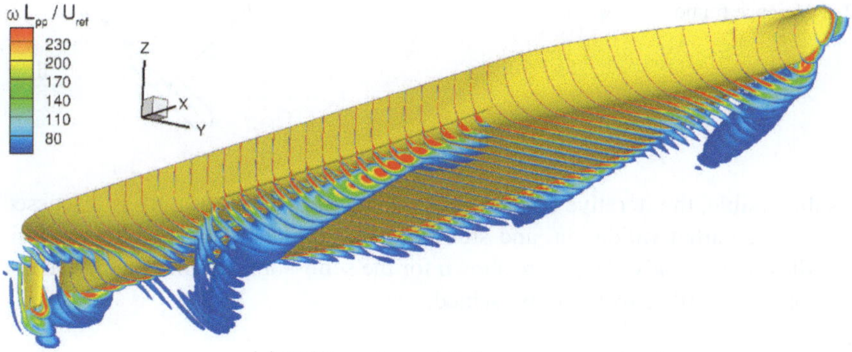

(a) LNG carrier at 140 degrees current heading.

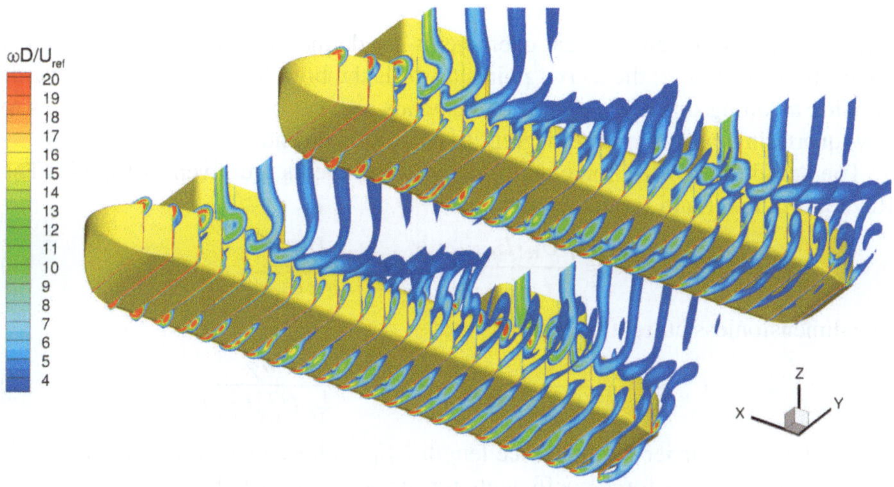

(b) Semi-submersible at 150 degrees current heading.

Fig. 1 Impression of the flow field around the LNG carrier and semi-submersible illustrated by the vorticity distribution around the structures

An LNG carrier appended with bilge-keels and rudder (streamlined body), and a semi-submersible constituted by four rounded-square columns mounted on large block-coefficient ship-shaped pontoons (blunt body) are considered, since they are typical offshore constructions. MARIN's in-house URANS code ReFRESCO [3] is used. Figure 1 presents the geometries and illustrates the calculated flow field for a specific current angle. Previous work done on these structures [1, 4, 5] is here extended, and the lessons learned from the Current Affairs JIP, see [1, 4], are considered in order to improve the accuracy of the results.

The paper is organized as follows. After this introduction, definitions and details on the structures and measurements are presented, followed by the numerical settings used for the calculations. Afterwards, and for both the LNG carrier and

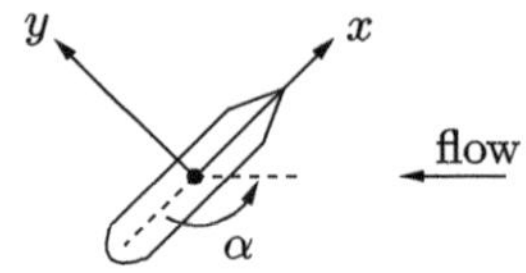

Fig. 2 Reference frame

semi-submersible, the iterative convergence and numerical uncertainty is discussed followed by detailed validation, and study of scale effects. Additionally, preliminary studies on unsteady effects are shown for the semi-submersible. Finally, major conclusions and further work are presented.

2 Definitions

The reference frame and current coefficients are defined following the OCIMF [6] convention, see Fig. 2: the x-axis points towards the bow, the y-axis points towards portside meaning that 180 degrees corresponds to head-on current, 135 degrees to bow-quartering current and 90 degrees to beam-on current.

The definitions of the force and moment coefficients are given in Eq. (2). The Reynolds number and Froude number are:

$$\mathrm{Re} = \frac{\rho U_{\mathrm{ref}} L_{\mathrm{ref}}}{\mu}, \qquad \mathrm{Fr} = \frac{U_{\mathrm{ref}}}{\sqrt{g L_{\mathrm{ref}}}}. \tag{1}$$

The dimensionless current coefficients are:

$$C_{X,Y} = \frac{F_{X,Y}}{\frac{1}{2}\rho U_{\mathrm{ref}}^2 L_{\mathrm{ref}} T}, \qquad C_M = \frac{M_Z}{\frac{1}{2}\rho U_{\mathrm{ref}}^2 L_{\mathrm{ref}}^2 T}. \tag{2}$$

For the LNG carrier the reference length L_{ref} is chosen equal to L_{pp}. For confidentiality reasons the force coefficients for the semi-submersible are scaled by the maximum value found in the wind tunnel experiments. In Sect. 6 these scaled values are denoted by C_X^*, C_Y^* and C_M^*.

Table 1 Nomenclature

α	Current angle [deg]	$F = (F_X, F_Y, F_Z)$	Forces [$\mathrm{kg\,m\,s^{-2}}$]
L_{pp}	Length between perpendiculars [m]	$M = (M_X, M_Y, M_Z)$	Moments [$\mathrm{kg\,m^2\,s^{-2}}$]
T	Draft [m]	ρ	Density [$\mathrm{kg\,m^{-3}}$]
W_D	Water depth [m]	μ	Dynamic viscosity [$\mathrm{kg\,m^{-1}\,s^{-1}}$]
U_{ref}	Reference velocity [$\mathrm{m\,s^{-1}}$]	$\vec{\omega} = \nabla \times \vec{u}$	Vorticity vector [$\mathrm{s^{-1}}$]
L_{ref}	Reference length [m]	$\omega = \lvert\vec{\omega}\rvert$	Norm of the vorticity vector [$\mathrm{s^{-1}}$]
$T_{\mathrm{ref}} = L_{\mathrm{ref}}/U_{\mathrm{ref}}$	Reference time [s]		

Table 2 Main particulars of 135,000 m^3 LNG carrier

Description	Value
Length between perpendiculars	$L_{pp} = 274.0$ m
Draft	$T = 11.0$ m
Breadth	$B = 44.2$ m
Water-depth	$W_D = 53.0$ m
Capacity	135,000 m^3
Current velocity	$U_{ref} = 2.06\ \mathrm{m\,s^{-1}}$

3 Measurements

3.1 LNG Carrier

During the HAWAII JIP, current loads for a 135,000 m^3 LNG carrier have been measured in MARIN's shallow water basin by towing the model through otherwise calm water for flow angles between 0 and 180 degrees, see [7]. The model of the LNG carrier included bilge keels, propeller and rudder and the scale was 1:50. Studs were used on the bow and stern to trigger the boundary layer to become turbulent. From a Reynolds sensitivity check it was concluded that for the tested current velocity, the Reynolds dependency at model scale on the measured force coefficients was less than 5 %. The ratio of the water depth to the draft is $W_D/T = 4.8$, indicating that shallow-water effects might have an influence on the current forces, see also [4]. In order for the free-surface effects to be negligible the Froude number was set small to $\mathrm{Fr} = 0.04$. During the experiments no significant waves were observed, see [7]. The model-scale Reynolds number is equal to $1.6 \cdot 10^6$. At full scale the Reynolds number is equal to $5 \cdot 10^8$. The full-scale particulars of the LNG carrier are given in Table 2.

3.2 Semi-submersible

Wind-tunnel tests at Force Technology [8] have been carried out at scale 1:200. The current loads have been tested in an airflow corresponding to a vertically uniform current. The forces and moments were measured for angles in the range 0 to 360 degrees in increments of 10 degrees. The tests have been carried out with 8 thrusters placed under the pontoons modeled by a single ring. No roughness was applied on the hull. Later in the project the semi-submersible was also tested in MARIN's Offshore Basin and the length of the pontoons has been changed between the wind tunnel tests and the basin measurements. Therefore, the length of the pontoons for the wind tunnel tests was 4.5 % shorter than used in the Offshore Basin and CFD calculations. The comparison between the basin measurements, wind tunnel and CFD results can be found in [5].

Table 3 Information on computational grids for the LNG carrier and semi-submersible

	LNG carrier			Semi-submersible			
	Model scale		Full scale	Model scale		Full scale	
	# cells	max y^+	max y^+	# cells	max y^+	# cells	max y^+
Grid 1	0.70M	0.95	99.5	0.9M	1.1	1.2M	12
Grid 2	1.15M	0.77	78.5	1.7M	0.81	2.4M	9.0
Grid 3	1.85M	0.74	74.8	3.4M	0.62	4.7M	5.8
Grid 4	3.33M	0.59	68.1	7.0M	0.53	9.5M	5.0
Grid 5	5.72M	0.58	61.5	14.0M	0.36	19.3M	4.2

The Froude number based on the column diameter at water surface level is equal to $\mathrm{Fr} = 0.11$. The Reynolds number based on L_{pp} is equal to $2 \cdot 10^8$ at full scale and equal to $5 \cdot 10^5$ in the wind tunnel and model-scale CFD calculations. For confidentiality reasons the main particulars of the semi-submersible can not be shared in this paper.

4 Computational Setup

4.1 Computational Grids

For both geometries five consecutively refined block-structured grids have been constructed using the package GridPro [9], see Table 3 and Fig. 3. At model scale the maximum y^+ value is below 1 on all grids and no wall functions are used. However, for the LNG carrier at full scale, having y^+ values below 1 requires extremely thin cells, especially close to the bilge keels, which lead to severe numerical problems. Using the same clustering as on model scale, we obtain maximum y^+ values below 100 and wall functions are applied. For the semi-submersible the model-scale grids are adapted by refining the first element in the boundary layer to obtain maximum y^+ values below 11 and no wall functions are used.

Wall functions model the viscous sublayer near the wall. The use of wall functions effectively avoids numerical issues due to extremely thin cells but also introduces an additional modelling error in the computations. The effect of using wall functions is not addressed in this paper, but certainly deserves further investigation.

4.2 Boundary and Initial Conditions

Since the Froude number is very small, wave generation is neglected and a symmetry boundary condition is imposed on the water surface. The bottom surface of the

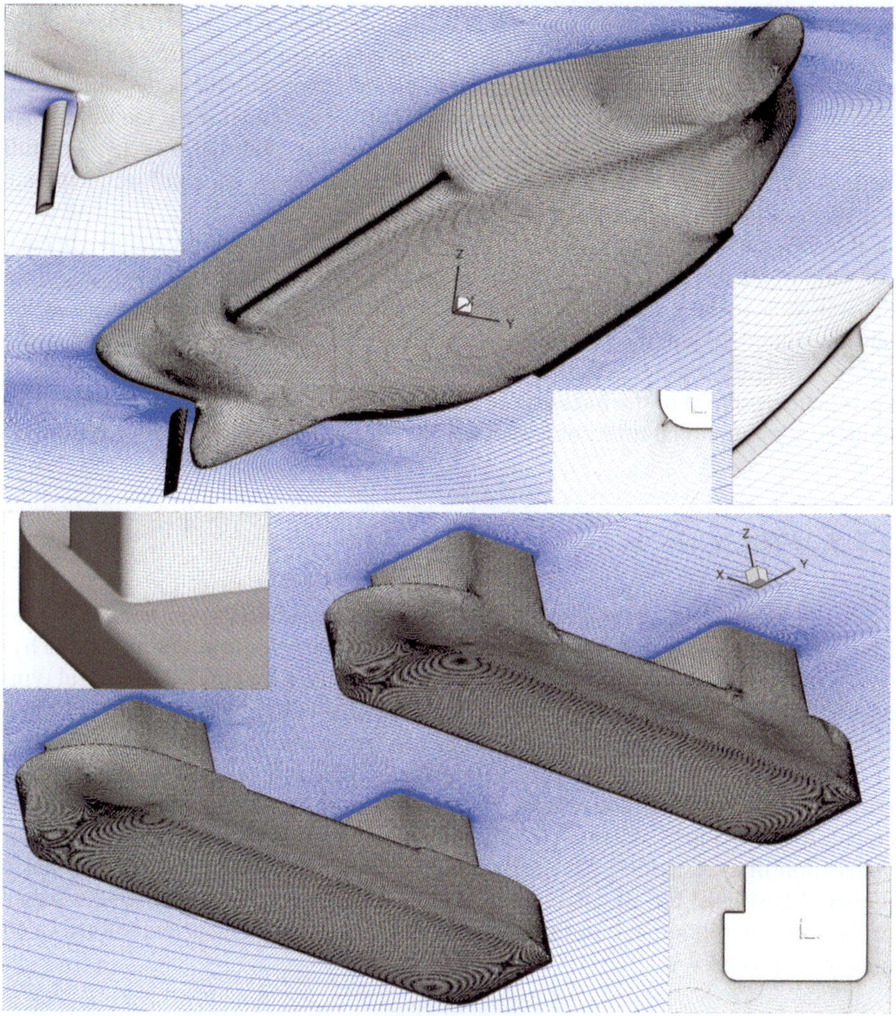

Fig. 3 Computational grid for the LNG carrier and semi-submersible. The *black lines* denote the grid on the surface of the carrier and the *blue lines* denote the grid on the water surface

domain is positioned at the same depth as in the measurements. For the LNG carrier the water depth to draft ratio is equal to 4.8. In [4] the effect of the distance of the bottom surface has been investigated and it was concluded that for this ratio the bottom surface should be taken into account. Therefore, a free-slip wall condition is prescribed at the bottom surface. For the semi-submersible the water depth to draft ratio is more than 25, so a constant-pressure boundary condition is prescribed.

In [4], it was shown that, for this waterdepth-to-draft ratio, the blockage effect of the basin side walls is negligible for the LNG carrier. Therefore, a cylindrical domain is chosen in order to use the same grid for all current angles. The cylinder is centered at the origin and has radius $3.5L_{pp}$ for the LNG carrier and $4L_{pp}$ for the semi-submersible. At the cylindrical boundary a constant uniform velocity is prescribed corresponding with the current angle, together with the eddy-viscosity to laminar-viscosity ratio and the turbulence intensity. For model scale the turbulence intensity is chosen equal to 1 % and the eddy-viscosity ratio is set to 1.0. For full scale these values are set to 10 % and 100.0, respectively. At the outflow, Neumann boundary conditions are applied for all variables.

For the coarsest grids the initial conditions for the calculations are defined in each computational cell by setting the velocity equal to the constant uniform velocity of the inflow boundary, the pressure is chosen equal to the reference pressure at the outflow boundary and the turbulence intensity and eddy-viscosity ratio equal to the inflow boundary settings. For the calculations on finer grids the solution on coarser grids is interpolated to the finer grid to serve as the initial condition. This procedure reduces computational time compared to calculations started from uniform flow.

4.3 ReFRESCO

The CFD calculations in this paper are carried out using MARIN's in-house viscous-flow URANS code ReFRESCO [3]. ReFRESCO is targeted and optimized for hydrodynamic applications exclusively, and it has already been applied to several typical offshore flows. In particular, current, wind and manoeuvring coefficients of semi-submersibles, submarines and ships have been successfully verified and validated, [1, 4, 5, 10, 11]. For all calculations here presented the following numerical settings have been used: (1) QUICK scheme for convection discretization of the momentum equations; (2) Central scheme for diffusion discretization; (3) Upwind scheme for convection discretization of the turbulence equations. The SST $k - \omega$ turbulence model [12] is used for all calculations. Parallelization has been employed because of the long computational times: some calculations have been carried out using 64 quad-core processors.

4.4 Verification and Validation Procedures

In any numerical calculation there are intrinsic errors which have to be controlled, and if possible quantified, e.g. iterative and discretization errors. However, for a complex CFD calculation this can be very time-consuming. Iterative errors are due to non-linear algorithms and iterative solvers utilized, and in principle should be of the same order as the round-off error. Previous studies with two different CFD codes, see [2], have shown that the iterative error should be at least two orders of magnitude lower than the discretization error, in order not to influence the accuracy of the results.

Several methodologies are available to determine the numerical uncertainty related to the discretization error [13, 14]. In this paper we follow the approach as described in [2]. The numerical uncertainty U_ϕ for any arbitrary flow quantity ϕ is determined using $U_\phi = F_s|\epsilon|$, where F_s represents a safety factor and ϵ denotes an estimate of the discretization error. These are determined by applying a least-squares fit of a error power law, αh_i^p, to the results obtained for grids with different densities or relative step size h_i. The choice of error estimator and safety factor depends on the apparent convergence condition (monotonic, oscillatory, non-convergent) and apparent order of convergence p (see for further details [2]).

Validation can only be done after verification and it involves numerical, experimental and parameter uncertainties. The aim of validation is to estimate the modelling error of a given mathematical model in relation to a given set of experimental data. If the validation is successful one cannot say that the code is validated, only that the model is valid for the problem at hand. A well-documented procedure [14] already applied for other ReFRESCO applications [11, 15, 16] is here employed. It compares the validation uncertainty U_{val} with the validation comparison difference E, which are defined by

$$U_{\text{val}} = \sqrt{U_\phi^2 + U_{\text{inp}}^2 + U_{\text{exp}}^2}, \qquad E = \phi_i - \phi_{\text{exp}}, \tag{3}$$

with U_ϕ the numerical uncertainty, U_{inp} the parameter uncertainty, i.e. uncertainties in the fluid properties, geometry and boundary conditions, and U_{exp} the experimental uncertainty. ϕ_i and ϕ_{exp} represent the numerical and experimental value, respectively. The outcome of the validation exercise is decided from the comparison of $|E|$ with U_{val}:

- If $|E| > U_{\text{val}}$, the comparison difference is probably dominated by the modelling error, which indicates that the model must be improved;
- If $|E| < U_{\text{val}}$, the modelling error is within the "noise level" imposed by the three uncertainties. This can mean two things: if E is considered sufficiently small, the model and its solution are validated (with U_{val} precision) against the given experiment; else the quality of the numerical solution and/or the experiment should be improved before conclusions can be drawn about the adequacy of the mathematical model.

For a precise validation, the experimental uncertainty U_{exp} is also needed. This is rarely assessed, and few experimental data for current loads exists in the open literature for which uncertainties are presented. In this paper, the experimental uncertainty U_{exp} is assumed to be equal to 5 % for the current loads obtained in MARIN's shallow-water basin and 10 % for those from the wind tunnel. These values are based on in-house studies taking into account reproducibility for different test runs, manufacturing tolerances and uncertainties of the sensors. The experimental uncertainty for the wind-tunnel experimental data is larger due to the measurement procedure: the model is placed on a flat splitter plate to position the model in an uniform airflow. However, along the splitter plate a boundary layer develops which has an influence on the forces on the hull.

5 Results for LNG Carrier

5.1 Iterative Convergence

The level and speed of the iterative convergence is dependent on the angle, grid resolution and scale of the calculation. Figure 4 shows three typical convergence histories: (1) for model scale and 180 degrees the convergence is fast, and a decrease in residuals more than 6 orders is obtained; (2) for model scale and 140 degrees the convergence is slower and the residuals stagnate at 5 orders of decrease; (3) for full scale at 90 degrees, the convergence history is the worst for all calculations presented in this paper, and the residuals stagnate at 3–4 orders. For all cases here presented, the force coefficients become constant after a few hundred iterations and no oscillations are seen, not even for current angles 140 degrees and 90 degrees. In general, full-scale calculations are more difficult to converge than for model scale, and for the same inflow angle one order less of residuals decrease is obtained.

5.2 Numerical Uncertainties

In Table 4 the current loads at model and full scale are presented for three headings using five consecutively refined grids, together with the numerical uncertainties for the finest grid. Figure 5 shows the results of the uncertainty procedure explained in Sect. 4.4, for three angles, both for model and full scale. For the 140 degrees model-scale case the convergence is not monotonic and the uncertainty procedure is not able to perform a fit to the error power law. Nevertheless, it is able to estimate an uncertainty value. One can see that the differences in the solution between two consecutive refinements are getting smaller for all cases, with already relatively small values for grid 3. However, this has no relation with the uncertainties calculated, which can still be large.

For model scale the numerical uncertainties are relatively small. In general, the uncertainties are larger for the coefficients with lower absolute values. At full scale the uncertainties are larger. Also, at full scale, the use of wall-functions adds an additional modeling error which could explain why C_X is more sensitive to grid density than at model scale. The largest sensitivity is found at 140 degrees where significant flow separation and recirculation contradict the assumptions underlying wall functions. Notice though, that the uncertainty value of 94.3 % for the full-scale C_X at 140 degrees, corresponds to a variation of $2 \cdot 10^{-3}$ in this coefficient, i.e. in absolute magnitude this uncertainty is not relevant.

5.3 Validation with Model-Scale Experiments

From the verification exercise presented in Sect. 5.2 the numerical uncertainty U_ϕ is known. Having in the mind the experimental accuracy stated in Sect. 4.4, the validation procedure can be employed, and the comparison error E and the validation uncertainty U_{val} can be calculated. Note that, the numerical uncertainties have been

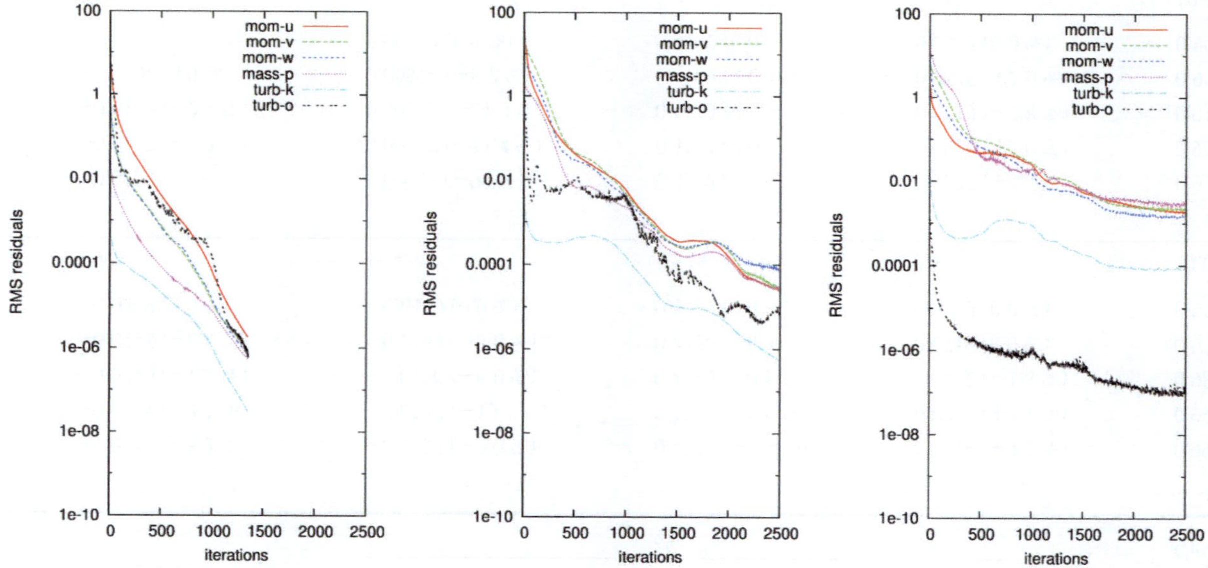

Fig. 4 LNG carrier: examples of iterative convergence for model scale at 180 degrees (*left*); model scale at 140 degrees (*middle*); full scale at 90 degrees (*right*)

Table 4 LNG carrier: results for model and full-scale current coefficients at 180, 140 and 90 degrees obtained using grids with different resolution. *Between brackets* the difference compared to the result on the finest grid is presented. The numerical uncertainty U_ϕ for the results of each coefficient for the finest grid is also presented

Grid	180 degrees	140 degrees			90 degrees
	C_X	C_X	C_Y	C_M	C_Y
Model scale					
Grid 1	−0.0241 (−3.3 %)	−0.0137 (−9.0 %)	0.4721 (−0.4 %)	0.0749 (+4.2 %)	0.9978 (+2.7 %)
Grid 2	−0.0244 (−2.1 %)	−0.0147 (−13.1 %)	0.4714 (−0.6 %)	0.0741 (+3.1 %)	0.9916 (+2.0 %)
Grid 3	−0.0246 (−1.2 %)	−0.0156 (−6.0 %)	0.4716 (−0.5 %)	0.0732 (+1.8 %)	0.9855 (+1.4 %)
Grid 4	−0.0248 (−0.4 %)	−0.0163 (−3.6 %)	0.4726 (−0.3 %)	0.0719 (0.0 %)	0.9781 (+0.6 %)
Grid 5	−0.0249 (0.0 %)	−0.0169 (0.0 %)	0.4739 (0.0 %)	0.0719 (0.0 %)	0.9716 (0.0 %)
U_ϕ	2.9 %	16.4 %	1.5 %	3.8 %	13.0 %
Full scale					
Grid 1	−0.0120 (−15.4 %)	0.00043 (−90 %)	0.4178 (+1.8 %)	0.0527 (+4.4 %)	0.924 (+1.9 %)
Grid 2	−0.0112 (−7.7 %)	0.00200 (−53.4 %)	0.4141 (+0.9 %)	0.0520 (+3.0 %)	0.920 (+1.5 %)
Grid 3	−0.0107 (−2.9 %)	0.00312 (−27.3 %)	0.4123 (+0.5 %)	0.0519 (+2.8 %)	0.916 (+1.1 %)
Grid 4	−0.0104 (0.0 %)	0.00394 (−8.2 %)	0.4109 (+0.1 %)	0.0515 (+2.0 %)	0.910 (+0.4 %)
Grid 5		0.00429 (0.0 %)	0.4102 (0.0 %)	0.0505 (0.0 %)	0.906 (0.0 %)
U_ϕ	25.7 %	94.3 %	2.0 %	12.8 %	10.3 %

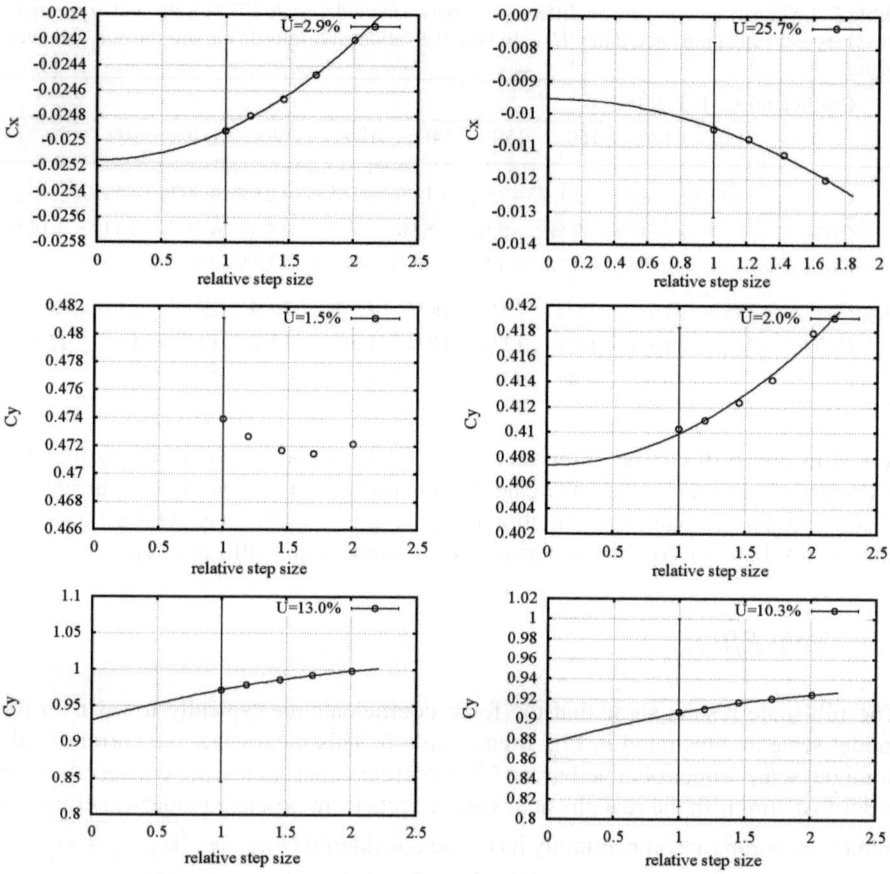

Fig. 5 LNG carrier: examples of numerical uncertainty analysis for C_X and C_Y coefficient. Model scale (*left*); full scale (*right*); 180 degrees (*top*); 140 degrees (*middle*); 90 degrees (*bottom*)

only computed for the angles 180, 140 and 90 degrees. For the sake of the validation exercise we consider the maximum value of the uncertainties, for each coefficient, for the remaining angles. This will probably lead to larger validation uncertainties, but also to more validated results. For the LNG test-case alone around 40 calculations have been performed in total. In order to obtain U_ϕ for all angles and both in model and full-scale situations, more than 100 calculations should have been done. Table 5 shows the final results of the validation exercise. Most of the results for C_Y are validated within ± 15 % of uncertainty. It can be observed that for angles where the coefficients have small values the CFD results are not validated. The value for C_X for 130 degrees shows the use of relative differences: the comparison difference E is equal to 165 %, which is due to the fact that, coincidentally for this angle, the C_X coefficient is almost zero.

An extra outcome from the validation procedure is that the graphical comparison between numerical and experimental results can be done using error bars as pre-

Table 5 LNG carrier: comparison difference E between model-scale CFD results and experimental values and validation uncertainty U_{val}. In *bold* the validated results, in *cursive* the non-validated results

	Coefficient	Angle [deg]									
		180	170	160	150	140	130	120	110	100	90
E	C_X	−11 %	−11 %	−19 %	−25 %	+11 %	+165 %	+23 %	+40 %	+49 %	–
	C_Y	–	−20 %	−8 %	−5 %	−8 %	−7 %	−4 %	+10 %	+11 %	+14 %
	C_M	–	−7 %	+4 %	−4 %	−16 %	−21 %	−17 %	+0.5 %	+27 %	–
U_{val}	C_X	**15 %**	**15 %**	*13 %*	*12 %*	**18 %**	*43 %*	*20 %*	*23 %*	*25 %*	–
	C_Y	–	*10 %*	**12 %**	**12 %**	**12 %**	**12 %**	**12 %**	**14 %**	**14 %**	**15 %**
	C_M	–	*4 %*	**4 %**	*4 %*	*4 %*	*3 %*	*3 %*	**4 %**	*5 %*	–

sented in Fig. 6. It can be observed that the global trend for the three coefficients is correctly captured by the CFD calculations for the complete range of current angles. Using this graphical presentation of the results, validation is obtained once the uncertainty-bars for both the experiments and numerical results overlap.

5.4 Scale Effects

The full-scale results show that the force coefficients are typically lower than for model scale as illustrated in Fig. 6 and Table 6. This means that the current loads at model scale, either obtained using CFD or from experiments, are conservative. In order to distinguish the real physical scale effects from possible numerical effects, a numerical comparison uncertainty has to be considered $U_{comp} = \sqrt{U^2_{\phi,MS} + U^2_{\phi,FS}}$. If the difference between model and full-scale results is larger than this comparison uncertainty then one can say that the scale effect has been captured correctly. If not, no conclusions must be made since the numerical uncertainties taint the real physical effects. As presented in Table 6 the comparison uncertainties are too large for C_X at 140 degrees and C_Y at 90 degrees. Nevertheless, we can still, with some carefulness, observe several trends:

- Scale effects are largest when the friction component is dominant. For example, the C_X coefficient at 180 degrees is 42 ± 26 % lower at full scale than at model scale;
- The scale effects for C_Y are small. For C_X the scale effects are larger. For C_M it depends on the angle, but they are also clearly visible;
- The scale effect for the friction component is larger than for the pressure component for both C_X and C_Y. The ratio of the friction component to the pressure component does not remain constant between model and full scale;
- Having in mind that the pressure component of the calculated coefficients, except for C_X at 180 degrees, is larger than the friction component, it is not straightforward to apply the extrapolation techniques used for ship resistance based on the form-factor hypothesis, see for instance [17].

Fig. 6 LNG carrier: model and full-scale current coefficients. CFD results versus experimental results

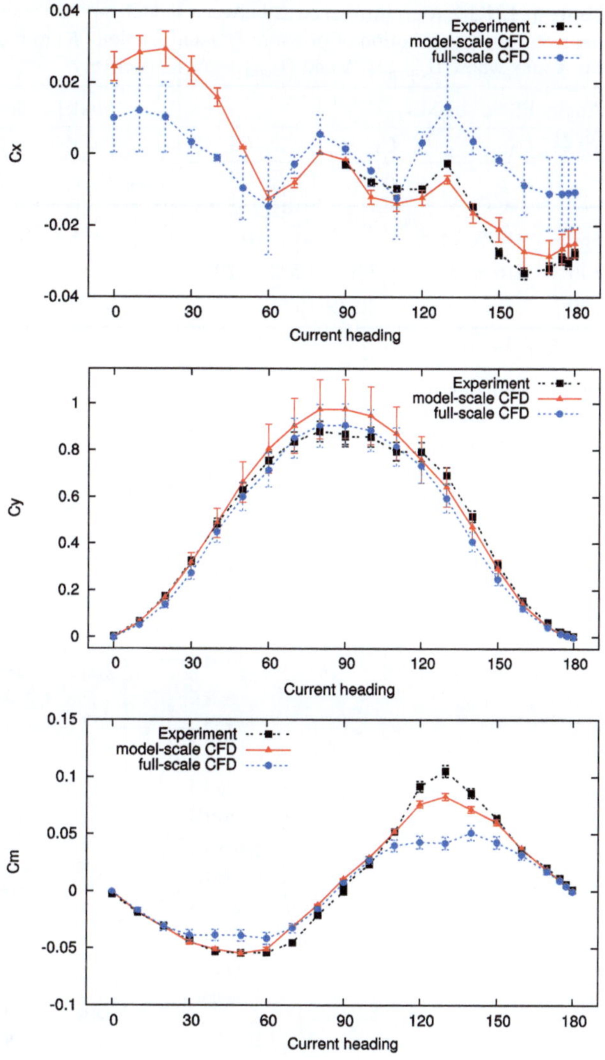

6 Results for Semi-submersible

6.1 Iterative Convergence

For the semi-submersible the level and speed of the iterative convergence are also dependent on the angle, grid resolution and scale of the calculation. Good convergence is often obtained, i.e. a decrease in residuals of more than 5 orders is achieved as illustrated in Fig. 7. However, it also occurs that the iterative convergence stagnates at 4 orders. In general, full-scale calculations are more difficult to converge than for model scale, and for the same inflow angle one order less is obtained. For

Table 6 LNG carrier: difference Δ between model and full-scale values with comparison uncertainty U_{comp}. Contribution of pressure (P) and friction (F) to the total force. In **bold** and *cursive* the results where $U_{\text{comp}} \leq \Delta$ and $U_{\text{comp}} > \Delta$, respectively

Angle [deg]	FS versus MS						Model scale				Full scale			
	C_X		C_Y		C_M		C_X		C_Y		C_X		C_Y	
	Δ	U_{comp}	Δ	U_{comp}	Δ	U_{comp}	P	F	P	F	P	F	P	F
180	**−42** %	26 %	–	–	–	–	14 %	86 %	–	–	14 %	86 %	–	–
140	*−16* %	96 %	**−7.8** %	2.5 %	**−29** %	13 %	74 %	26 %	98 %	2 %	57 %	43 %	99 %	1 %
90	–	–	*−3.2* %	16 %	–	–	–	–	99 %	1 %	–	–	99 %	1 %

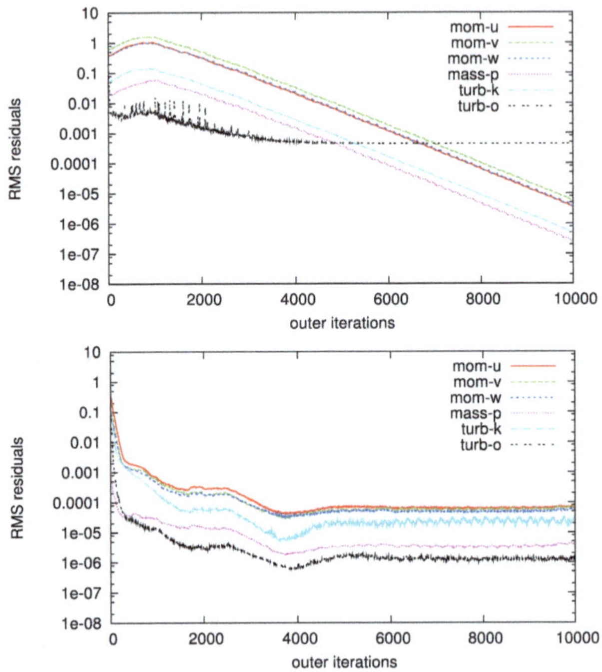

Fig. 7 Semi-submersible: examples of iterative convergence for model scale at 90 degrees (*top*) and full scale at 180 degrees (*bottom*)

all cases here presented, the force coefficients become constant after a few hundred iterations and no oscillations are seen. Compared to the LNG carrier the convergence for the semi-submersible is slower and more difficult to obtain due to the unsteadiness of the flow and complex flow with large separated flow regions.

6.2 Numerical Uncertainties

In Table 7 the results for the semi-submersible without thrusters are presented using subsequently refined grids for model and full scale, all for the angles 180, 150 and

Table 7 Semi-submersible without thrusters: results for model and full-scale current at 180, 150 and 90 degrees obtained using grids with different resolution. *Between brackets* the difference compared to the result on the finest grid is presented. The numerical uncertainty U_ϕ for the results of each coefficient for the finest grid is also presented

Grid	180 degrees	150 degrees			90 degrees
	C_X^*	C_X^*	C_Y^*	C_M^*	C_Y^*
Model scale					
Grid 1	0.651 (+14.8 %)	0.810 (+3.1 %)	0.439 (−5.8 %)	0.725 (+28.9 %)	0.677 (+0.9 %)
Grid 2	0.642 (+13.2 %)	0.799 (+1.7 %)	0.444 (−4.7 %)	0.693 (+23.1 %	0.670 (−0.1 %)
Grid 3	0.612 (+7.9 %)	0.787 (+0.1 %)	0.459 (−1.5 %)	0.614 (+9.1 %)	0.664 (−1.0 %)
Grid 4	0.590 (+4.1 %)	0.786 (0.0 %)	0.467 (+0.2 %)	0.565 (+0.4 %)	0.672 (+0.1 %)
Grid 5	0.566 (0.0 %)	0.786 (0.0 %)	0.466 (0.0 %)	0.563 (0.0 %)	0.671 (0.0 %)
U_ϕ	29.5 %	2.2 %	5.5 %	31.6 %	3.7 %
Full scale					
Grid 1	0.484 (+5.0 %)	0.782 (+8.0 %)	0.390 (+16.8 %)	0.946 (+35.4 %)	0.623 (+11.8 %)
Grid 2	0.490 (+6.3 %)	0.771 (+6.6 %)	0.385 (+15.3 %)	0.890 (+27.3 %)	0.598 (+7.5 %)
Grid 3	0.476 (+3.3 %)	0.748 (+3.3 %)	0.373 (+12.0 %)	0.735 (+5.1 %)	0.580 (+4.1 %)
Grid 4	0.471 (+2.2 %)	0.732 (+1.1 %)	0.348 (+4.2 %)	0.712 (+1.9 %)	0.567 (+1.8 %)
Grid 5	0.461 (0.0 %)	0.724 (0.0 %)	0.334 (0.0 %)	0.699 (0.0 %)	0.557 (0.0 %)
U_ϕ	12.4 %	21.8 %	33.1 %	19.7 %	6.2 %

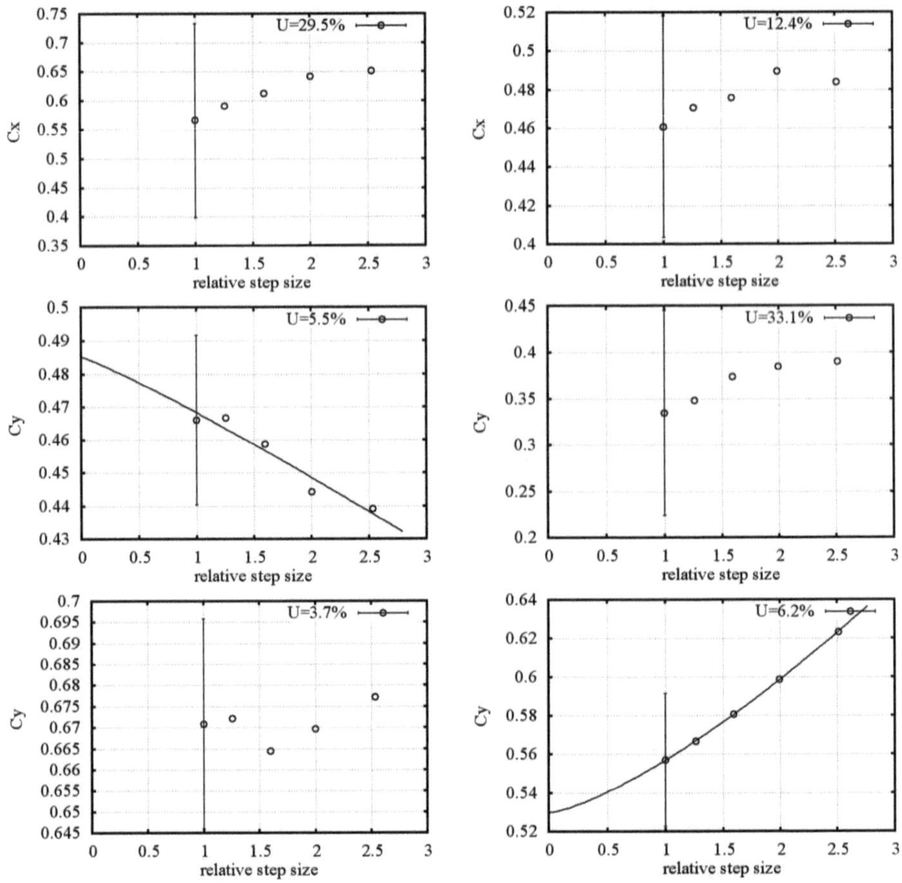

Fig. 8 Semi-submersible: examples of numerical uncertainty analysis for C_X and C_Y coefficient. (*left*) Model scale (*right*) Full scale; 180 degrees (*top*); 150 degrees (*middle*); 90 degrees (*bottom*)

90 degrees. In Fig. 8 the results of the uncertainty procedure is presented for three angles both for model and full scale. It can be observed that the force coefficients converge when using finer grids for all angles both for model and full scale. Also, for all cases the results for grid 4 are at most 4 % different from the results on grid 5.

The numerical uncertainties U_ϕ presented in Table 7 show that for model scale the uncertainties are small except for the C_X coefficient for 180 degrees and for the C_M coefficient for 150 degrees. For 180 degrees the drag force is very small, leading to a higher uncertainty similar to as found for the LNG carrier. The moment on the semi-submersible is sensitive to the precise location of the flow separation. When refining the grid this location changes slightly leading to a change in the moment. The numerical uncertainties for the full-scale calculations are higher than for model scale.

Table 8 Semi-submersible with thrusters: comparison difference E between model-scale CFD results and experimental values and validation uncertainty U_{val}. In *bold* the validated results and in *cursive* the non-validated results

	Coefficient	Angle [deg] 180	170	160	150	140	130	120	110	100	90
E	C_X^*	+9 %	+16 %	+4 %	−2 %	−12 %	−9 %	−35 %	−46 %	−73 %	–
	C_Y^*	–	−27 %	−27 %	−15 %	−8 %	−12 %	−10 %	−11 %	−11 %	−11 %
U_{val}	C_X^*	**35 %**	**37 %**	**33 %**	**13 %**	**29 %**	**30 %**	*22 %*	*19 %*	*13 %*	*10 %*
	C_Y^*	*10 %*	*12 %*	*12 %*	*13 %*	**13 %**	**13 %**	**13 %**	**13 %**	**13 %**	**13 %**

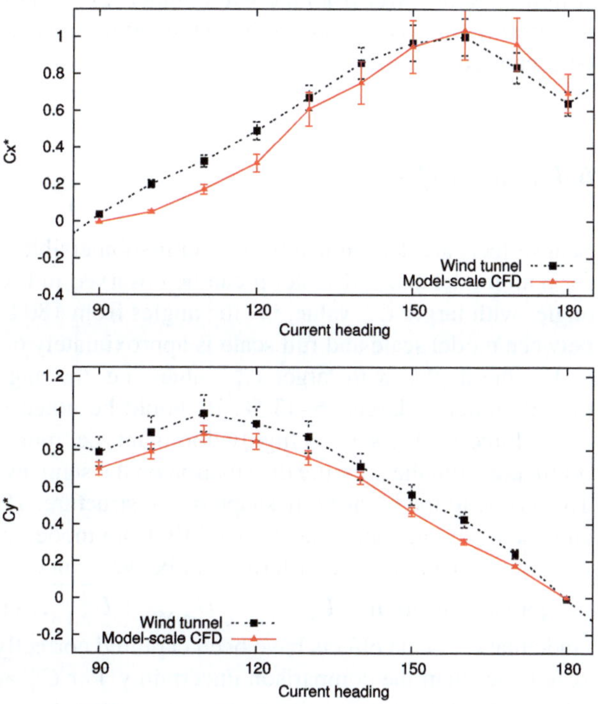

Fig. 9 Semi-submersible with thrusters: comparison of CFD results with experimental results

6.3 Validation with Model-Scale Experiments

In the wind-tunnel experiments eight thrusters were modelled under the hull of the semi-submersible. For validation purposes it would have been better to test and calculate the bare hull of the semi-submersible. In [5] the effect of modeling the thrusters has been investigated and it was concluded that for certain current headings the thrusters have a significant effect on the calculated results. Therefore, for the comparison between the CFD and wind tunnel results, following [5], the thrusters have been taken into account in the CFD calculations presented in Fig. 9 and Table 8. However, in the wind tunnel simple rings were used for the thrusters. In the CFD calculations the exact geometry of the thrusters has been taken into account but the thrusters have been closed. This leads to an additional uncertainty when comparing the results between wind tunnel and CFD, which can be taken into account through the parameter uncertainty U_{inp} in Eq. (3). We assume that this uncertainty due to the thrusters is 4 %. Furthermore, as explained in Sect. 3.2, the pontoons in the wind tunnel were 4.5 % shorter than in the CFD calculations. This leads to approximately 1 % larger drag forces and 3 % larger side forces in the CFD calculations. Therefore, we assume that the total value for the parameter uncertainty U_{inp} is equal to 8 %.

From Fig. 9 it appears that good agreement between the wind tunnel and CFD is obtained. For the angles with larger forces, i.e. the range 180 to 130 degrees for C_X^* and the range 140 to 90 degrees for C_Y^*, the CFD results are validated and within 12 % or lower from the wind tunnel results. However, for C_X^* the validation uncertainty U_{val} is large and both the numerical and experimental uncertainty should be decreased.

6.4 Scale Effects

Scale effects are determined for the semi-submersible without thrusters using grid 4 for both model and full scale. It can be observed in Fig. 10 and Table 9 that for the angles with larger C_X^* value, i.e. the angles from 180 to 120 degrees, the difference between model scale and full scale is approximately 6–30 %.

For the angles with larger C_Y^* values, i.e. the angles from 160 to 90 degrees, the difference is larger: 6–43 %. It should be noted that the pressure component of the force for the semi is highly dominant, i.e. more than 90 % of the total force originates from the pressure distribution on the semi, even for the C_X^* at 180 degrees. This is due to the blunt-body shape of the structure. The ratio between the pressure and friction component changes slightly from model to full scale.

When considering the difference Δ between model and full-scale results and the comparison uncertainty $U_{\mathrm{comp}} = \sqrt{U_{\phi,\mathrm{MS}}^2 + U_{\phi,\mathrm{FS}}^2}$, as presented in Table 9, we conclude that the scale effects have been captured correctly for C_Y^* since the difference Δ is larger than the comparison uncertainty. For C_X^* and C_M^* the numerical uncertainty of the full-scale results are large indicating that one should be careful to draw strong conclusions based on these full-scale results.

6.5 Unsteady Calculations

To investigate the change in forces due to unsteady effects preliminary unsteady URANS calculations are carried out. These calculations require large CPU time, and for practical applications they are usually not performed. The results here presented are for both model and full scale for current headings 180 and 150 degrees. For 180 degrees the flow is very unsteady due to vortices being shed from the first

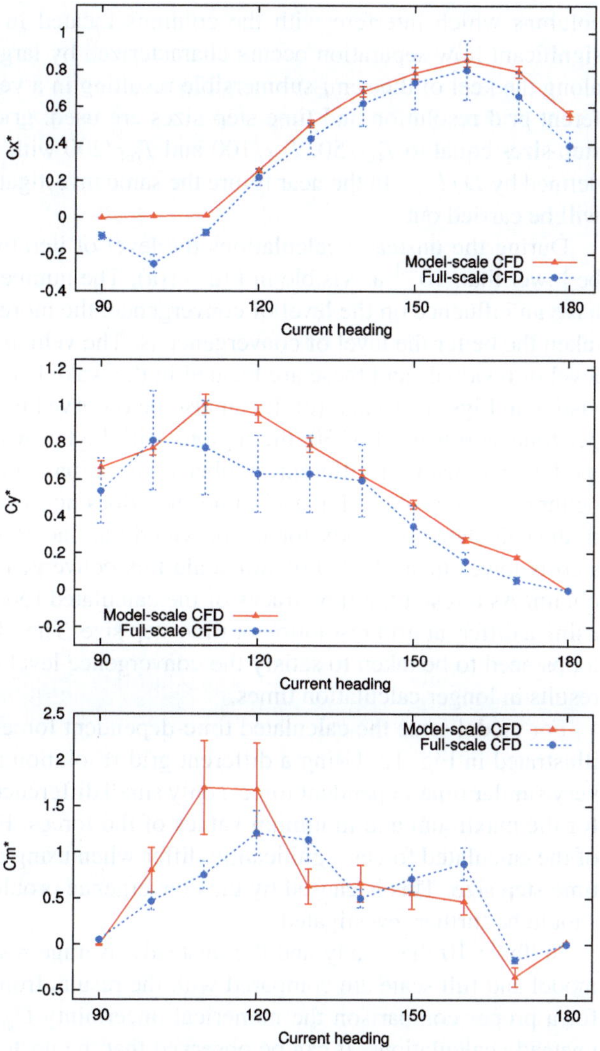

Fig. 10 Semi-submersible without thrusters: model and full-scale current coefficients

Table 9 Semi-submersible without thrusters: difference between model and full-scale values and contribution of pressure (P) and friction (F) to the total force. Results obtained on fine grid. In **bold** and *cursive* the results where $U_{\text{comp}} \leq \Delta$ and $U_{\text{comp}} > \Delta$, respectively

Angle [deg]	FS versus MS						Model scale				Full scale			
	C_X^*		C_Y^*		C_M^*		C_X^*		C_Y^*		C_X^*		C_Y^*	
	Δ	U_{comp}	Δ	U_{comp}	Δ	U_{comp}	P	F	P	F	P	F	P	F
180	*−20 %*	32 %	–	–	–	–	90 %	10 %	–	–	95 %	5 %	–	–
150	*−9 %*	22 %	**−36 %**	33 %	*+20 %*	37 %	92 %	8 %	98 %	2 %	97 %	3 %	99 %	1 %
90	–	–	**−19 %**	7 %	–	–	–	–	98 %	2 %	–	–	99 %	1 %

columns which interfere with the columns located in the wake. At 150 degrees significant flow separation occurs characterized by large, steady, coherent vortices along the keel of the semi-submersible resulting in a very steady flow pattern. Different grid resolution and time step sizes are used: grid 3 to 5 together with time step sizes equal to $T_{ref}/50$, $T_{ref}/100$ and $T_{ref}/200$ with T_{ref} the reference time here defined by D/U_{ref}. In the near future the same investigations for heading 90 degrees will be carried out.

During the unsteady calculations the level of iterative convergence is aimed to be lower than 10^{-4} as visible in Fig. 11(a). The number of iterations per time step have an influence on the level of convergence: the more iterations per time step are taken the better the level of convergence is. The velocity variables have the highest level of residuals and these are located in the wake behind the semi-submersible as visible in Figs. 11(b) and (c). It can also be observed that when using 100 iterations per time step instead of 50, the regions with higher residuals decrease in size. For model scale using a different number of iterations per time step, i.e. aiming at a better convergence level at each time step, does not result in significant differences in the calculated unsteady forces, provided that the convergence level at each time step is lower than 10^{-4}. For full scale this convergence level is more difficult to obtain. As a result the time traces of the calculated forces differ significantly when using a different grid resolution or time step size. Therefore, at full scale more outer loops need to be taken to satisfy the convergence level of 10^{-4}. However, this also results in longer calculation times.

For model scale the calculated time-dependent forces become nicely periodic as illustrated in Fig. 12. Using a different grid resolution and time step size results in very similar time-dependent forces, only small differences of less than 2 % are found for the maximum and minimum values of the forces. For full scale the time traces of the calculated forces significantly differ when using a different grid resolution or time step size. This is caused by the convergence problems as described above and should be further investigated.

In Table 10 the steady and the unsteady average results obtained with CFD for model and full scale are compared with the results from the wind tunnel. Note that for a proper comparison the numerical uncertainty U_ϕ should be obtained for the unsteady calculations. It can be observed that the unsteady average is almost equal to the steady-state result for all cases. The model-scale values obtained with CFD are less than 11 % different from the wind tunnel. The full-scale CFD values are 7–29 % different from the model-scale CFD values. Contrary to the obtained results in [1] we do not find significant unsteady effects on the calculated average force coefficients. However, only two current angles have been investigated in this paper and only one angle in [1]. One major issue is the iterative convergence for unsteady calculations, especially at full scale. The other point may be that the grid topology and its refinement here used might not be adequate for unsteady calculations, where highly refined grids are needed in the shear-layers and in particular in the wakes of all structure components. Furthermore, the numerical uncertainty U_ϕ should be obtained for these unsteady calculations using different grid resolution and time step size.

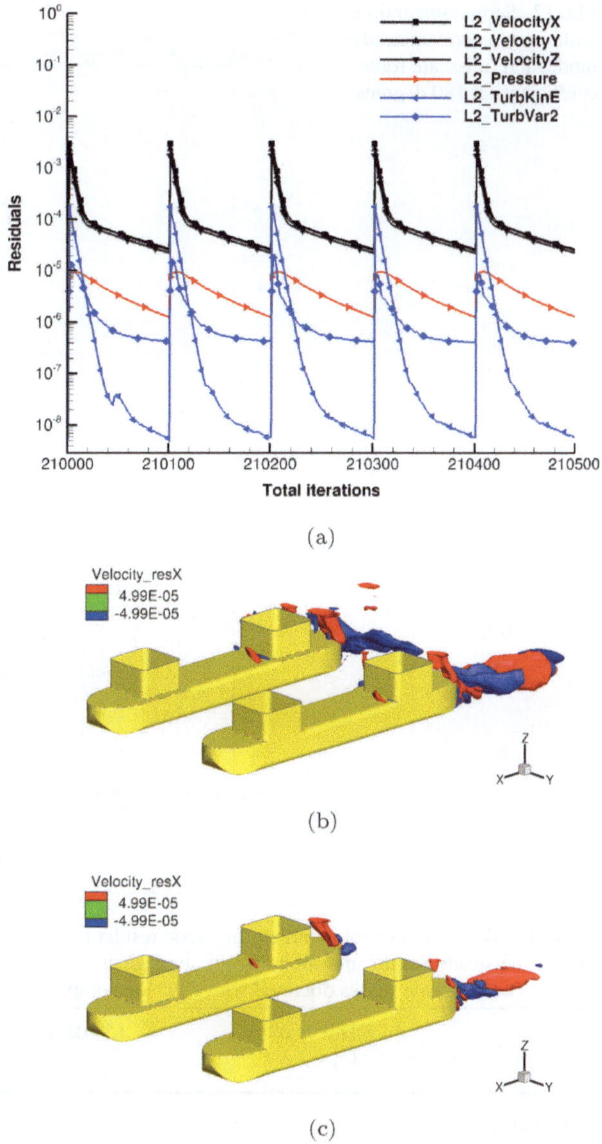

Fig. 11 Semi-submersible without thrusters: unsteady model-scale calculations at 180 degrees using grid 4 and $\Delta t = T_{\text{ref}}/50$. (**a**) Residuals during time steps. Highest residuals in the flow field at the last iteration of a certain time step using (**b**) 50 or (**c**) 100 iterations per time step

Therefore, we recommend to investigate the unsteady effects further despite the long calculation times required.

7 Conclusions and Future Work

In this paper, CFD calculations for current loads on an LNG carrier and a semi-submersible are presented, both for model and full-scale situations, for current angles ranging from 180 to 0 degrees. MARIN's in-house URANS code ReFRESCO

Fig. 12 Semi-submersible without thrusters: unsteady model and full-scale force coefficients at 180 degrees

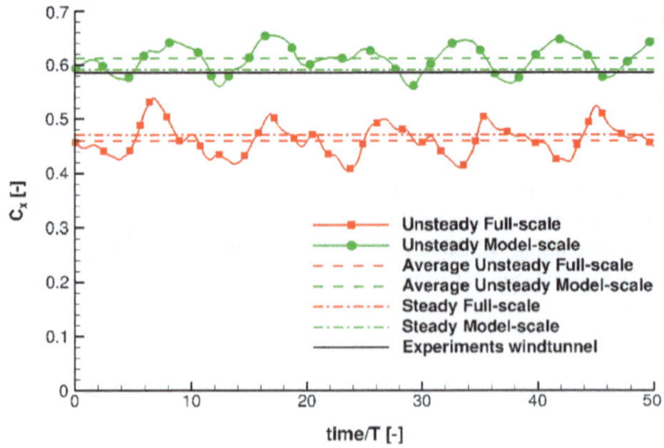

(a) C_X^*

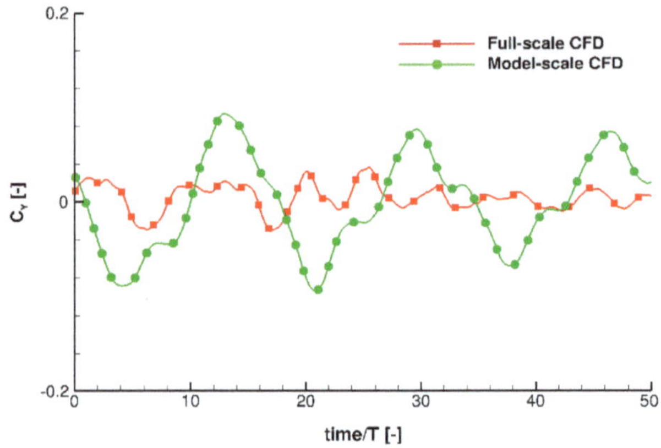

(b) C_Y^*

Table 10 Semi-submersible without thrusters: results for current coefficients at 180 and 150 degrees. The steady results and the unsteady average are compared with the results from the wind tunnel. The full-scale values obtained with CFD are compared to the model-scale CFD results

	180 degrees	150 degrees		
	C_X^*	C_X^*	C_Y^*	C_M^*
Windtunnel	–	–	–	–
Steady MS	+1 %	−11 %	−9 %	+3 %
Unsteady MS	+4 %	−11 %	−9 %	+6 %
Steady FS	−20 %	−7 %	−26 %	+29 %
Unsteady FS	−22 %	−7 %	−24 %	+22 %

is used. Numerical studies are carried out concerning iterative convergence and grid refinement. In total, more than 100 calculations have been performed. Detailed verification analysis is carried out using modern techniques, and numerical uncertainties are calculated. Afterwards, quantitative validation for model-scale Reynolds number is done. Scale effects on the current coefficients are investigated, having in mind the estimated numerical uncertainties, and unsteady effects are briefly studied.

Good iterative convergence is obtained in most calculations, i.e. a decrease in residuals of more than 5 orders is achieved. The level and speed of the iterative convergence is dependent on the current angle, grid resolution and scale of the calculation. Nicely streamlined flows are easier to solve than flows with large separated flow regions. Full-scale calculations are more difficult to converge than for model scale and for the same flow angle one order less is obtained. For unsteady model-scale calculations a better iterative convergence level does not result in significant changes in the calculated unsteady forces, provided that the convergence level at each time step is lower than 10^{-4}. For full scale this convergence level is more difficult to obtain and more iterations per time step should be taken to satisfy this convergence level. However, the authors emphasize that an adequate absolute value for iterative convergence is very much dependent on the employed linear solvers, residual normalization and numerical tool.

The sensitivity to grid resolution at model and full scale has been investigated for both cases using five consecutively refined grids and for 3 current headings. The differences in the solution between two consecutive refinements converge for all cases. The grid 4 results, i.e. 3 million cells for the LNG carrier and 7 million cells for the semi-submersible, are at most 4 % different from the results on the finest grid used with 6 million cells for the LNG carrier and 20 million cells for the semi-submersible. However, this does not mean that the numerical uncertainties are low. The numerical uncertainties are larger for angles with small values of the loads, which is also expected for the experimental results. In some cases, such as C_X for 180 degrees current heading at model-scale, the numerical uncertainty value is too large, 29.5 %. In order to further decrease the numerical uncertainties, better iterative convergence should be achieved and even finer grids should be used. In general, for full-scale situations the numerical uncertainties are higher. Also, for full-scale situations no wall-functions should be used, since this adds an additional modelling inaccuracy, and possibility of numerical scattering due to different boundary conditions for different grids.

Comparison with experiments shows that ReFRESCO provides good quantitative prediction of the current loads at model scale. Taking into account the numerical and experimental uncertainties, it is found that for angles with larger forces the CFD results are validated with 15 % of uncertainty. Nevertheless, for the semi-submersible, for some validated situations, the validation uncertainties are too large due to the numerical uncertainties, but also due to the large experimental and input-parameters uncertainties. This should be further investigated.

To determine scale effects the numerical uncertainties must be considered in order to prevent wrong conclusions drawn on basis of numerical differences rather than on physical differences. When the difference between model and full-scale results is smaller than the comparison uncertainty these values should be considered with care. For the full-scale results larger numerical uncertainties are found than for model scale and for absolute values for scale effects this uncertainty should be improved. For the LNG carrier significant scale effects, i.e. more than 40 %, have been obtained for current angles where the friction component is dominant. For these cases the numerical uncertainty is relatively low. For the other current angles differences of 8–30 % between model and full scale can be observed, but here larger uncertainties are found. For the semi-submersible the numerical uncertainties for the full-scale results are larger than for the LNG carrier. For the semi-submersible the pressure component of the force is highly dominant, i.e. larger than 90 % of the total force. On average the full-scale current coefficients are 20 % lower than at model scale, but larger differences for a number of angles can be observed. For both the semi-submersible and the LNG carrier it is found that the ratio between the pressure contribution and friction contribution to the force does not remain constant comparing model scale to full scale. For the angles where the pressure component is larger than the friction component, it is not straightforward to apply extrapolation methods as used for ship resistance.

Lastly, a preliminary study into the unsteady effects on the current loads has been carried out. These calculations require much CPU time and are therefore only presented for 180 and 150 degrees current heading. Contrary to the obtained results from [1] we do not find significant unsteady effects on the average of the calculated force coefficients. However, only two current angles have been investigated in this paper and only one angle in [1]. More calculations for different headings should be carried out before a valid conclusion on unsteady effects can be drawn. Unsteady problems can be found in many offshore applications such as Vortex Induced Motions (VIM), illustrating the importance to accurately calculate the unsteady loads. Therefore, we recommend to further investigate the unsteady effects despite the long calculation times required.

References

1. Vaz G, Waals O, Fathi F, Ottens H, Le Souef T, Kwong K (2009) Current affairs—model tests, semi-empirical predictions and CFD computations for current coefficients of semi-submersibles. In: Proceedings of OMAE2009, Honolulu, Hawaii, USA
2. Eça L, Vaz G, Hoekstra M (2010) A verification and validation exercise for the flow over a backward facing step. In: Proceedings of ECCOMAS-CFD2010, Lisbon, Portugal
3. Vaz G, Jaouen F, Hoekstra M (2009) Free-surface viscous flow computations. Validation of URANS code FreSCo. In: Proceedings of OMAE2009, Honolulu, Hawaii, USA
4. Fathi F, Klaij CM, Koop A (2010) Predicting loads on a LNG carrier with CFD. In: Proceedings of OMAE2010, Shanghai, China
5. Koop A, Bereznitski A (2011) Model-scale and full-scale CFD calculations for current loads on semi-submersible. In: Proceedings of OMAE2011, Rotterdam, The Netherlands
6. Oil Companies International Marine Forum (1994) Prediction of wind and current loads on VLCCs, 2nd edn
7. Waals OJ (2007) Current force measurements on a 135,000 m^3 and 224,000 m^3 LNG carrier—HAWAII JIP current coefficients. Technical report 19436-1-BT, MARIN
8. http://www.force.dk
9. http://www.gridpro.com/
10. Koop A, Klaij CM, Vaz G (2010) Predicting wind shielding for FPSO tandem offloading using CFD. In: Proceedings of OMAE2010, Shanghai, China

11. Vaz G, Toxopeus SL, Holmes S (2010) Calculation of manoeuvring forces on submarines using two viscous-flow solvers. In: Proceedings of OMAE2010, Shanghai, China
12. Menter F (1994) Two-equation eddy viscosity turbulence models for engineering applications. AIAA J 32:1598–1605
13. American Institute for Aeronautics and Astronautics (1998) Guide for verification and validation of computational fluid dynamics simulations. Technical report AIAA-G-077-1998, AIAA
14. American Society of Mechanical Engineers (2008) ASME guide on verification and validation in computational fluid dynamics and heat transfer. Technical report ASME Committee PTC-61, ANSI standard V&V-20
15. Rijpkema D, Vaz G (2011) Viscous flow computations on propulsors: verification, validation and scale effects. In: Proceedings of RINA-CFD2011, London, UK
16. Toxopeus SL (2011) Practical application of viscous-flow calculations for the simulation of manoeuvring ships. PhD thesis, Delft University of Technology, Faculty Mechanical, Maritime and Materials Engineering
17. Raven HC, van der Ploeg A, Starke AR, Eça L (2008) Towards a CFD-based prediction of ship performance. Progress in predicting full-scale resistance and scale effects. In: Proceedings of RINA-CFD-2008, London, UK

An Absorbing Boundary Condition for Regular and Irregular Wave Simulations

Bülent Düz, Rene H.M. Huijsmans, Arthur E.P. Veldman, Mart J.A. Borsboom, and Peter R. Wellens

Abstract In this paper, the use of an absorbing boundary condition (ABC) is investigated for the numerical simulation of regular and irregular waves in three dimensional computational domains where Navier-Stokes equations describe the motion of the fluid. The numerical implementation of the ABC using a staggered grid arrangement is explained in detail. All of the numerical modifications are incorporated into the CFD simulation tool ComFLOW which employs a volume-of-fluid (VOF) method. Numerical examples are provided to demonstrate the performance of the ABC. The reflection character of the ABC is observed and the results of the computations are discussed and compared.

Keywords Absorbing boundary condition (ABC) · Regular and irregular waves

B. Düz (✉) · R.H.M. Huijsmans
Department of Ship Hydrodynamics, Technical University of Delft, Mekelweg 2, 2628 CD Delft, The Netherlands
e-mail: b.duz@tudelft.nl
url: www.tudelft.nl

R.H.M. Huijsmans
e-mail: r.h.m.huijsmans@tudelft.nl

A.E.P. Veldman
Institute for Mathematics and Computer Science, University of Groningen, P.O. Box 407, 9700 AK Groningen, The Netherlands
e-mail: a.e.p.veldman@rug.nl
url: www.rug.nl

M.J.A. Borsboom · P.R. Wellens
Deltares, P.O. Box 177, 2600 MH Delft, The Netherlands

M.J.A. Borsboom
e-mail: mart.borsboom@deltares.nl
url: www.deltares.nl

P.R. Wellens
e-mail: peter.wellens@deltares.nl

L. Eça et al. (eds.), *MARINE 2011, IV International Conference on Computational Methods in Marine Engineering*, Computational Methods in Applied Sciences 29,
DOI 10.1007/978-94-007-6143-8_2,

1 Introduction

The CFD simulation of ocean waves remains a challenge even today. Although highly capable numerical features are at the disposal of researchers, particular aspects of numerically solving wave problems in unbounded domains cause various bottlenecks. Typically the phenomena of interest are local but embedded in a vast spatial domain. At this point, the infinite domain, although sometimes it may not be truly unbounded, is truncated via artificial boundaries, thus introducing a finite computational domain and a residual infinite domain. One of the aforementioned bottlenecks is developing a robust and efficient boundary condition to be imposed on these artificial boundaries.

The Sommerfeld boundary condition [1] was the cornerstone of non-reflecting boundary conditions. Engquist and Majda [2] presented a method to develop the first hierarchy of absorbing boundary conditions. Higdon [3] generalized this theory and showed that Engquist and Majda boundary condition is a subset of the Higdon operators. Since high order boundary operators include high order derivatives both in time and space, Collino and Joly [4] introduced the use of auxiliary variables to circumvent this difficulty. This idea has found widespread interest and has been used by Grote and Keller [5], Givoli and Neta [6], and Hagstrom and Warburton [7] among others. For a general review regarding high order local non-reflecting boundary conditions, see [8].

In this paper, we present the derivation of an absorbing boundary condition (ABC) [9, 10] along with the numerical implementation of the analytical operator. The ABC is applied in three dimensional computational domains where a regular Stokes wave and an irregular JONSWAP spectrum wave are traveling under an angle of incidence. Here, we focus our attention specifically on the reflection behavior of the ABC for the duration of the simulations. We end the paper with some concluding remarks.

2 Statement of the Problem

If we consider water as a homogeneous, incompressible, viscous fluid, we can describe fluid motion in a three-dimensional domain Ω (see Fig. 1) by the continuity equation and the Navier-Stokes equations in a conservative form as,

$$\oint_{\Gamma} \mathbf{u} \cdot \mathbf{n} \, d\Gamma = 0, \tag{1}$$

$$\oint_{\Omega} \frac{\partial \mathbf{u}}{\partial t} d\Omega + \oint_{\Gamma} \mathbf{u}\mathbf{u}^T \cdot \mathbf{n} \, d\Gamma = -\frac{1}{\rho} \oint_{\Gamma} (p\mathbf{n} - \mu \nabla \mathbf{u} \cdot \mathbf{n}) \, d\Gamma + \oint_{\Omega} \mathbf{F} \, d\Omega. \tag{2}$$

In (1) and (2), Ω denotes a volume with boundary Γ and normal vector $\mathbf{n}$, $\mathbf{u} = (u, v, w)^T$ is the flow velocity, ρ is the fluid density, p is the pressure, μ is the dynamic viscosity, ∇ is the gradient operator and $\mathbf{F} = (F_x, F_y, F_z)^T$ represents external body forces acting on the fluid such as gravity.

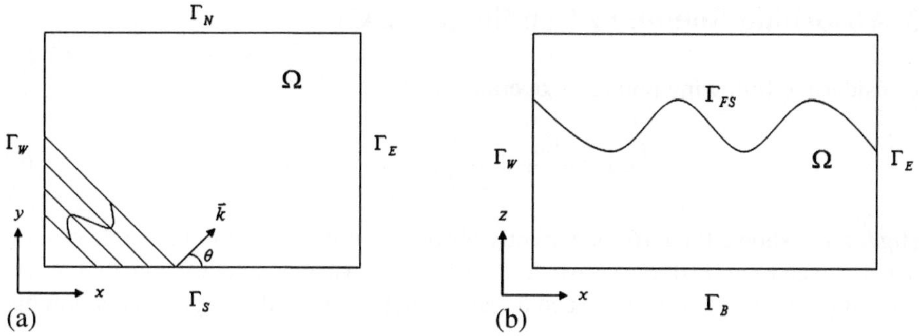

Fig. 1 A computational domain with Γ_N and Γ_E as artificial boundaries

To solve the above equations in Ω, we impose four types of boundary conditions: a free surface, a wall, an inflow and an absorbing boundary condition. On the west and south boundaries Γ_W and Γ_S, the incoming wave is prescribed. The incoming wave propagating at an angle θ $(0 < \theta < \pi/2)$ with the x-axis can be simulated by using either a regular wave such as Airy wave or Stokes wave, or an irregular wave such as a superposition of Airy waves. In either case, at every time step starting from $t = 0$, free surface elevations, values of the velocity components and the pressure are provided on Γ_W and Γ_S. At the bottom Γ_B, we specify a no-slip no-penetration condition which is simply the Dirichlet condition, i.e. $\mathbf{u} = 0$. At the free surface Γ_{FS}, resulting from the continuity of normal and tangential stresses, the following conditions are implemented for the velocity components and the pressure,

$$\mu\left(\frac{\partial u_n}{\partial t} + \frac{\partial u_t}{\partial n}\right) = 0, \tag{3}$$

$$-p + 2\mu\frac{\partial u_n}{\partial n} = -p_0 + \sigma\kappa, \tag{4}$$

where u_n and u_t correspond to the normal and tangential component of the velocity, respectively, p_0 is the atmospheric pressure, σ is the surface tension and κ is the total curvature of the free surface. If we describe the position of the free surface by $s(\mathbf{x}, t) = 0$, the displacement of the free surface can be computed via,

$$\frac{Ds}{Dt} = \frac{\partial s}{\partial t} + (\mathbf{u} \cdot \nabla)s = 0. \tag{5}$$

We now introduce two artificial boundaries Γ_N and Γ_E, see Fig. 1. To complete the statement of the problem, we shall employ an ABC on these artificial boundaries. In this study, we will restrict ourselves for a discussion about the behavior of an ABC in wave simulations where the Navier-Stokes are implemented as the governing equations.

3 Absorbing Boundary Condition (ABC)

Consider the following boundary operator on Γ_E:

$$\left(\cos\alpha\frac{\partial}{\partial t}+c\frac{\partial}{\partial x}\right)\Phi=0 \quad \text{on } \Gamma_E. \tag{6}$$

Higdon [3] showed that (6) is perfectly absorbing if α is equal to the angle of incidence θ (see Fig. 1(a)) for a wave described by the wave or velocity potential Φ and traveling with phase speed c. Such a wave which satisfies the Laplace equation has the form

$$\Phi=\left(C_1e^{+kz}+C_2e^{-kz}\right)\sin(kx\cos\theta+ky\sin\theta-\omega t+\psi), \tag{7}$$

where k is the wave number, ω is the wave frequency and ψ is its phase. The unknowns C_1 and C_2 can be determined via the boundary conditions imposed on Γ_{FS} and Γ_B.

If we replace c in (6) by the dispersion relation, namely,

$$c=\sqrt{gh}\sqrt{\frac{\tanh(kh)}{kh}}, \tag{8}$$

we can rewrite (6) as

$$\left(\cos\alpha\frac{\partial}{\partial t}+\sqrt{gh}\sqrt{\frac{\tanh(kh)}{kh}}\frac{\partial}{\partial x}\right)\Phi=0. \tag{9}$$

The boundary condition (9) is perfectly absorbing for this single component but recall that any solution to the Laplace equation for the velocity potential can be represented by a linear superposition of waves which will be referred to as an *irregular* wave here and elsewhere. Each individual component of this irregular wave has its own frequency, amplitude, wave number and phase. Therefore, the boundary condition (9) cannot annihilate all these wave components simply because it is evidently designed for only one of them.

The corresponding velocity potential of such an irregular wave can be written as

$$\Phi=\sum_{j=1}^{N}\left(C_{1_j}e^{+k_jz}+C_{2_j}e^{-k_jz}\right)\sin(k_jx\cos\theta+k_jy\sin\theta-\omega_jt+\psi_j), \tag{10}$$

where N denotes the number of modes or components. All flow variables can be calculated by taking derivatives of (10). At this point a question crosses one's mind: Is it possible to develop a boundary condition which has the feature of allowing reflection only to an acceptable threshold for all the wave components which all together form an irregular wave? One can deduce from the way this question is asked that we expect some amount of reflection for such a boundary condition but it will be restricted within certain limits.

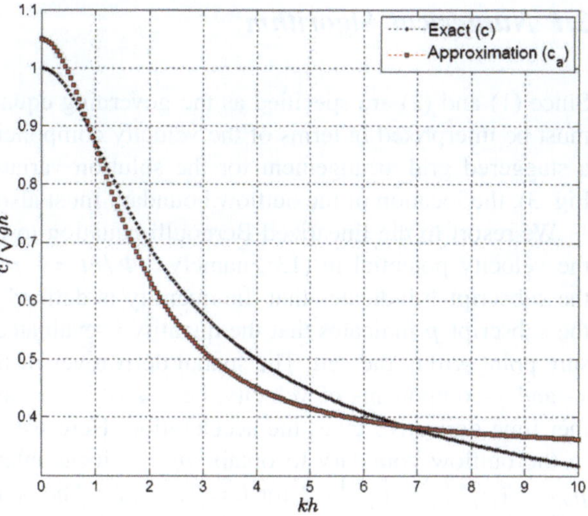

Fig. 2 Approximation of the dispersion relation

Now we introduce the following rational expression which approximates the dispersion relation (8),

$$c_a \approx \sqrt{gh}\frac{a_0 + a_1(kh)^2}{1 + b_1(kh)^2}, \tag{11}$$

where a proper choice of coefficients a_0, a_1 and b_1 would lead to a close approximation for the largest possible range of kh values, see Fig. 2. Thus, reflection from the boundary will be minimized over that specific range of kh values.

As a result of strong effect of dispersion especially in deep water, any wave behaves as the sum of a large number of wave components, each traveling at its own dispersive phase speed. To compute these *local* velocities we will exploit the exponential behavior of (7) and (10) in z direction. After straightforward algebraic manipulations, one can derive the following relation

$$k^2\Phi = \frac{\partial^2}{\partial z^2}\Phi. \tag{12}$$

By employing (12) the dependency of the boundary condition on the wave number is removed since it is calculated using the velocity potential Φ.

Finally we substitute (12) and (11) in (6) to reach the final form of the absorbing boundary condition to be applied on Γ_E

$$\left(1 + b_1 h^2 \frac{\partial^2}{\partial z^2}\right)\cos\alpha \frac{\partial \Phi}{\partial t} + \sqrt{gh}\left(a_0 + a_1 h^2 \frac{\partial^2}{\partial z^2}\right)\frac{\partial \Phi}{\partial x} = 0 \quad \text{on } \Gamma_E. \tag{13}$$

Following the same method, it is rather easy to write the ABC on Γ_N.

3.1 Numerical Algorithm

Since (1) and (2) are specified as the governing equations, the ABC given in (13) must be interpreted in terms of the velocity components and pressure. As we have a staggered grid arrangement for the solution variables inside volume cells (see Fig. 3), the location of the outflow boundary must also be specified appropriately.

We resort to the linearized Bernoulli equation to replace the time derivative of the velocity potential in (13), namely, $\partial\Phi/\partial t = -p_b - gz_p$. Here and elsewhere the subscript b indicates that the quantity is defined at the outflow boundary and the subscript p indicates that the quantity is evaluated at the elevation of the pressure point within the cell. The spatial derivatives of the velocity potential give the x- and y-components of velocity, i.e. $\partial\Phi/\partial x = u_b$ and $\partial\Phi/\partial y = v_b$, while a further time derivative gives the acceleration. Here we will use mirror cells adjacent to the outflow boundary to obtain p_b by linear interpolation (see Fig. 3(a)), i.e. $p_{b,k} = (p^{n+1}_{I+1,k} + p^{n+1}_{I,k})/2$ for $k = 1, \ldots, K$. The shaded area contains the mirror cells which have indices $(I+1,k)$ for $k = 1, \ldots, K$. The outflow boundary is situated at the same position along x-direction as u, therefore we can impose the following, $u_b = u^{n+1}_{I,k}$. Note that velocity components and pressure are discretized at the same position on the boundary and also at the same instant in time.

Utilizing the momentum equation (2), the velocity component at the new time step u^{n+1} can be written in terms of the pressure p^{n+1} and the intermediate velocity $\tilde{u}^n$ which includes convective and diffusive effects [11]. This modification is necessary to easily plug the ABC into the pressure Poisson equation which is solved inside the computational domain for the pressure at the new time step p^{n+1}. As a result, the ABC has the same temporal character as the pressure Poisson equation. Consequently, we obtain the discrete form of the ABC to be prescribed on Γ_E as follows

$$\begin{aligned}
&\left[\frac{1}{2}\cos\alpha + a_0\sqrt{gh}\frac{\Delta t}{\Delta x_{p(I+1,k)}}\right.\\
&\quad\left. + \left(\frac{1}{2}b_1h^2\cos\alpha + a_1h^2\sqrt{gh}\frac{\Delta t}{\Delta x_{p(I+1,k)}}\right)\frac{\partial^2}{\partial z^2}\right]p^{n+1}_{I+1,k}\\
&\quad + \left[\frac{1}{2}\cos\alpha - a_0\sqrt{gh}\frac{\Delta t}{\Delta x_{p(I+1,k)}}\right.\\
&\quad\left. + \left(\frac{1}{2}b_1h^2\cos\alpha - a_1h^2\sqrt{gh}\frac{\Delta t}{\Delta x_{p(I+1,k)}}\right)\frac{\partial^2}{\partial z^2}\right]p^{n+1}_{I,k}\\
&= \left(a_0\sqrt{gh} + a_1h^2\sqrt{gh}\frac{\partial^2}{\partial z^2}\right)\tilde{u}^n_{I,k} - gz_{p(I+1,k)}\cos\alpha \quad \text{on } \Gamma_E,
\end{aligned} \tag{14}$$

where

$$\Delta x_{p(I+1,k)} = x_{p(I+1,k)} - x_{p(I,k)}. \tag{15}$$

Fig. 3 Discretization of the ABC in space

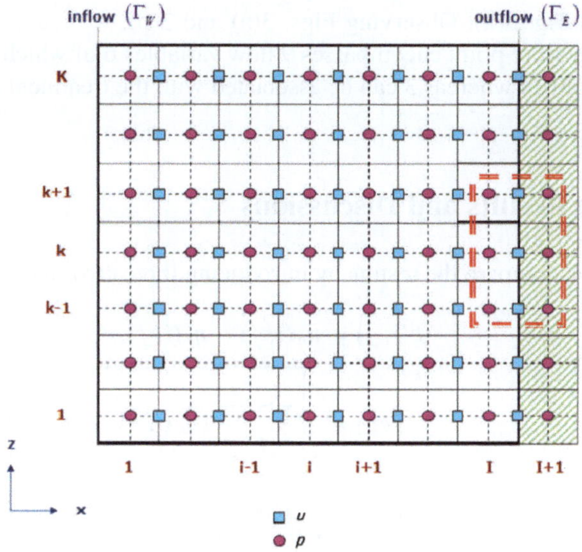

(a) Stencil of the ABC in $x - z$ domain.

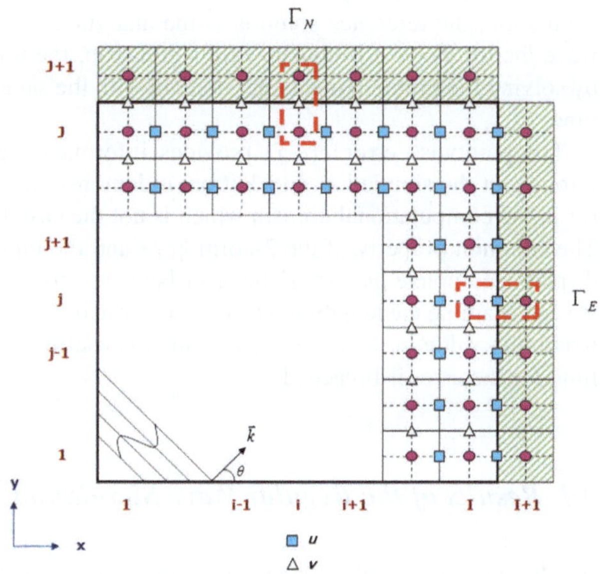

(b) Stencil of the ABC in $x - y$ domain (top view).

Following the same steps, one can easily derive the ABC on Γ_N. The discrete ABCs on Γ_E and Γ_N are equations for the pressure values in the mirror cells outside the domain, see Fig. 3(b). The stencil for $p_{I+1,k}^{n+1}$ is plotted by a double dashed line in Fig. 3(a). Observing Figs. 3(a) and 3(b), we realize that a typical stencil for a pressure point encompasses 9 flow variables 6 of which reside in the computational domain whereas 3 can be associated with the treatment of the boundary condition.

4 Results and Discussions

We compare the results by introducing three error measures,

$$e(i,j) = \eta_n(i,j) - \eta_r(i,j), \tag{16}$$

$$\|e\|_2 = \sqrt{\sum_{i=1}^{I}\sum_{j=1}^{J}\big(\eta_n(i,j) - \eta_r(i,j)\big)^2}, \tag{17}$$

$$\|e\|_\infty = \max_{i=1,2,\ldots,I}\ \max_{j=1,2,\ldots,J}\big\{\big|\eta_n(i,j) - \eta_r(i,j)\big|\big\}, \tag{18}$$

where η is the free surface elevation. Here the subscript n indicates the numerical results and the subscript r indicates the reference solution. For the regular wave simulation, the reference solution is the analytical results arising from the Stokes wave theory. For the irregular wave simulation, the reference solution is obtained by solving the problem in a larger domain with the same discretization in space and time.

The pointwise error $e(i,j)$ provides information at particular time instances throughout the simulation. In addition, it demonstrates the exact location of the error in the computational domain which is not the case for the other error measures. The common property of the 2-norm $\|e\|_2$ and the infinity norm $\|e\|_\infty$ is that they display a complete picture of the error behavior in a single plot. More particularly, we can examine the length of the error vector using $\|e\|_2$ whereas $\|e\|_\infty$ captures the maximum value in the error vector which is useful especially to check if a certain limit for the error is breached.

4.1 Results of the Regular Wave Simulation

A fully developed fifth-order Stokes wave is generated and initialized everywhere in the computational domain at $t = 0$ as depicted in Fig. 4. Since we know the exact values of the solution variables for a fifth-order Stokes wave [12], we can compare the numerical results with the theoretical results. The fifth-order Stokes wave with wave height $H = 9$ m, wave period $T = 10$ s, wave length $\lambda = 161$ m, phase speed $c = 16.1$ m/s is simulated by performing 7143 time-steps at $\Delta t = 0.007$ s. The length and the width of the computational domain is the same, $l_x = l_y = 340$ m, and its depth is $l_z = 179$ m with the water depth of $h = 170$ m. The grid resolution is

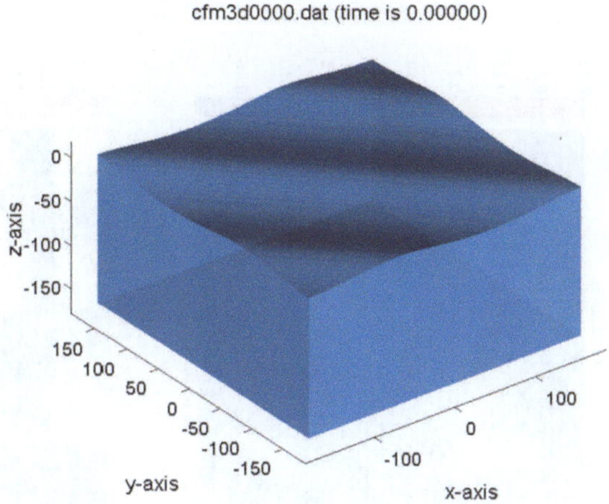

Fig. 4 Initial condition for the simulation of the fifth-order Stokes wave. Angle of incidence $\theta = 45°$

$\Delta x \times \Delta y \times \Delta z = 2.26$ m $\times$ 2.26 m $\times$ 0.95 m with 6 % vertical stretching at the free surface.

Figure 5 shows the absolute pointwise errors $e(i, j)$ corresponding to times $t = 8.589$ s, $t = 16.667$ s, $t = 25.256$ s, $t = 33.334$ s, $t = 42.427$ s and $t = 49.994$ s. The amplitudes of the maximum errors increase in time although not substantially. The reflected waves from the outflow boundaries travel back and perturb the solution in the entire computational domain. Figure 6 demonstrates the 2-norm $\|e\|_2$ and the infinity norm $\|e\|_\infty$ both of which are normalized by the wave height. Observing Fig. 5, we notice that large errors are very local whereas in the major part of the domain we have relatively small errors. This is consistent with the fact that $\|e\|_2$ has an oscillating character below the maximum value of 2.5 % throughout 5 wave periods. Evidently, $\|e\|_\infty$ shows a similar behavior but it oscillates generally between the values of 9 % and 4 %. For the maximum values of $\|e\|_\infty$, we believe that we are encountering the effects of reconstruction of the free surface in the VOF algorithm.

4.2 Results of the Irregular Wave Simulation

We apply the discrete ABCs to a problem in a three dimensional computational domain where an irregular wave is traveling under an angle of incidence, $\theta = 45°$. The initial condition for the simulation of the irregular wave with 537 Fourier components is shown in Fig. 7(a). The domain length in x- and y-direction is the same, $l_x = l_y = 70$ m whereas $l_z = 8$ m with the water depth $h = 5$ m. The grid resolution is $\Delta x \times \Delta y \times \Delta z = 0.28$ m $\times$ 0.28 m $\times$ 0.23 m. A JONSWAP spectrum wave with $T_p = 10$ s and $H_s = 1.0$ m is simulated by performing 3964 time-steps at $\Delta t = 0.007$ s.

As mentioned before, we compute the reference solution by solving the problem in a large domain Ω_L which is twice the size of the small domain Ω_S in x- and

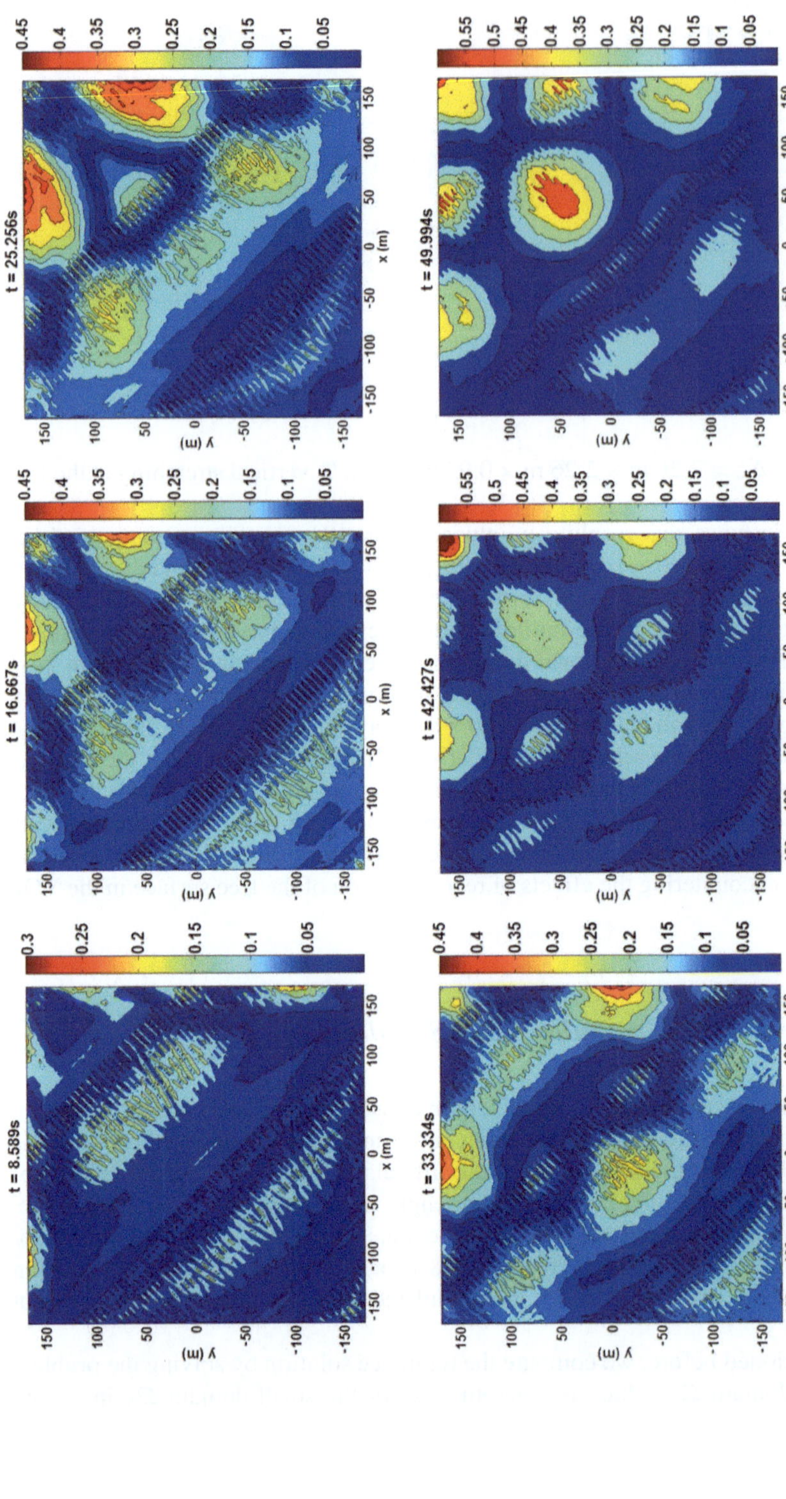

Fig. 5 The absolute pointwise errors $e(i, j)$ between the numerical and theoretical results for the simulation of the fifth-order Stokes wave. Errors are shown at times: $t = 8.589$ s, $t = 16.667$ s, $t = 25.256$ s, $t = 33.334$ s, $t = 42.427$ s and $t = 49.994$ s

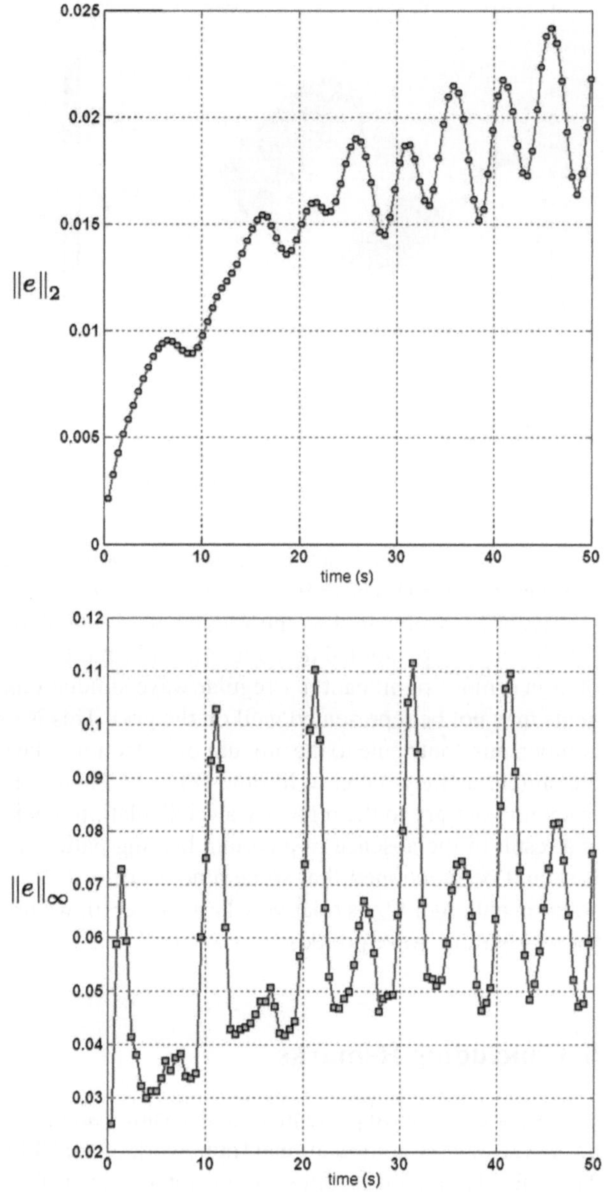

Fig. 6 The relative 2-norm $\|e\|_2$ and the infinity norm $\|e\|_\infty$ in space as a function of time for the regular wave simulation. Both error measures are normalized by the wave height $H = 9$ m

y-direction, see Fig. 7(b) for the illustration of the problem. For each time step, the computational solution in Ω_S is compared to the reference solution in Ω_L. In both Ω_S and Ω_L the numerical parameters are the same. Since the flow behavior is highly nonlinear, the linear theory fails to produce correct results under the current circumstances. Therefore, it is not possible to make a comparison with the analytical solution for the irregular wave simulation.

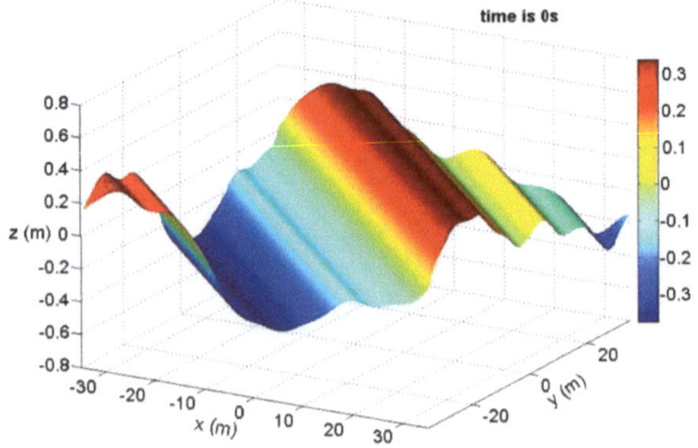

(a) The initial condition for the irregular wave simulation.

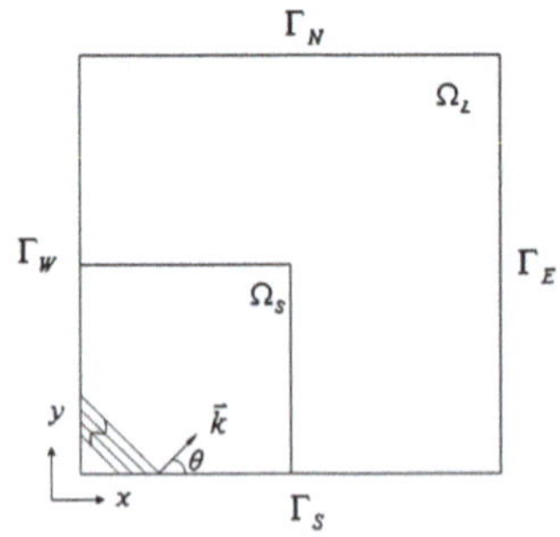

(b) Large domain Ω_L is considered as the reference domain (top view).

Fig. 7 The setup for the irregular wave simulations

In Fig. 8, we demonstrate the absolute pointwise errors $e(i, j)$ corresponding to times $t = 4.504$ s, $t = 9.008$ s, $t = 13.504$ s, $t = 18.008$ s, $t = 23.008$ s and $t = 27.752$ s. Compared to the regular wave simulation, large errors cover wider parts in the computational domain, thus $\|e\|_2$ has higher values, see Fig. 9. Grid resolution is more significant in irregular wave simulations because short wave components may not be represented well on the grid. This is certainly undesirable as these components contribute to the amount of reflection. The characters of $\|e\|_\infty$ and $\|e\|_2$ are similar as they increase in time (Fig. 9). The error norms show an exponential behavior contrary to the regular wave calculation in which they are oscillatory. This is a result of the absence of a certain beating pattern in irregular waves. Moreover, it should be mentioned that some amount of error is also present in the reference solution although Ω_L is relatively large. Overall, we find deviations of less than 9 % for nearly three wave periods.

5 Concluding Remarks

In this paper, we have presented the derivation and the numerical implementation of an ABC using the computational framework of the CFD simulation tool ComFLOW. The ABC is applied in three dimensional free surface simulations of regular and irregular waves propagating under an angle of incidence. For this purpose, a fifth-order Stokes wave and a JONSWAP spectrum wave are generated at the inflow boundaries of the computational domains. The results of the numerical computations are compared to various reference solutions to provide sufficient information regarding the performance of the proposed boundary condition. Additionally, the reflection character of the ABC is monitored throughout the calculations and different error measures are exploited to deliver a comprehensive picture for the error behavior.

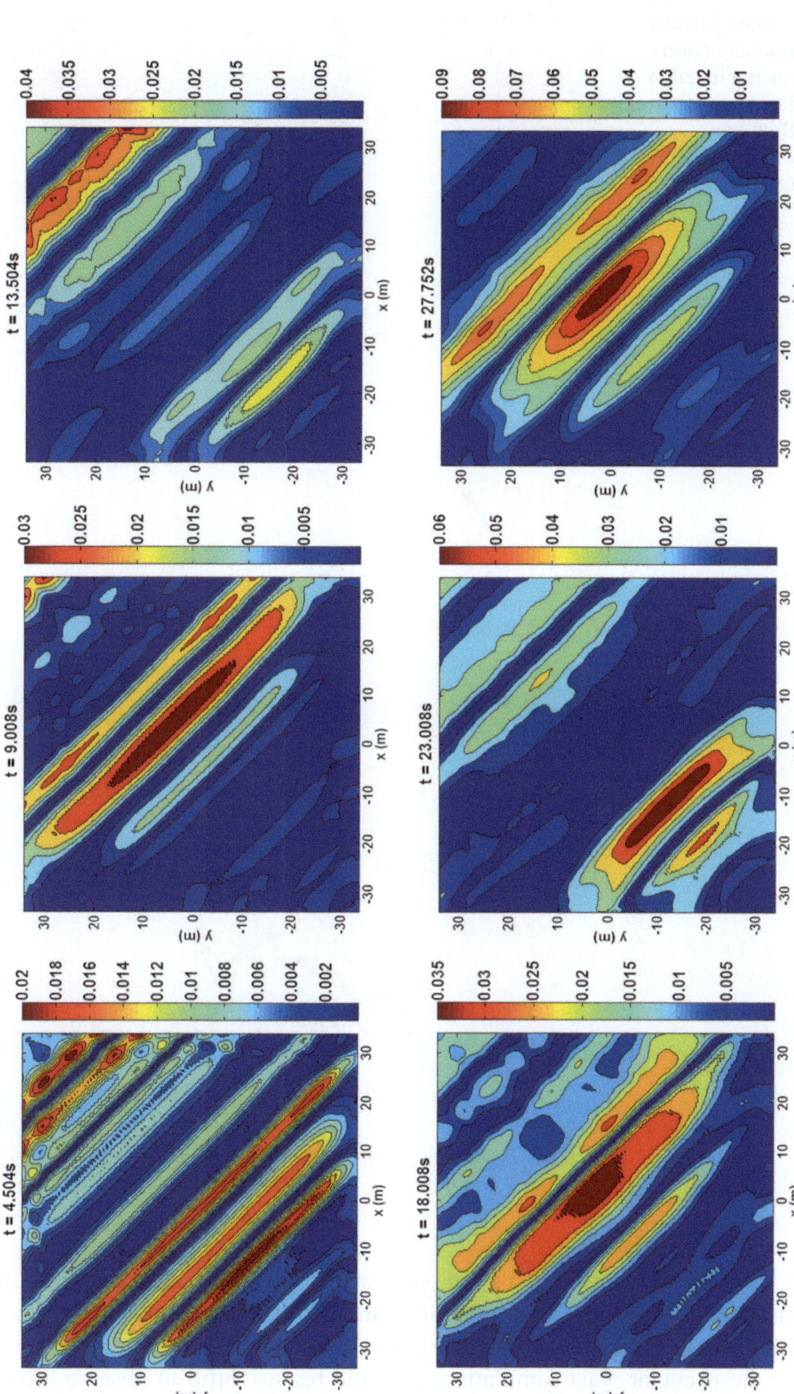

Fig. 8 The absolute pointwise errors $e(i, j)$ between the numerical results in large and small domains for the irregular wave simulation. Errors are shown at times: $t = 4.504$ s, $t = 9.008$ s, $t = 13.504$ s, $t = 18.008$ s, $t = 23.008$ s and $t = 27.752$ s

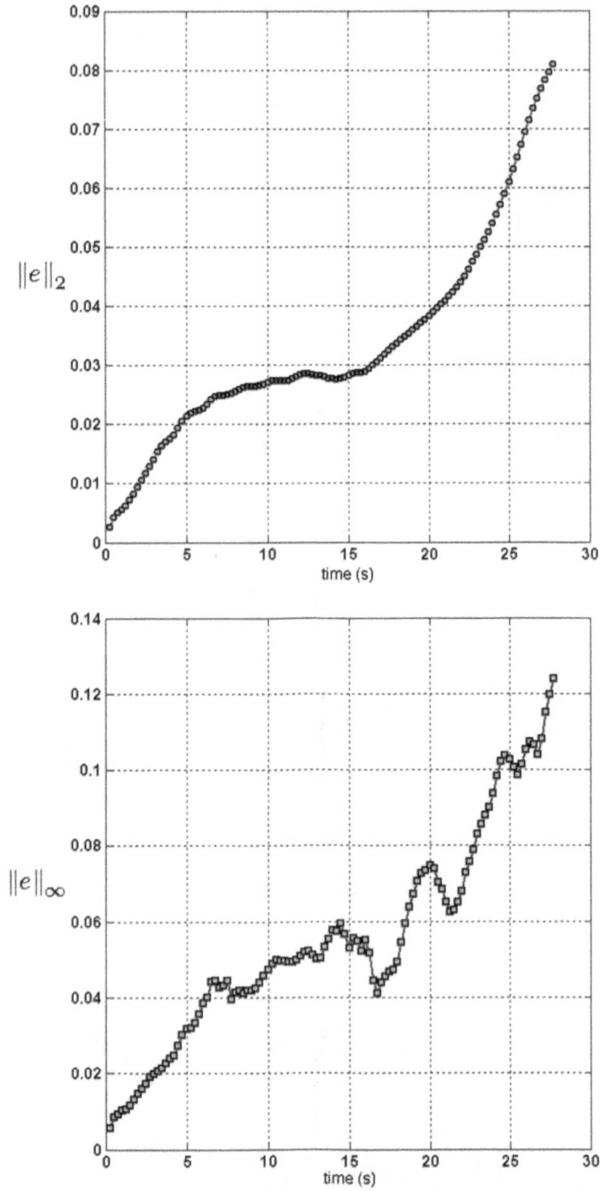

Fig. 9 The relative 2-norm $\|e\|_2$ and the infinity norm $\|e\|_\infty$ in space as a function of time for the irregular wave simulation. Both error measures are normalized by the wave height $H_s = 1.0$ m

Overall, the ABC demonstrated a good performance. In both regular and irregular wave simulations, we notice that reflections are less than acceptable thresholds. The numerical results are in reasonable agreement with the reference solutions. Particularly for the irregular wave simulation, it would be insightful to observe error behaviors for a longer duration of simulation which will be the subject of the future work.

Acknowledgements This research is supported by the Dutch Technology Foundation STW, applied science division of NWO and the technology programme of the Ministry of Economic Affairs in The Netherlands (contracts GWI.6433 and 10475).

References

1. Sommerfeld A (1949) Partial differential equations in physics. Academic Press, San Diego
2. Engquist B, Majda A (1977) Absorbing boundary conditions for the numerical simulation of waves. Math Comput 31:629–651
3. Higdon RL (1987) Numerical absorbing boundary conditions for the wave equation. Math Comput 49:65–90
4. Collino F, Joly P (1995) New absorbing boundary conditions for the finite element solution of 3D Maxwell's equations. IEEE Trans Magn 31(3):1696–1701
5. Grote MJ, Keller JB (1996) Nonreflecting boundary conditions for time-dependent scattering. J Comput Phys 127(1):52–65
6. Givoli D, Neta B (2003) High-order non-reflecting boundary conditions for dispersive waves. Wave Motion 37(3):257–271
7. Hagstrom T, Warburton T (2004) A new auxiliary variable formulation of high-order local radiation boundary conditions: corner compatibility conditions and extensions to first-order systems. Wave Motion 39(4):327–338
8. Givoli D (2004) High-order local non-reflecting boundary conditions: a review. Wave Motion 39:319–326
9. Wellens P (2011) Wave simulations in truncated domains for offshore applications. PhD thesis, Technical University of Delft, The Netherlands
10. Duz B, Huijsmans RHM, Wellens PR, Borsboom MJA, Veldman AEP (2011) Towards a general purpose open boundary condition for wave simulations. In: Proc 30th conf on ocean, offshore and arctic engineering OMAE, OMAE2011-49979
11. Kleefsman KMT, Fekken G, Veldman AEP, Iwanowski B, Buchner B (2005) A volume-of-fluid based simulation method for wave impact problems. J Comput Phys 206:363–393
12. Skjelbreia L, Hendrickson J (1961) Fifth order gravity wave theory. In: Proc 7th conf of coastal engineering, pp 184–196

Part II
Fluid-Structure Interaction

Free-Surface Flow and Fluid-Object Interaction

I. Akkerman, K. Benner, and Y. Bazilevs

Abstract We present our free-surface flow and fluid-object interaction computational framework. The framework is an instantiation of the Mixed Interface-Tracking/Interface-Capturing Technique (MITICT) (Akin et al. in Comput. Fluids 36:2–11, 2007; Cruchaga et al. in Int. J. Numer. Methods Fluids 54:1021–1031, 2007; Tezduyar in Arch. Comput. Methods Eng. 8:83–130, 2001) where the level-set method is used for the air-water interface description and the ALE (Hughes et al. in Comput. Methods Appl. Mech. Eng. 29:329–349, 1981) technique is employed to track the moving fluid-object interface. We discuss the definition of the local mesh size used in the level-set formulation, which is an important aspect of this work. We show two example computations, the dam break and Fridsma hull, and validate our methodology using the experimental data available for these cases.

Keywords Free-surface flow · Level set method · ALE-VMS · Fluid-object interaction · Planing hulls

1 Introduction

In this work we present our efforts to develop a robust and accurate methodology for simulating air-water free-surface flow and its interaction with rigid objects [2, 4, 5, 29]. We are interested in a class of large-scale applications in coastal, marine, and offshore engineering. Therefore, we need a unified numerical methodology that is able to accurately represent the entire range of free-surface behavior, from the smooth, quasi-steady free surface, as in the case of surface ship wakes at low-to-moderate Froude number, to the violent free-surface motions with topological changes, as in the case of liquids sloshing in tanks or waves impacting offshore

I. Akkerman
School of Engineering and Computing Sciences, Durham University, South Road, Durham, UK

K. Benner · Y. Bazilevs (✉)
Department of Structural Engineering, University of California, San Diego, 9500 Gilman Drive, La Jolla, CA 92093, USA
e-mail: yuri@ucsd.edu

L. Eça et al. (eds.), *MARINE 2011, IV International Conference on Computational Methods in Marine Engineering*, Computational Methods in Applied Sciences 29,
DOI 10.1007/978-94-007-6143-8_3,

structures. As a result, the methods we developed for free-surface flow are based on the level set technique [33, 34, 36], which is an interface-capturing methodology [39], and which puts no limitations on the complexity of free-surface flow behavior.

To account for the motion of floating objects, and the corresponding changes in the geometry of the free-surface flow domain, an ALE-VMS approach [2, 12, 23, 26] is adopted. This results in a so-called Mixed Interface-Tracking/Interface-Capturing Technique (MITICT) [1, 19, 38], where the computational mesh follows (or "tracks") the fluid-object interface, while the air-water interface is "captured" on that mesh. The attractive feature of this approach is that the quality of the boundary-layer discretization near the solid object is maintained in the simulation, even as the object moves through space.

To enhance the accuracy of the free-surface fluid-object simulations, weakly enforced no-slip boundary conditions are employed [13–15] at the fluid-object interface, an approach which has similarities to wall-function modeling [20].

The ALE-VMS formulation is suitable for discretization using variable-order finite elements [29] as well as Isogeometric Analysis (IGA) [17, 25]. The results in [4] illustrate that on a per-degree-of-freedom basis, IGA based on non-Uniform Rational B-Splines (NURBS) is more accurate for free-surface flow than lower-order FEM.

This book chapter is outlined as follows. We briefly recall our free-surface fluid-object interaction numerical formulation and point the reader to the appropriate references for more details. We give a detailed discussion of the functions and mesh-dependent parameters used to distribute the fluid properties in the air and water subdomains. We feel this is an important aspect of free-surface flow formulations on unstructured meshes where the element sizes in the problem domain may differ by several orders of magnitude. We then present two numerical test cases that illustrate the accuracy and versatility of the proposed computation framework.

The first test case is the dam break problem, which involves large motions of the free surface accompanied by violent wave breaking. This test case is well known [21, 30, 31], and has good experimental data available, which allows one to perform validation studies on one's free-surface flow solver. The ability of the same ALE-VMS formulation to deliver accurate results for very different underlying interpolation functions is demonstrated by employing an unstructured mesh consisting of linear tetrahedral elements and a block-structured mesh consisting of quadratic NURBS. The second test case is the Fridsma planing hull [5, 22]. This problem is not as well known in the community as the dam break problem; it involves a lab-scale planing hull, and likewise has a very well documented experimental data set, allowing one to perform fluid-object interaction validation studies.

2 Formulation

In the following, we provide a brief description of the free-surface fluid-object interaction formulation. The reader is referred to [2, 4, 29] for additional details.

To model free-surface flow, we use the level set method [32, 34, 36]. The level-set function is used to separate the air and water subdomains, and provide the spatial distribution of the fluid density and viscosity. Using variable density and viscosity distribution, the Navier–Stokes equations of incompressible flows are solved everywhere in the fluid domain and the level-set function is convected with the computed flow velocity. We also impose the requirement that the level-set function satisfies the signed-distance property in order to control the size of the air-water interface in the computations.

The Navier–Stokes equations of incompressible flows are employed in both air and water subdomains. The Navier–Stokes and level-set equations are written in the Arbitrary Lagrangian–Eulerian (ALE) frame [26], which is well suited for moving domain problems. The Navier–Stokes and level-set equations are discretized using the ALE-VMS formulation [12]. The ALE-VMS formulation is a moving-domain extension of the Residual-Based Variational Multiscale (RBVMS) formulation, originally proposed in [6] in the context of LES turbulence modeling, and successfully applied to simulate turbulent flow and fluid-structure interaction in [3, 7–11, 14, 15, 24]. The level-set function is prescribed to satisfy the signed-distance property (or "redistanced") by solving the Eikonal partial differential equation, which is discretized on the same mesh as the fluid mechanics equations, using the SUPG formulation [16]. Mass conservation, which is important in some cases, is achieved by adding a small perturbation to the redistanced level-set function. This small perturbation may be either a field [4], which results in local mass conservation, or a global constant [2, 4, 5], which results in global mass conservation.

The rigid object is described using balance equations of linear and angular momenta. The rigid-object formulation is written in terms of the linear and angular velocities of the object's center-of-mass. The rotation matrix is also treated as an unknown governed by an appropriate evolution equation. The rotation-matrix equation is integrated in time using a midpoint method, which preserves the orthonormal property of the columns of the rotation matrix, as shown in [27].

The fluid domain mesh is deformed using the equations of linear elastostatics subject to the kinematic boundary conditions coming from the motion of the rigid object. Jacobian-based stiffening is employed to protect the small elements in the boundary layers near the solid surfaces [28].

The proposed methodology is an instantiation of the more general Mixed Interface-Tracking/Interface-Capturing Technique (MITICT) where the fluid-solid interface is tracked with the mesh, while the air-water interface is captured on that mesh. While boundary-layer meshing is employed near solid surfaces, a factor that is important for boundary-layer accuracy, resolution of the air-water interface depends on the mesh size in the neighborhood of where the interface is. The MITICT [38] was introduced primarily for fluid-object interactions with multiple fluids (see, for example, [37]), and was successfully tested in [1], where the interface-tracking technique used was a space–time formulation [40, 41], and the interface-capturing method was the Edge-Tracked Interface Locator Technique (ETILT) [38]. It was also tested in [19] by using a moving Lagrangian interface technique [18] for interface tracking and the ETILT for interface capturing.

In the next subsection we give a short discussion of how the air and water material parameters are represented in our method, and give some useful definitions of the local mesh size. We note that the free-surface flow solutions are sensitive to the choice of the local mesh size, much like the solutions using stabilized and multiscale methods are sensitive to the definition of the stabilization parameters.

2.1 Mesh-Dependent Parameters

Let ϕ^h be the discrete level-set function. At the air-water interface, $\phi^h = 0$. For this two-fluid case, the density and viscosity are defined as

$$\rho(\phi^h) = \rho_w H_\varepsilon(\phi^h) + \rho_a(1 - H_\varepsilon(\phi^h)) \tag{1}$$

$$\mu(\phi^h) = \mu_w H_\varepsilon(\phi^h) + \mu_a(1 - H_\varepsilon(\phi^h)), \tag{2}$$

where ρ_w and ρ_a are the densities of water and air, respectively, μ_w and μ_a are the dynamic viscosities of water and air, respectively, and

$$H_\varepsilon(\phi) = \begin{cases} 0 & \text{if } \phi \le -\varepsilon; \\ \frac{1}{2}(1 + \frac{\phi}{\varepsilon} + \frac{1}{\pi}\sin(\frac{\phi\pi}{\varepsilon})) & \text{if } |\phi| < \varepsilon; \\ 1 & \text{if } \phi \ge \varepsilon, \end{cases} \tag{3}$$

is a smoothed or regularized Heaviside function. The width of the air-water interface, ε, is given by $\varepsilon = \tilde{\varepsilon} h_\phi$, where h_ϕ is the local mesh size. To define the local mesh size we use the element metric tensor given by

$$\mathbf{G} = \frac{\partial \boldsymbol{\xi}}{\partial \mathbf{x}}^T \frac{\partial \boldsymbol{\xi}}{\partial \mathbf{x}}, \tag{4}$$

where $\partial \mathbf{x}/\partial \boldsymbol{\xi}$ is the Jacobian of the mapping between the element in the parent and physical domains, and $\partial \boldsymbol{\xi}/\partial \mathbf{x}$ is its inverse. The parameter $\tilde{\varepsilon}$ is defined by the user and is interpreted as the number of elements in the interface layer. Smaller values of $\tilde{\varepsilon}$ give a sharper interface, while larger values of $\tilde{\varepsilon}$ lead to a more diffuse interface. In our computations we typically choose $\tilde{\varepsilon} \in [1.0, 2.0]$.

Two types of h_ϕ are considered: *isotopic* and *directional*. In the isotopic case, h_ϕ may be computed as

$$h_\phi = \frac{C}{\sqrt{\operatorname{tr} \mathbf{G}}} \tag{5}$$

or

$$h_\phi = C\sqrt{\operatorname{tr} \mathbf{G}^{-1}}, \tag{6}$$

where C is a positive constant that depends on the element topology and parametric domain. For example, we take $C = 2$ for a hexahedron whose parametric domain is a bi-unit cube. In Eq. (5), the element's shorter dimension is favored, while Eq. (6) favors the element's longer dimension.

In the directional case, we use

$$h_\phi = \sqrt{\frac{\nabla\phi^h \cdot \nabla\phi^h}{\nabla\phi^h \cdot \mathbf{G}\nabla\phi^h}}. \tag{7}$$

Equation (7) essentially gives the element size in the direction of the level-set gradient, which is normal to the air-water interface.

Equations (1) and (2) are typically invoked at mesh quadrature points. On unstructured grids, rapidly changing local mesh size h_ϕ may lead to non-smooth and non-monotone definitions of ρ and μ. To improve on this, we propose to project the mesh size h_ϕ to the finite-element space. Any of the above definitions of h_ϕ may be used. This automatically gives a continuous mesh size, which is more likely to produce a monotone transition in the material parameters from water to air. In this case, the fluid density and viscosity are given by

$$\rho = \sum_{A=1}^{N_n} \rho_A N_A \tag{8}$$

$$\mu = \sum_{A=1}^{N_n} \mu_A N_A, \tag{9}$$

where N_A's are the basis functions used in the discretization, N_n is their total number, and ρ_A and μ_A are the nodal values of the density and viscosity, respectively, computed using Eqs. (1) and (2) evaluated with h_A and ϕ_A, the nodal values of the mesh size and level-set function, respectively.

3 Computational Results

In this section we present two test cases: the dam break problem and Fridsma planing hull.

3.1 The MARIN Dam Break Problem

The setup of the dam break problem, initially proposed by the Maritime Research Institute Netherlands (MARIN) [30], is depicted in Fig. 1. The problem consists of a column of water, initially at rest, that collapses under the action of gravity and impacts a fixed rectangular container. We compute the problem using two types of the spatial discretization: linear tetrahedral finite elements and NURBS. The tetrahedral mesh, which is completely unstructured, consists of 2,999,780 elements and

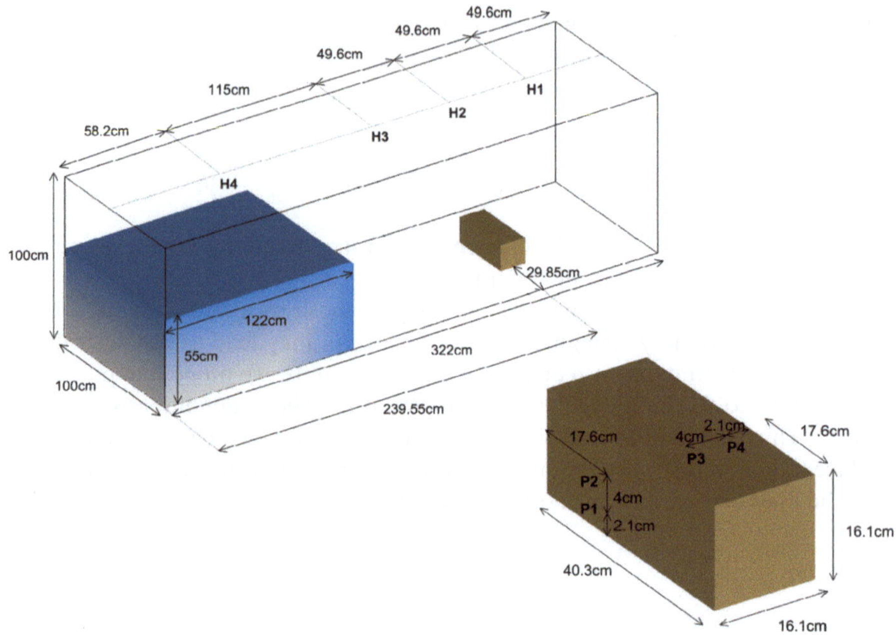

Fig. 1 The MARIN dam break problem. Geometry definition. The computational domain is a rectangular box with dimensions 3.22 m × 1 m × 1 m. The object has dimensions 0.2 m × 0.2 m × 0.4 m and is placed at the back end of the tank. The water column, initially at rest, has dimensions 1 m × 1 m × 0.55 m. The locations where pressure and water height are sampled are also depicted

517,791 nodes. The quadratic NURBS mesh is significantly more coarse. It is block-structured and consists of only 65,356 elements and 85,464 control points. Free-slip and no-penetration boundary conditions are applied on all surfaces, including the top of the tank. The problem is run until $T = 6$ s, using 2,400 time steps of equal size. We would like to note that significant computational savings may be obtained for this problem when using time-step adaptivity (see, e.g., [35, 42]). Snapshots comparing the solutions coming from tetrahedral FEM and NURBS computations are given in Fig. 2. Large-scale features of the solution are very similar in the two simulations, however the details of the small-scale features are better represented on a much finer tetrahedral grid, as expected.

Time series of the pressure at different locations on the obstacle are shown in Fig. 3. The first wave hits the block at approximately $t = 0.5$ s, and the second, much smaller wave arrives at the block at about $t = 5$ s. The wave impact times and pressure peaks are predicted very well with both linear elements and quadratic NURBS. Given that the NURBS mesh has about half of the degrees-of-freedom of the linear FEM mesh in each Cartesian direction, the accuracy of NURBS results is remarkable; linear FEM is not capable of attaining such accuracy at this level of resolution (see [4]), and requires a finer mesh for comparable accuracy.

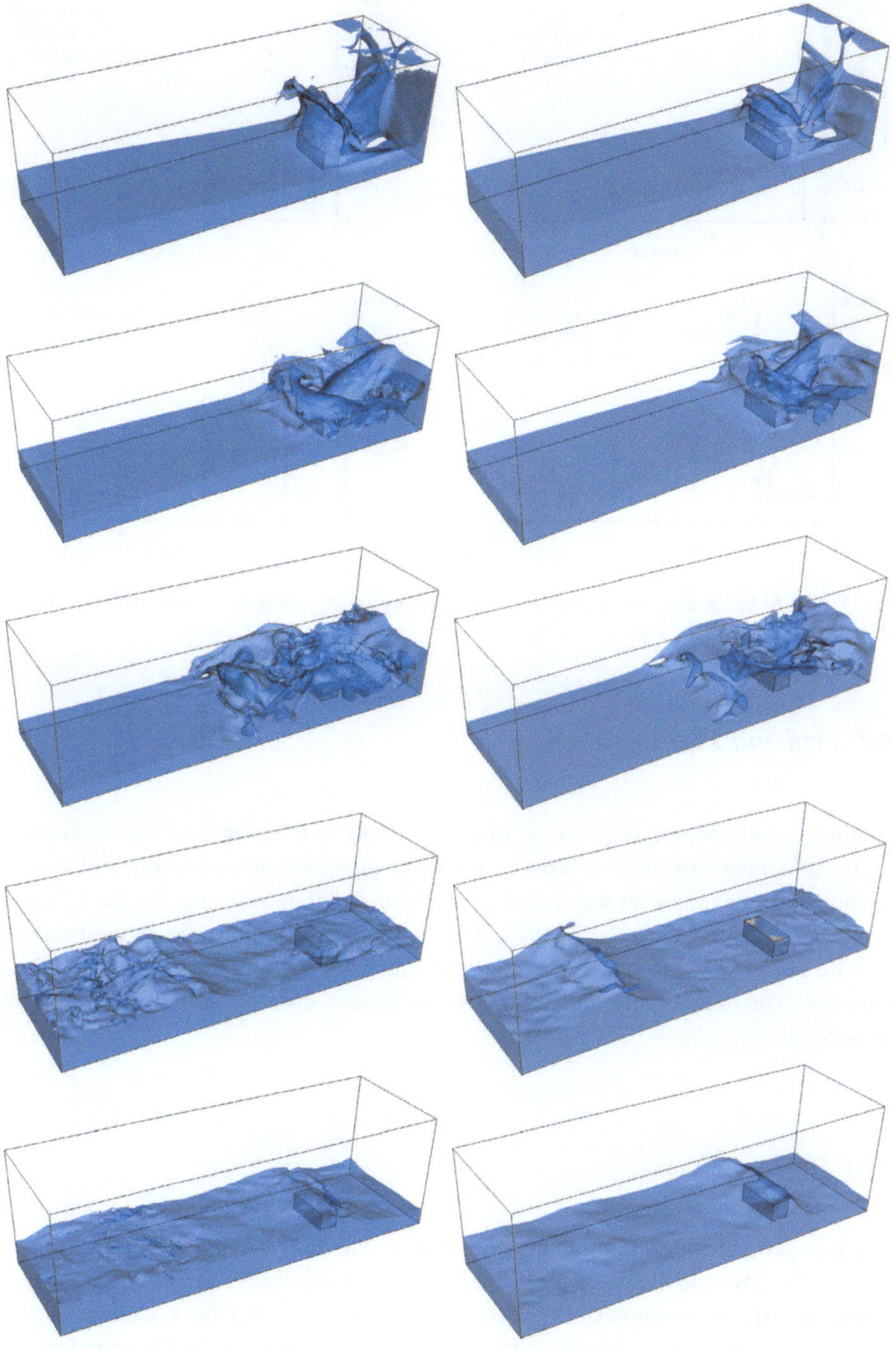

Fig. 2 The MARIN dam break problem. Snapshots of the free surface solution on the tetrahedral (*left*) and NURBS (*right*) meshes at $t = 1.0$, 1.5, 2.0, 4.0, and 5.0 s

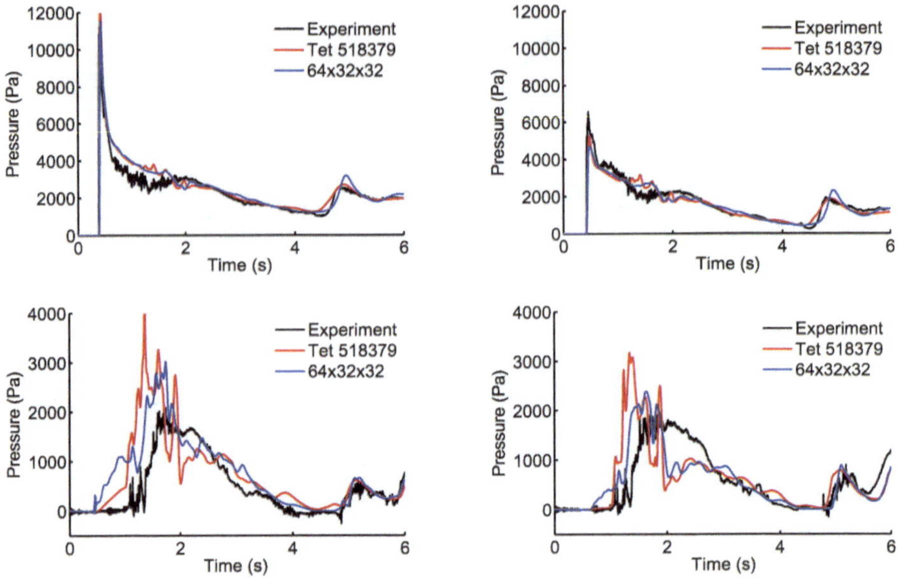

Fig. 3 The MARIN dam break problem. Time history of the pressure at four locations on the obstacle. Experimental data is from [30]

3.2 *Fridsma Planing Hull*

Planing vessels are of importance in civilian and military marine transportation. In the associated free-surface flow problem air plays an important role and should be modeled together with water. We present our simulation results for the Fridsma planing hull from [5] in this section. We give a detailed definition of the hull geometry, present a mesh refinement study, and assess the effect of hull speed on drag force and trim angle. Only flat-water (i.e., no waves), constant-hull-speed cases are considered.

3.2.1 Hull Definition

The hull geometry definition is given in Fig. 4. The hull is comprised of idealized shapes: the bow consisting of four ruled surfaces and a wedge-shaped straight section with a deadrise angle of 20°. Analytical expressions for the bounding curves for the ruled surfaces are provided in the figure and the relevant hull parameters are summarized in Table 1. The data pertains to the center of gravity located at 70 % of the hull length measured from the aft [22].

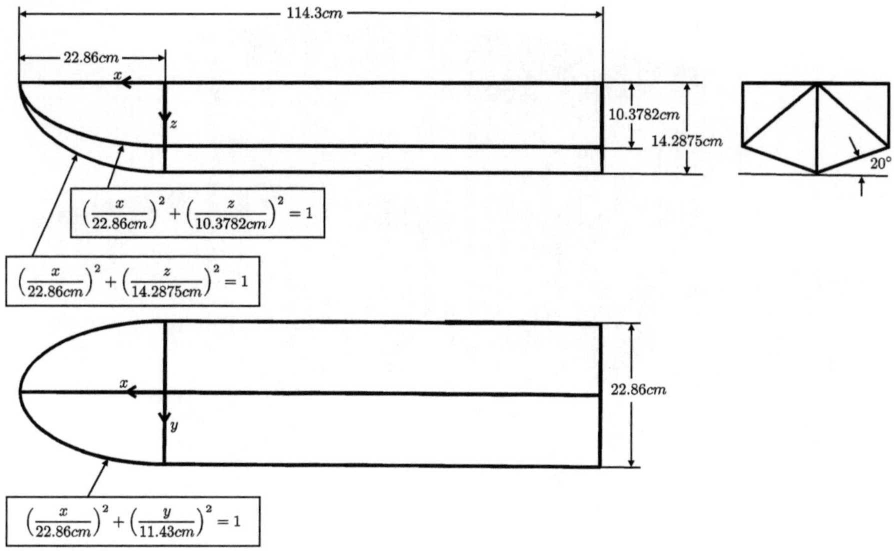

Fig. 4 Fridsma hull. Geometry definition. *Top left*: side view. *Top right*: aft view. *Bottom left*: top view

Table 1 Fridsma hull. Global geometric and inertial parameters

Geometric parameters		Inertial parameters	
Length (L)	114.3 cm	Mass (m)	7.257 kg
Beam (b)	22.86 cm	x_{cg}	80.01 cm
Height	14.2875 cm	z_{cg}	6.721 cm
Deadrise	20°	Gyradius (r)	25 %L
		$I_{yy} = m(rL)^2$	0.6165 kg m^2

3.2.2 Mesh Convergence Study

We perform a convergence study at Froude number Fr $= 0.8950$.[1] The Froude number is defined as

$$\mathrm{Fr} = \frac{u}{\sqrt{gL}}, \tag{10}$$

where u is the hull speed, g is the magnitude of the gravitational acceleration, and L is the hull length. The Froude number represents the significance of inertial forces relative to gravity. At this chosen Froude number, according to [22], the trim angle

[1] In [22] the results were reported in terms of the Speed-Length Ratio (SLR), $u/\sqrt{L}$, which is a dimensional quantity. Here report the results in terms of the Froude number, which is non-dimensional.

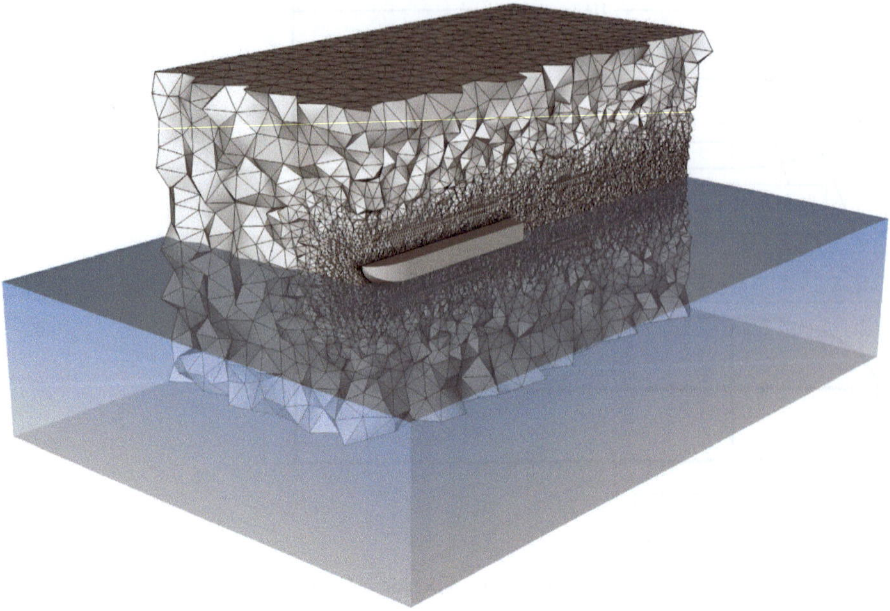

Fig. 5 Fridsma hull. Coarsest mesh with water and air domains depicted

Table 2 Fridsma hull. Mesh parameters for the four meshes used in this work

Mesh	Nodes	Elements
1	111,596	615,932
2	211,429	1,165,064
3	488,419	2,721,768
4	963,731	5,373,290

is at a maximum. We use a sequence of four meshes (see Table 2 for the mesh parameters). The coarsest mesh is shown in Fig. 5, which also shows the water and air subdomains in the undisturbed configuration. The mesh density is increased in the boundary layer near the hull surface and in the wake. The hull is fixed in the direction of travel, and the corresponding velocity is set at the inflow of the computational domain together with the level-set function. The hull is allowed to pitch and displace in the vertical direction. At the outflow a hydrostatic pressure profile is imposed as a traction boundary condition. On the side, bottom, and top boundaries of the computational domain free-slip boundary conditions are imposed. Figure 6 shows the deformed free surface colored by the flow speed relative to the hull speed corresponding to the 2,721,768-element mesh simulation. The hull rises up and develops a trim angle such that the bow is higher than the aft. Note the presence of the "rooster tail" feature, which is typical for planing hulls. Figure 7 shows convergence of the drag force and trim angle with the drag force non-dimensionalized by the gravitational force. From the results we see that the drag force converges quickly

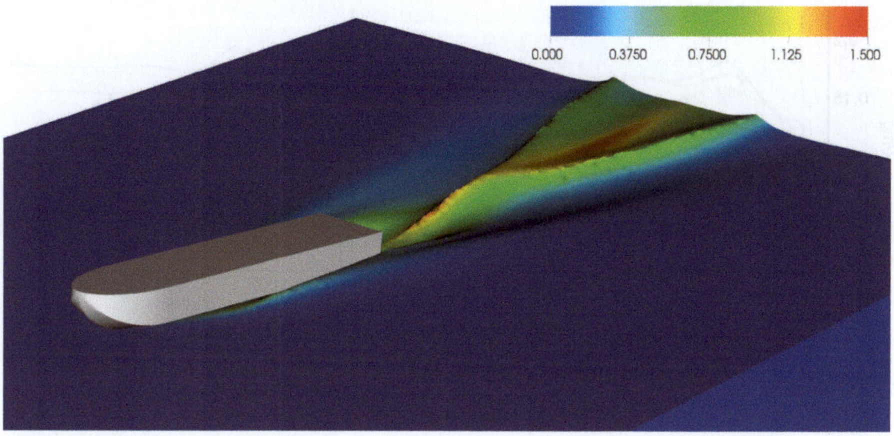

Fig. 6 Fridsma hull. Free surface colored by flow speed relative to hull speed in m/s

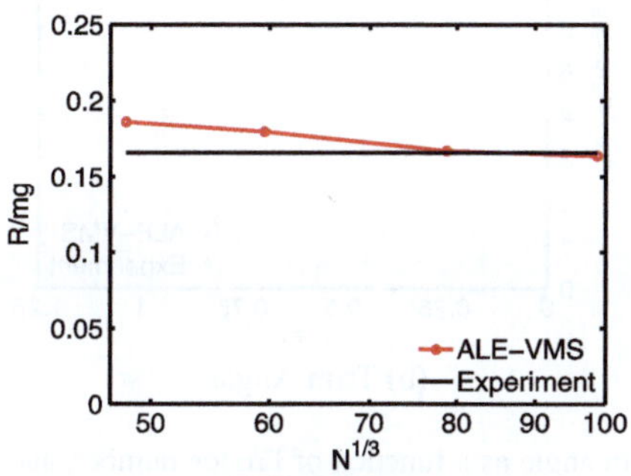

(a) Drag Force

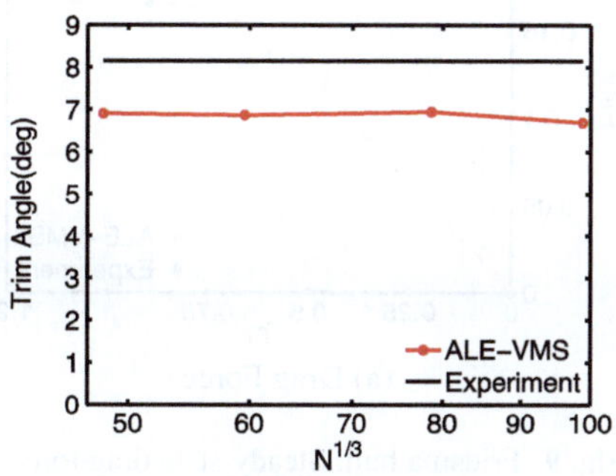

(b) Trim Angle

Fig. 7 Fridsma hull. Convergence of the drag force and trim angle with mesh refinement and comparison with experimental data from [22]

to the experimental value. The trim angle, on the other hand, is underestimated by 12 % with respect to the experimental data, and does not improve with mesh refinement. Possible causes for this may be the choice of the location of downstream, lateral, and bottom boundaries. Errors in the experimental data are also possible.

3.2.3 Effect of Hull Speed

In this section the effect of hull speed on the drag force and trim angle is studied. For this purpose the mesh of 2,721,768 tetrahedral elements is used. In addition to the $\mathrm{Fr} = 0.8950$ case, we consider $\mathrm{Fr} = 0$, $\mathrm{Fr} = 0.5925$, and $\mathrm{Fr} = 1.190$. The simulations are started impulsively in the configuration depicted in Fig. 5. In the case of $\mathrm{Fr} = 0$, although the hull speed is zero, a non-zero trim angle develops such that the hull is in equilibrium with the hydrostatic forces. In all other cases, however,

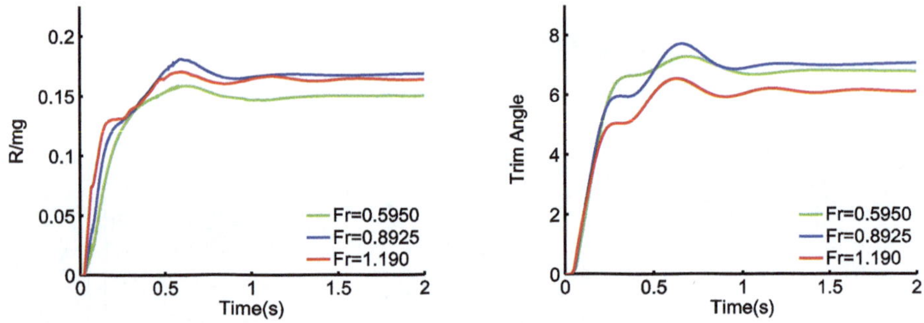

(a) Drag Force (b) Trim Angle

Fig. 8 Fridsma hull. Transient behavior of the drag force and trim angle

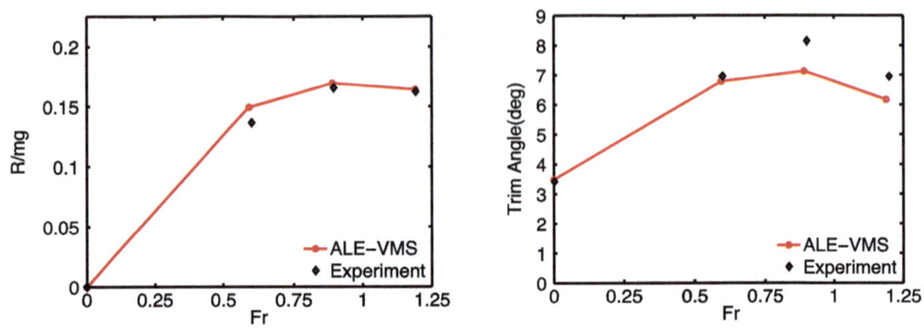

(a) Drag Force (b) Trim Angle

Fig. 9 Fridsma hull. Steady-state drag force and trim angle as a function of Froude number, and comparison with experimental data from [22]

there is a rapid transient followed by a largely steady-state response. Figure 8 depicts this transient behavior, and shows that the steady state for all cases is attained in about 1 s. For higher hull speeds the decay in the transient response is slower. The steady-state drag force and trim angle are plotted as a function of Froude number, and compared to the experimental results in Fig. 9. Accurate prediction of the drag force is attained in all cases. The trim angle is predicted very well for the first two Froude number cases, and the 10–12 % deviation from the experiment is seen in the remaining two cases.

4 Conclusions

A free-surface flow and fluid-object interaction computational framework is presented that can be used to compute free-surface motions of arbitrary complexity. The numerical formulation is able to accommodate a variety of discretizations, including lower- and higher-order FEM and IGA. Both air and water are modeled, which enables the application of the proposed methodology to planing hulls as well as other engineering systems, such as offshore wind turbines. The framework is applied to the simulation of the dam break and Fridsma hull problems, where very good comparison with the experimental data was observed. In the case of the dam break problem, IGA showed remarkable per-degree-of-freedom accuracy.

Acknowledgements This research was supported through ARO Award W911NF-10-1-0247. This support is gratefully acknowledged. We also wish to thank the Texas Advanced Computing Center (TACC) at the University of Texas at Austin and San Diego Supercomputer Center (SDSC) at the University of California, San Diego for providing HPC resources that contributed to the research results reported in this paper.

References

1. Akin JE, Tezduyar TE, Ungor M (2007) Computation of flow problems with the mixed interface-tracking/interface-capturing technique (MITICT). Comput Fluids 36:2–11
2. Akkerman I, Bazilevs Y, Benson DJ, Farthing MW, Kees CE (2012) Free-surface flow and fluid-object interaction modeling with emphasis on ship hydrodynamics. J Appl Mech 79:010905. doi:10.1115/1.4005072
3. Akkerman I, Bazilevs Y, Calo VM, Hughes TJR, Hulshoff S (2008) The role of continuity in residual-based variational multiscale modeling of turbulence. Comput Mech 41:371–378
4. Akkerman I, Bazilevs Y, Kees C, Farthing M (2011) Isogeometric analysis of free-surface flow. J Comput Phys 230:4137–4152. doi:10.1016/j.jcp.2010.11.044
5. Akkerman I, Dunaway J, Kvandal J, Spinks J, Bazilevs Y (2012) Toward free-surface modeling of planing vessels: simulation of the Fridsma hull using ALE-VMS. Comput Mech. doi:10.1007/s00466-012-0770-2
6. Bazilevs Y, Calo VM, Cottrel JA, Hughes TJR, Reali A, Scovazzi G (2007) Variational multiscale residual-based turbulence modeling for large eddy simulation of incompressible flows. Comput Methods Appl Mech Eng 197:173–201
7. Bazilevs Y, Calo VM, Hughes TJR, Zhang Y (2008) Isogeometric fluid-structure interaction: theory, algorithms, and computations. Comput Mech 43:3–37
8. Bazilevs Y, Calo VM, Zhang Y, Hughes TJR (2006) Isogeometric fluid-structure interaction analysis with applications to arterial blood flow. Comput Mech 38:310–322
9. Bazilevs Y, Gohean JR, Hughes TJR, Moser RD, Zhang Y (2009) Patient-specific isogeometric fluid-structure interaction analysis of thoracic aortic blood flow due to implantation of the Jarvik 2000 left ventricular assist device. Comput Methods Appl Mech Eng 198:3534–3550
10. Bazilevs Y, Hsu M-C, Akkerman I, Wright S, Takizawa K, Henicke B, Spielman T, Tezduyar TE (2011) 3D simulation of wind turbine rotors at full scale. Part I: Geometry modeling and aerodynamics. Int J Numer Methods Fluids 65:207–235
11. Bazilevs Y, Hsu M-C, Kiendl J, Wuechner R, Bletzinger K-U (2011) 3D simulation of wind turbine rotors at full scale. Part II: Fluid-structure interaction. Int J Numer Methods Fluids 65:236–253
12. Bazilevs Y, Hsu M-C, Takizawa K, Tezduyar TE (2012) ALE-VMS and ST-VMS methods for computer modeling of wind-turbine rotor aerodynamics and fluid–structure interaction. Math Models Methods Appl Sci 22:1230002. doi:10.1142/S0218202512300025
13. Bazilevs Y, Hughes TJR (2007) Weak imposition of Dirichlet boundary conditions in fluid mechanics. Comput Fluids 36:12–26

14. Bazilevs Y, Michler C, Calo VM, Hughes TJR (2007) Weak Dirichlet boundary conditions for wall-bounded turbulent flows. Comput Methods Appl Mech Eng 196:4853–4862
15. Bazilevs Y, Michler C, Calo VM, Hughes TJR (2010) Isogeometric variational multiscale modeling of wall-bounded turbulent flows with weakly-enforced boundary conditions on unstretched meshes. Comput Methods Appl Mech Eng 199(13–16):780–790
16. Brooks AN, Hughes TJR (1982) Streamline upwind/Petrov-Galerkin formulations for convection dominated flows with particular emphasis on the incompressible Navier-Stokes equations. Comput Methods Appl Mech Eng 32:199–259
17. Cottrell JA, Hughes TJR, Bazilevs Y (2009) Isogeometric analysis: toward integration of CAD and FEA. Wiley, Chichester
18. Cruchaga M, Celentano D, Tezduyar T (2001) A moving Lagrangian interface technique for flow computations over fixed meshes. Comput Methods Appl Mech Eng 191:525–543
19. Cruchaga MA, Celentano DJ, Tezduyar TE (2007) A numerical model based on the mixed interface-tracking/interface-capturing technique (mitict) for flows with fluid-solid and fluid-fluid interfaces. Int J Numer Methods Fluids 54:1021–1031
20. Eca L, Hoekstra M, Hay A, Pelletier D (2007) Verification of RANS solvers with manufactured solutions. Eng Comput 23:253–270
21. Elias RN, Coutinho ALGA (2007) Stabilized edge-based finite element simulation of free-surface flows. Int J Numer Methods Fluids 54:965–993
22. Fridsma G (1968) A systematic study of the rough-water performance of planing boats. Davidson Laboratory report 1275
23. Hsu M-C, Akkerman I, Bazilevs Y (2012) Wind turbine aerodynamics using ALE–VMS: validation and the role of weakly enforced boundary conditions. Comput Mech 50:499–511
24. Hsu M-C, Bazilevs Y, Calo VM, Tezduyar TE, Hughes TJR (2010) Improving stability of multiscale formulations of fluid flow at small time steps. Comput Methods Appl Mech Eng 199:828–840
25. Hughes TJR, Cottrell JA, Bazilevs Y (2005) Isogeometric analysis: CAD, finite elements, NURBS, exact geometry, and mesh refinement. Comput Methods Appl Mech Eng 194:4135–4195
26. Hughes TJR, Liu WK, Zimmermann TK (1981) Lagrangian-Eulerian finite element formulation for incompressible viscous flows. Comput Methods Appl Mech Eng 29:329–349
27. Hughes TJR, Winget J (1980) Finite rotation effects in numerical integration of rate constitutive equations arising in large-deformation analysis. Int J Numer Methods Eng 15:1862–1867
28. Johnson AA, Tezduyar TE (1994) Mesh update strategies in parallel finite element computations of flow problems with moving boundaries and interfaces. Comput Methods Appl Mech Eng 119:73–94
29. Kees C, Akkerman I, Farthing M, Bazilevs Y (2011) A conservative level set method suitable for variable-order approximations and unstructured meshes. J Comput Phys 230:4536–4558. doi:10.1016/j.jcp.2011.02.030
30. Kleefsman KMT, Fekken G, Veldman AEP, Iwanowski B, Buchner B (2005) A volume-of-fluid based simulation method for wave impact problems. J Comput Phys 206:363–393
31. Lins EF, Elias RN, Rochinha FA, Coutinho ALGA (2010) Residual-based variational multiscale simulation of free surface flows. Comput Mech 46:545–557
32. Nagrath S, Jansen KE, Lahey RT (2005) Computation of incompressible bubble dynamics with a stabilized finite element level set method. Comput Methods Appl Mech Eng 194:4565–4587
33. Osher S, Fedkiw R (2003) Level set methods and dynamic implicit surfaces. Applied mathematical sciences, vol 153. Springer, New York
34. Sethian JA (1999) Level set methods and fast marching methods. Cambridge University Press, Cambridge
35. Söderlind G (2002) Automatic control and adaptive time-stepping. Numer Algorithms 31:281–310
36. Sussman M, Smereka P, Osher SJ (1994) A level set approach for computing solutions to incompressible two-phase flows. J Comput Phys 114:146–159

37. Takizawa K, Tanizawa K, Yabe T, Tezduyar TE (2007) Ship hydrodynamics computations with the CIP method based on adaptive Soroban grids. Int J Numer Methods Fluids 54:1011–1019
38. Tezduyar TE (2001) Finite element methods for flow problems with moving boundaries and interfaces. Arch Comput Methods Eng 8:83–130
39. Tezduyar TE (2003) Computation of moving boundaries and interfaces and stabilization parameters. Int J Numer Methods Fluids 43:555–575
40. Tezduyar TE, Behr M, Liou J (1992) A new strategy for finite element computations involving moving boundaries and interfaces—the deforming-spatial-domain/space–time procedure: I. The concept and the preliminary numerical tests. Comput Methods Appl Mech Eng 94(3):339–351
41. Tezduyar TE, Behr M, Mittal S, Liou J (1992) A new strategy for finite element computations involving moving boundaries and interfaces—the deforming-spatial-domain/space–time procedure: II. Computation of free-surface flows, two-liquid flows, and flows with drifting cylinders. Comput Methods Appl Mech Eng 94(3):353–371
42. Valli AMP, Carey GF, Coutinho ALGA (2005) Control strategies for timestep selection in finite element simulation of compressible flows and coupled reaction-convection-diffusion processes. Int J Numer Methods Fluids 47:201–231

The Particle Finite Element Method (PFEM). An Effective Numerical Technique for Solving Marine, Naval and Harbour Engineering Problems

E. Oñate, S.R. Idelsohn, M.A. Celigueta, and B. Suárez

Abstract We present some developments in the Particle Finite Element Method (PFEM) for the solution of complex coupled problems in marine, naval and harbour engineering involving fluid-soil-structure interaction (FSSI). The PFEM uses an updated Lagrangian description to model the motion of nodes (particles) in a continuum domain containing fluid, soil/rock and structures subdomains. A mesh connects the nodes defining the discretized domain where the governing equations for each of the constituent materials are solved with the FEM. The stabilization for dealing with an incompressibility material is introduced via the finite calculus (FIC) method. An incremental iterative scheme for solving the non linear transient FSSI problem is described. The procedure to model frictional contact conditions and material erosion at fluid-solid and solid-solid interfaces is described. We present examples of application of the PFEM to solve FSSI problems in marine, naval and harbour engineering such as the motion of rocks by water streams, the stability of breakwaters and constructions under sea waves, the sinking of ships and the collision of a ship with ice blocks.

Keywords Particle finite element method · Marine engineering · Naval engineering · Harbour engineering

1 Introduction

The analysis of problems involving the interaction of fluids, soil/rocks and structures is relevant in many areas of marine, naval and harbour engineering. Some representative examples are the study of off-shore and harbour structures under large

E. Oñate (✉) · M.A. Celigueta
International Center for Numerical Methods in Engineering (CIMNE), Barcelona, Spain
e-mail: onate@cimne.upc.edu

S.R. Idelsohn
ICREA Research Professor at CIMNE, Barcelona, Spain

B. Suárez
Universitat Politècnica de Catalunya (UPC), Barcelona, Spain

L. Eça et al. (eds.), *MARINE 2011, IV International Conference on Computational Methods in Marine Engineering*, Computational Methods in Applied Sciences 29,
DOI 10.1007/978-94-007-6143-8_4, © Springer Science+Business Media Dordrecht 2013

waves, constructions hit by tsunamis, erosion of sea bed and landscape adjacent to sea shore, motion of ships under severe sea conditions, the simulation of the sinking of a ship, the collision of a ship with ice blocks, excavation and drilling problems in the sea in petroleum and gas engineering, etc.

The authors have developed in previous works a particular class of Lagrangian formulation for solving problems involving complex interactions between (free surface) fluids and solids. The so-called the particle finite element method (PFEM, www.cimne.com/pfem), treats the mesh nodes in the fluid and solid domains as particles which can freely move and even separate from the main fluid domain representing, for instance, the effect of water drops. A mesh connects the nodes discretizing the domain where the governing equations are solved using a stabilized FEM.

An advantage of the Lagrangian formulation is that the non-linear and non symmetric convective terms disappear from the fluid equations [1, 2]. The difficulty is however transferred to the problem of adequately (and efficiently) moving the mesh nodes. The theory and applications of the PFEM are reported in [1–18].

The solution of a FSSI problem requires solving the momentum and mass balance equations for the underlying continuum that may include incompressible domains (as it is the case for a fluid and some soils). In our work we use a stabilized mixed FEM based on Finite Calculus (FIC) which allows us to solve incompressible continua with a linear approximation for the velocity and pressure variables [6, 9, 18, 20–23].

The layout of the chapter is as follows. In the next section the key ideas of the PFEM are outlined. Next the basic equations for a general continuum using a Lagrangian description and the FIC formulation are schematically presented. Then an algorithm for the transient solution is briefly described. The treatment of the coupled FSSI problem and the method for mesh generation and for identification of the free surface nodes are outlined. The procedure for treating the frictional contact interaction between fluid, soil and structure interfaces is explained. We present several examples of application of the PFEM to solve FSSI problems in marine, naval and harbour engineering such as the motion of rocks by water streams, the stability of breakwaters and constructions hit by sea waves, the study of the sinking of ships and the collision of a ship with ice blocks.

2 The Basis of the Particle Finite Element Method

Let us consider a domain containing both fluid and solid subdomains (the solid subdomain may include soil/rock materials and/or structural elements). The moving fluid particles interact with the solid boundaries, thereby inducing the deformation of the solid which in turn affects the flow motion and, therefore, the problem is fully coupled.

In the PFEM both the fluid and the solid domains are modelled using an *updated Lagrangian formulation* [24]. That is, all variables are assumed to be known in the *current configuration* at time t. The new set of variables in both domains is sought for in the *next or updated configuration* at time $t + \Delta t$. The finite element method (FEM) is used to solve the equations of continuum mechanics for each of the subdomains. Hence a mesh discretizing these domains must be generated in order to solve the governing equations for each subdomain in the standard FEM fashion [24–27].

The quality of the numerical solution depends on the discretization chosen as in the standard FEM. Adaptive mesh refinement techniques can be used to improve the solution.

2.1 Basic Steps of the PFEM

For clarity purposes we will define the *collection or cloud of nodes* (C) pertaining to the fluid and solid domains, the *volume* (V) defining the analysis domain for the fluid and the solid and the *mesh* (M) discretizing both domains.

A typical solution with the PFEM involves the following steps.

1. The starting point at each time step is the cloud of points in the fluid and solid domains. For instance ${}^nC$ denotes the cloud at time $t = t_n$ (Fig. 1).
2. Identify the boundaries for both the fluid and solid domains defining the analysis domain ${}^nV$ in the fluid and the solid. This is an essential step as some boundaries (such as the free surface in fluids) may be severely distorted during the solution, including separation and re-entering of nodes. The Alpha Shape method [19] is used for the boundary definition.
3. Discretize the fluid and solid domains with a finite element mesh ${}^nM$.We use an effect mesh generation scheme based on the extended Delaunay tessellation [3, 4].
4. Solve the coupled Lagrangian equations of motion for the overall continuum. Compute the state variables in at the next (updated) configuration for $t + \Delta t$: velocities, pressure and viscous stresses in the fluid and displacements, stresses and strains in the solid.
5. Move the mesh nodes to a new position ${}^{n+1}C$ where $n + 1$ denotes the time $t_n + \Delta t$, in terms of the time increment size. This step is typically a consequence of the solution process of step 4.
6. Go back to step 1 and repeat the solution for the next time step to obtain ${}^{n+2}C$ (Fig. 1).

3 FIC/FEM Formulation for a Lagrangian Continuum

3.1 Governing Equations

The equations to be solved are the standard ones in Lagrangian continuum mechanics [24].

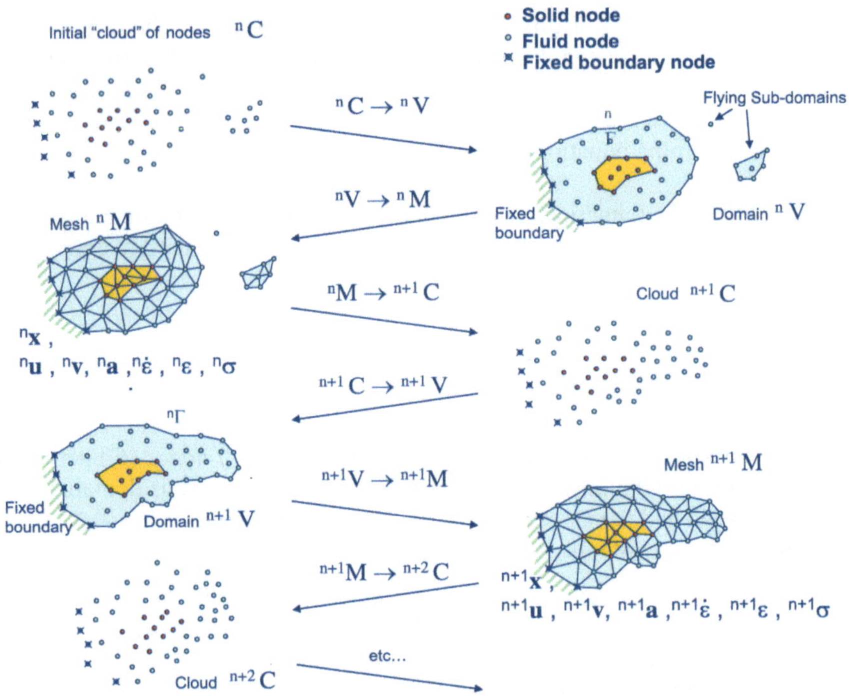

Fig. 1 Sequence of steps to update a "cloud" of nodes representing a domain containing a fluid and a solid part from time n ($t = t_n$) to time $n + 2$ ($t = t_n + 2\Delta t$)

Momentum

$$\rho \frac{\partial v_i}{\partial t} = \frac{\partial \sigma_{ij}}{\partial x_j} + b_i \quad \text{in } V \tag{1}$$

Pressure-velocity relationship

$$\frac{1}{K}\frac{\partial p}{\partial t} - \frac{\partial v_i}{\partial x_i} = 0 \quad \text{in } V \tag{2}$$

In above equations v_i is the velocity along the ith global (Cartesian) axis, p is the pressure (assumed to be positive in tension) ρ and K are the density and bulk modulus of the material, respectively, b_i and σ_{ij} are the body forces and the (Cauchy) stresses. Equations (1) and (2) are completed with the constitutive relationships [12]:

Incompressible continuum

$$^{t+1}\sigma_{ij} = 2\mu\dot{\varepsilon}_{ij} + {}^{t+1}p\delta_{ij} \tag{3}$$

Compressible/quasi-incompressible continuum

$$^{t+1}\sigma_{ij} = {}^{t}\hat{\sigma}_{ij} + 2\mu\dot{\varepsilon}_{ij} + \lambda\dot{\varepsilon}_{ii}\delta_{ij} \tag{4a}$$

where $\hat{\sigma}_{ij}$ are the component of the Cauchy stress tensor $[\hat{\sigma}]$

$$[\hat{\sigma}] = \frac{1}{J}\mathbf{F}^T\mathbf{S}\mathbf{F} \tag{4b}$$

where $\mathbf{S}$ is the second Piola-Kirchhoff stress tensor, $\mathbf{F}$ is the deformation gradient tensor and $J = \det \mathbf{F}$ [24]. Parameters μ and λ take the following values for a fluid or solid material:

Fluid μ: viscosity; $\lambda = \Delta t K - \frac{2\mu}{3}$

Solid $\mu = \frac{\Delta t G}{J}$; $\lambda = \frac{2G\nu\Delta t}{J(1-2\nu)}$, where ν is the Poisson ratio, G is the shear modulus and Δt the time increment.

In Eqs. (3) and (4a), (4b), $\dot{\varepsilon}_{ij}$ is the rate of deformation and δ_{ij} is the Kronecker delta. ${}^t(\cdot)$ denotes values at time t.

Note that ${}^t\widehat{\sigma}_{ij} = 0$ for a fluid in Eq. (4a), as the stresses depend on the rates of deformation only.

Indexes in Eqs. (1)–(4b) range from $i, j = 1, n_d$, where n_d is the number of space dimensions. These equations are completed with the standard boundary conditions of prescribed velocities and surface tractions in the mechanical problem [8–13].

3.2 Discretization of the Equations

A key problem in the numerical solution of Eqs. (1)–(4b) is the satisfaction of the mass balance condition for the fully incompressible case (i.e. $K = \infty$ in Eq. (2)). A number of procedures to solve this problem exists in the finite element literature [25, 27]. In our approach we use a stabilized formulation based in the so-called finite calculus (FIC) procedure [9, 14, 20–23]. The essence of this method is the solution of a *modified mass balance* equation which is written as

$$\frac{1}{K}\frac{\partial p}{\partial t} - \frac{\partial v_i}{\partial x_i} + \sum_{i=1}^{3} \tau \frac{\partial q}{\partial x_i}\left[\frac{\partial p}{\partial x_i} + \pi_i\right] = 0 \tag{5}$$

where q are weighting functions, τ is a stabilization parameter given by [9, 14, 21–23]

$$\tau = \left(\frac{2\rho|v|}{h} + \frac{8\mu}{3h^2}\right)^{-1} \tag{6}$$

In the above, h is a characteristic length of each finite element and $|v|$ is the modulus of the velocity vector. In Eq. (5) π_i are auxiliary pressure projection variables chosen so as to ensure that the second term in Eq. (5) can be interpreted as weighted sum of the residuals of the momentum equations and therefore it vanishes for the exact solution. The set of governing equations is completed by adding the following constraint equation [9]

$$\int_V \tau w_i \left(\frac{\partial p}{\partial x_i} + \pi_i \right) dV = 0 \quad i = 1, n_d \quad \text{(no sum in } i) \tag{7}$$

where w_i are arbitrary weighting functions.

The rest of the integral equations are obtained by applying the weighted residual technique to the governing Eqs. (1), (2), (3) and (5) and the boundary conditions [9, 12, 14, 18, 22, 26].

We interpolate next in the finite element fashion the set of problem variables. For 3D problems these are the three velocities v_i, the pressure p and the three pressure gradient projections π_i. In our work we use equal order *linear interpolation* for all variables over meshes of 3-noded triangles (in 2D) and 4-noded tetrahedra (in 3D). The resulting set of discretized equations using the standard Galerkin technique has the following form

Momentum

$$\mathbf{M}\dot{\bar{\mathbf{v}}} + \mathbf{K}\bar{\mathbf{v}} + \mathbf{G}\bar{\mathbf{p}} = \mathbf{f} \tag{8}$$

Pressure-velocity relationship

$$\bar{\mathbf{M}}\dot{\bar{\mathbf{p}}} + \mathbf{L}\bar{\mathbf{p}} - \mathbf{G}\bar{\mathbf{v}} + \mathbf{Q}\bar{\pi} = 0 \tag{9}$$

Pressure gradient projection

$$\widehat{\mathbf{M}}\bar{\pi} + \mathbf{Q}^T\bar{\mathbf{p}} = \mathbf{0} \tag{10}$$

In Eqs. (8)–(10) $\overline{(\cdot)}$ denotes nodal variables, $\dot{\overline{(\cdot)}} = \frac{\partial}{\partial t}\overline{(\cdot)}$. The matrices and vectors are given in [14, 18].

The solution in time of Eqs. (8)–(10) can be performed using any time integration scheme typical of the updated Lagrangian FEM [24]. A basic algorithm following the conceptual process described in Sect. 2 is presented in Box 1.

4 Generation of a New Mesh

A key point for the success of the PFEM is the fast regeneration of a mesh at every time step on the basis of the position of the nodes in the space domain. In our work the mesh is generated using the so called extended Delaunay tessellation (EDT) presented in [4].

The CPU time required for meshing grows linearly with the number of nodes. The CPU time for solving the equations exceeds that required for meshing as the number of nodes increases. As a general rule for large 3D problems meshing consumes around 15 % of the total CPU time per time step, while the solution of the equations (with typically 3 iterations per time step) and the system assembly consume approximately 70 % and 15 % of the CPU time per time step, respectively.

1. LOOP OVER TIME STEPS, $t = 1$, NTIME

Known values

${}^t\bar{\mathbf{x}}, {}^t\bar{\mathbf{v}}, {}^t\bar{\mathbf{p}}, {}^t\bar{\boldsymbol{\pi}}, {}^t\bar{\mathbf{T}}, {}^t\mu, {}^t\mathbf{f}, {}^t\mathbf{q}, {}^tC, {}^tV, {}^tM$

2. LOOP OVER NUMBER OF ITERATIONS, $i = 1$, NITER

- Compute the nodal velocities by solving Eq. (8)

$$\left[\frac{1}{\Delta t}\mathbf{M}+\mathbf{K}\right]^{t+1}\bar{\mathbf{v}}^{i+1} = {}^{t+1}\mathbf{f} - \mathbf{G}^{t+1}\bar{\mathbf{p}}^i + \frac{1}{\Delta t}\mathbf{M}^t\bar{\mathbf{v}}$$

- Compute nodal pressures from Eq. (9)

$$\left[\frac{1}{\Delta t}\bar{\mathbf{M}}+\mathbf{L}\right]^{t+1}\bar{\mathbf{p}}^{i+1} = \mathbf{G}^{T\,t+1}\bar{\mathbf{v}}^{i+1} - \mathbf{Q}^{t+1}\bar{\boldsymbol{\pi}}^i + \frac{1}{\Delta t}\bar{\mathbf{M}}^t\bar{\mathbf{p}}$$

- Compute nodal pressure gradient projections from Eq. (10)

$${}^{n+1}\bar{\boldsymbol{\pi}}^{i+1} = -\hat{\mathbf{M}}_D^{-1}[\mathbf{Q}^T]^{t+1}\bar{\mathbf{p}}^{i+1}, \quad \hat{\mathbf{M}}_D = \text{diag}[\hat{\mathbf{M}}_D]$$

- Update position of analysis domain nodes:

$${}^{t+\Delta t}\bar{\mathbf{x}}^{i+1} = {}^t\mathbf{x}^i + {}^{t+\Delta t}\mathbf{v}^{i+1}\Delta t$$

  Define new "cloud" of nodes ${}^{t+\Delta t}C^{i+1}$

Check convergence → NO → Next iteration $i \to i+1$

↓ YES

Next time step $t \to t+1$

- Identify new analysis domain boundary: ${}^{t+1}V$
- Generate mesh: ${}^{t+1}M$
  Go to 1

Box 1 Basic PFEM algorithm for a Lagrangian continuum

These figures refer to analyses in a single processor Pentium IV PC and prove that the generation of the mesh has an acceptable cost in the PFEM. Indeed considerable speed can be gained using parallel computing techniques.

5 Identification of Boundary Surfaces

One of the main tasks in the PFEM is the correct definition of the boundary domain. Boundary nodes are sometimes explicitly identified. In other cases, the total set of nodes is the only information available and the algorithm must recognize the boundary nodes (Fig. 2).

In our work we use an extended Delaunay partition for recognizing boundary nodes [4]. Considering that the nodes follow a variable $h(x)$ distribution, where $h(x)$ is typically the minimum distance between two nodes. *All nodes on an empty sphere with a radius greater than* αh, *are considered as boundary nodes*. In practice α is a parameter close to, but greater than one. Values of α ranging around 1.3 have been found to be optimal in all examples analyzed. This criterion is coincident with the Alpha Shape concept [19].

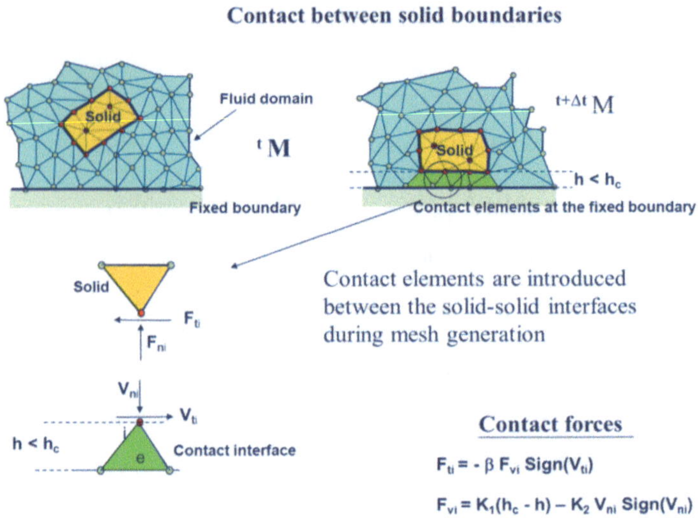

Fig. 2 Modelling of contact conditions at a solid-solid interface with the PFEM

Once a decision has been made concerning which nodes are on the boundaries, the boundary surface is defined by all the polyhedral surfaces (or polygons in 2D) having all their nodes on the boundary and belonging to just one polyhedron.

The method also allows one to identify isolated fluid particles outside the main fluid domain. These particles are treated as part of the external boundary where the pressure is fixed to the atmospheric value. We recall that each particle is a material point characterized by the density of the solid or fluid domain to which it belongs. The mass lost when a boundary element is eliminated due to departure of a node from the analysis domain is regained when the node falls down and a new boundary element is created by the Alpha Shape algorithm.

The boundary recognition method is useful for detecting contact conditions between the fluid domain and a boundary, as well as between different solids as detailed in the next section.

We emphasize that the key differences between the PFEM and the classical FEM are the remeshing technique and the identification of the domain boundary at each time step.

6 Treatment of Contact Conditions in the PFEM

6.1 Contact Between the Fluid and a Fixed Boundary

Known velocities at boundaries in the PFEM are prescribed in strong form to the boundary nodes. These nodes might belong to fixed external boundaries or to moving boundaries linked to the interacting solids. Contact between fluid particles and fixed boundaries is accounted for by the incompressibility condition which *naturally prevents fluid nodes to penetrate into the solid boundaries* [6, 9, 14].

6.2 Contact Between Solid-Solid Interfaces

The contact between two solid interfaces is treated by introducing a layer of *contact elements* between the two interacting solid interfaces. This layer is *automatically created during the mesh generation step* by prescribing a minimum distance (h_c) between two solid boundaries. If the distance exceeds the minimum value (h_c) then the generated elements are treated as fluid elements. Otherwise the elements are treated as contact elements where a relationship between the tangential and normal forces and the corresponding displacement is introduced (Fig. 2) [6, 10].

This algorithm allows us to identify and model complex frictional contact conditions between two or more interacting bodies moving in water in an extremely simple manner. The algorithm can also be used effectively to model frictional contact conditions between rigid or elastic solids in structural mechanics applications [16].

7 Modeling of Bed Erosion

Prediction of bed erosion and sediment transport in open channel flows are important tasks in river and environmental engineering. Bed erosion can lead to instabilities of the river basin slopes. It can also undermine the foundation of bridge piles thereby favouring structural failure. Modeling of bed erosion is also relevant for predicting the evolution of surface material dragged in earth dams in overspill situations. Bed erosion is one of the main causes of environmental damage in floods.

In recent works we have proposed an extension of the PFEM to model bed erosion [10, 14]. The erosion model is based on the frictional work at the bed surface originated by the shear stresses in the fluid. The resulting erosion model resembles Archard law typically used for modeling abrasive wear in surfaces under frictional contact conditions [28].

The algorithm for modeling bed erosion is the following:

1. Compute at the bed surface the resultant tangential stress induced by the fluid motion.
2. Compute the frictional work ${}^nW_f$ originated by the tangential stresses at the bed surface.
3. The onset of erosion at a bed point occurs when ${}^nW_f$ exceeds a critical threshold value W_c.
4. If ${}^nW_f > W_c$ at a bed node, then the node is detached from the bed region and it is allowed to move with the fluid. Also, the mass of the patch of bed elements surrounding the bed node is transferred to the new fluid node. This mass is subsequently transported with the fluid.

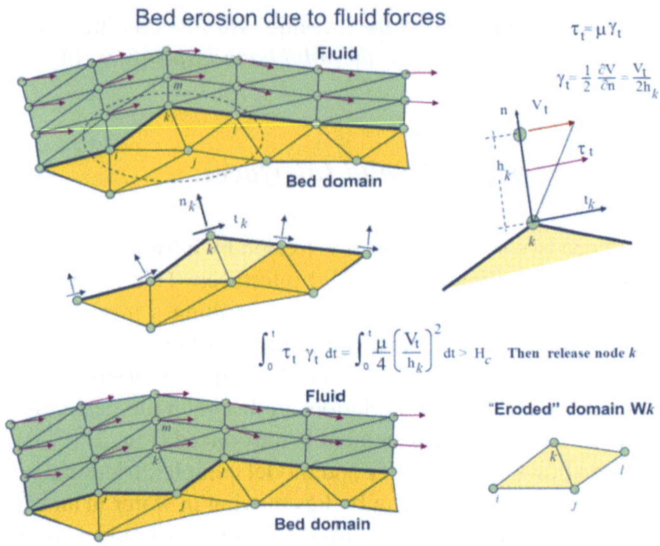

Fig. 3 Modeling of bed erosion with the PFEM by dragging of bed material

Figure 3 shows an schematic view of the bed erosion algorithm described.

Sediment deposition can be modeled by an inverse process. Hence, a suspended node adjacent to the bed surface with a velocity below a threshold value is attached to the bed surface.

Examples of the bed erosion algorithm for modeling excavation and rock cutting problems are presented in [16].

8 Examples

8.1 Dragging of Rocks by a Water Stream

Predicting the critical speed at which a rock will be dragged by a water stream is of great importance in many problems in hydraulic, harbour, civil and environmental engineering.

The PFEM has been successfully applied to the study of the motion of a 1 Tn quasi-spherical rock due to a water stream. The rock lays on a collection of rocks that are kept rigid.

Frictional conditions between the analyzed rock and the rest of the rocks have been assumed.

Figure 4a shows that a water stream of 1 m/s is not able to displace the individual rock. An increase of the water speed to 2 m/s induces the motion of the rock as shown in Fig. 4b.

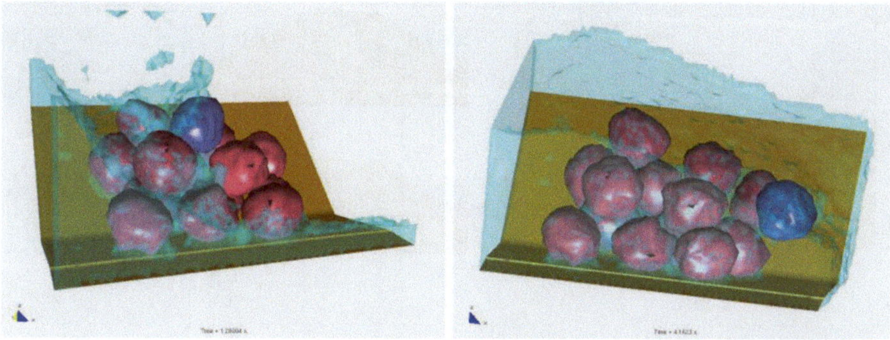

(a) Water speed of 1m/s. The individual rock can not be dragged by the stream

(b) Water speed of 2m/s. The individual rock is dragged by the stream

Fig. 4 Drag of a 1 Tn rock under a water stream at speeds of (**a**) 1 m/s and (**b**) 2 m/s

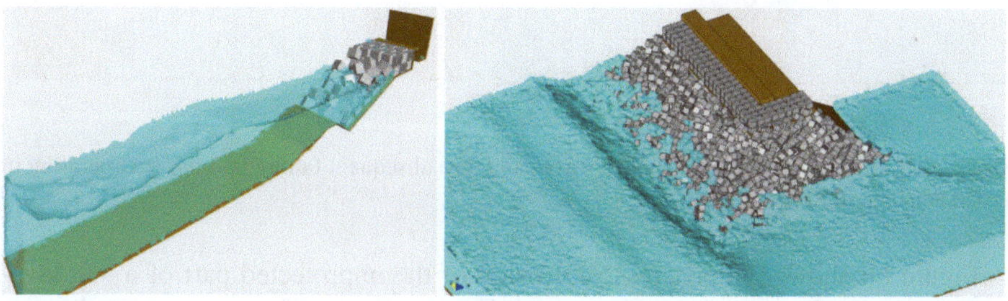

Fig. 5 Breaking waves on breakwater slopes containing reinforced concrete blocks

8.2 Impact of Sea Waves on Piers and Breakwaters

Figure 5 shows the analysis of the effect of breaking waves on two different sites of a breakwater containing reinforced concrete blocks (each one of $4 \times 4 \times 4$ mts). The figures correspond to the study of Langosteira harbour in A Coruña, Spain using PFEM.

8.3 Soil Erosion Problems

Figure 6a shows the capacity of the PFEM for modelling soil erosion, sediment transport and material deposition in a river bed. The soil particles are first detached from the bed surface under the action of the jet stream. Then they are transported by the flow and eventually fall down due to gravity forces into the bed surface at a downstream point.

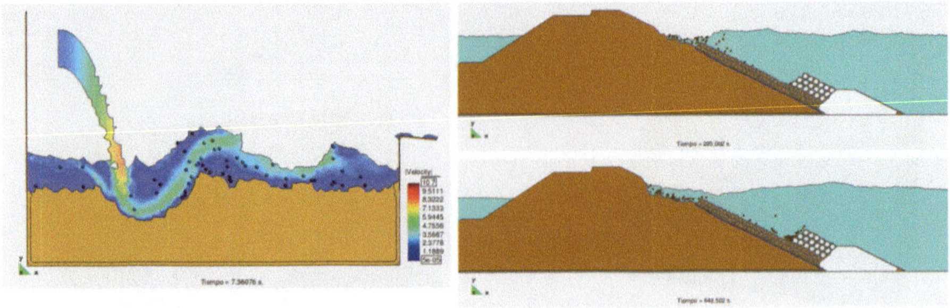

(a) (b)

Fig. 6 (**a**) Erosion, transport and deposition of soil particles at a river bed due to an impacting jet stream (**b**) Erosion of an unprotected shoulder of a breakwater due to sea waves

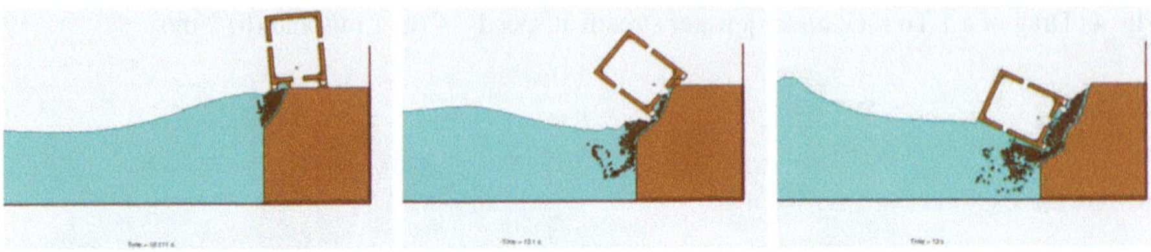

Fig. 7 Erosion of a soil mass due to sea waves and the subsequent falling into the sea operating in a road adjacent to the sea lorry

Figure 6b shows the progressive erosion of the unprotected part of a breakwater slope in the Langosteira harbour in A Coruña, Spain. The non-protected upper shoulder zone is progressively eroded as it is hit by the sea waves.

8.4 Falling of a Lorry into the Sea by Sea Wave Erosion of the Road Slope

Figure 7 shows a representative example of the progressive erosion of a soil mass adjacent to the shore due to sea waves and the subsequent falling into the sea of a 2D object representing the section of a lorry. The object has been modeled as a rigid solid.

This example and the previous ones, although still quite simple and schematic, show the possibilities of the PFEM for modeling complex FSSI problems involving soil erosion, free surface waves and rigid/deformable structures.

8.5 Impact of Waves over Transport Vehicles in Harbour

Figures 8 and 9 show two examples of the study of the impact of large waves on lorries operating in roads adjacent to a breakwater.

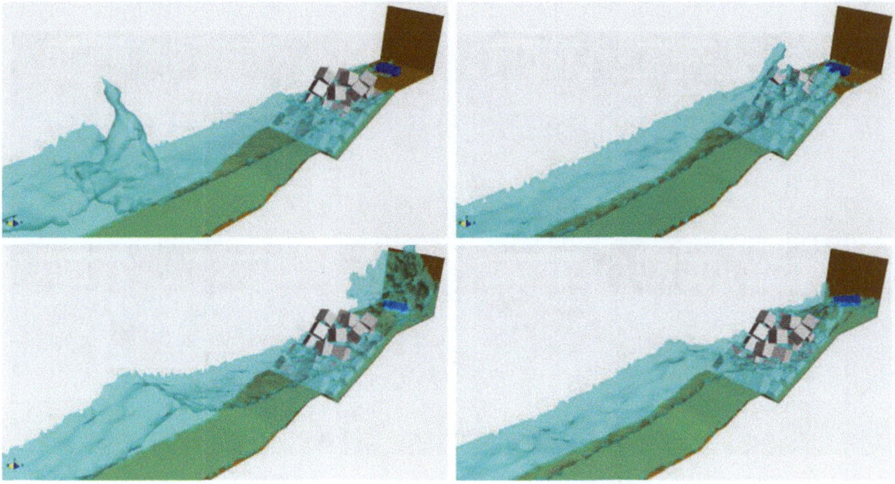

Fig. 8 20 mts amplitude wave hitting a 20 Tn lorry placed close to a breakwater slope

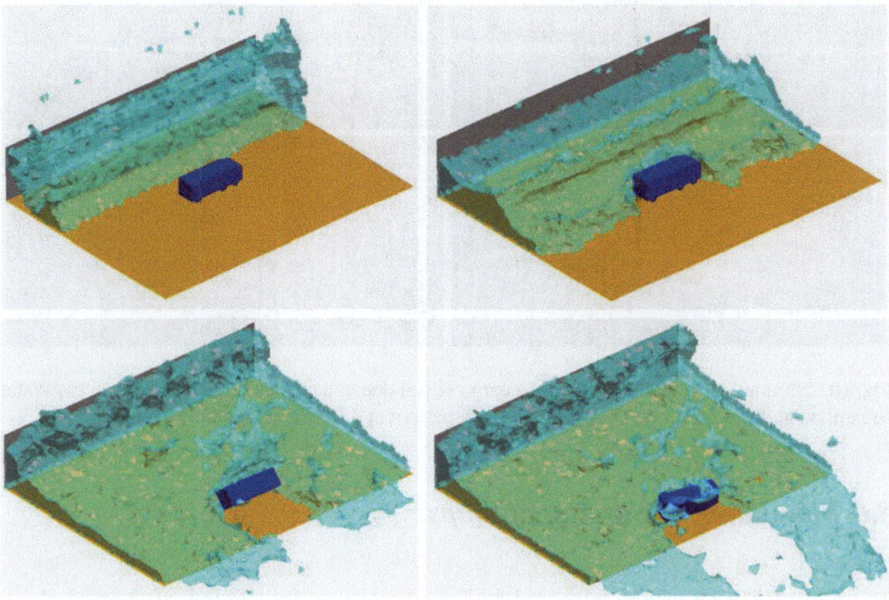

Fig. 9 Effect of an overtopping wave on a 20 Tn lorry in a road adjacent to a breakwater

Figure 8 shows the effect of a wave of 20 mts amplitude on a 20 Tns lorry adjacent to a breakwater shoulder.

Figure 9 displays the impact of an overtopping wave on the same lorry placed on a road behind a breakwater.

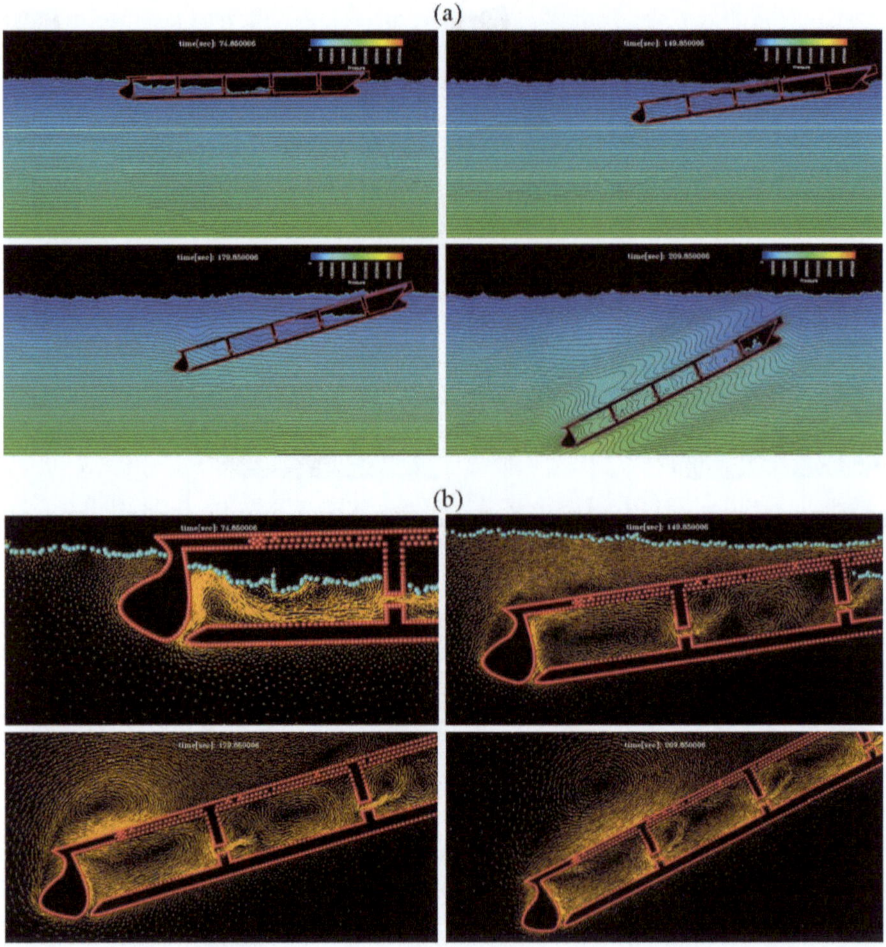

Fig. 10 2D simulation of the sinking of a cargo vessel due to a breach in the bow region. (**a**) Water streamline at different times. (**b**) Water velocity pattern at different times during sinking

8.6 Simulation of Sinking of Ships

The PFEM can be effectively applied for simulating the sinking of ships under a variety of scenarios.

Figure 10 shows images of the 2D simulation of the sinking of a cargo vessel induced by a breach in the bow region.

Figure 11 displays a 3D simulation of the sinking of a simple fisherman boat induced by a hole in the side of the hull.

These examples evidence the potential of PFEM for the study of the sinking of ships.

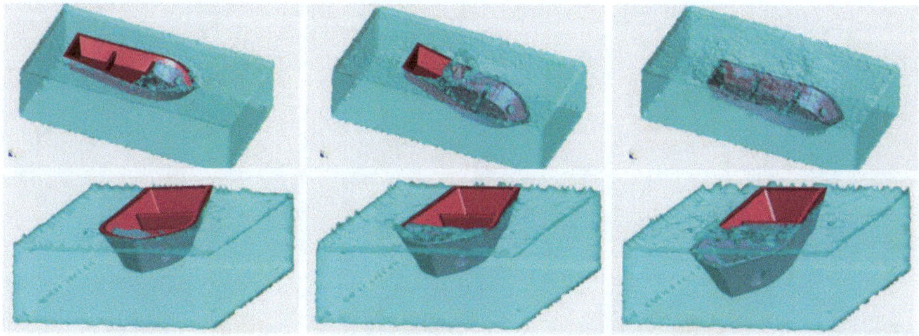

Fig. 11 3D simulation of the sinking of a boat induced by a hole in the side of the hull. Figures show different views of the water flow inside the boat during sinking

Fig. 12 3D simulation of a boat colliding with five ice blocks

8.7 Collision of Boat with Ice Blocks

Figures 12 and 13 show two examples of the application of PFEM to the study of the collision of a ship with ice blocks.

Figure 12 shows snapshots of the motion of a boat which collides with several ice blocks.

Figure 13 displays the interaction between a boat and two ice slabs that trap the boat in their motion.

We note that the boat and the ice blocks have been modelled as rigid bodies in these examples. Indeed, the deformation of the ship structure due to the ice-ship interaction forces can be accounted for in the analysis.

9 Conclusions

The particle finite element method (PFEM) is a promising numerical technique for solving fluid-soil-structure interaction (FSSI) problems in naval, marine and harbour engineering involving large motion of fluid and solid particles, surface waves, water splashing, frictional contact situations between fluid-solid and solid-solid interfaces and bed erosion, among other complex phenomena. The success of the PFEM lies in the accurate and efficient solution of the equations of an incompressible continuum using an updated Lagrangian formulation and a stabilized finite element method allowing the use of low order elements with equal order interpolation for all

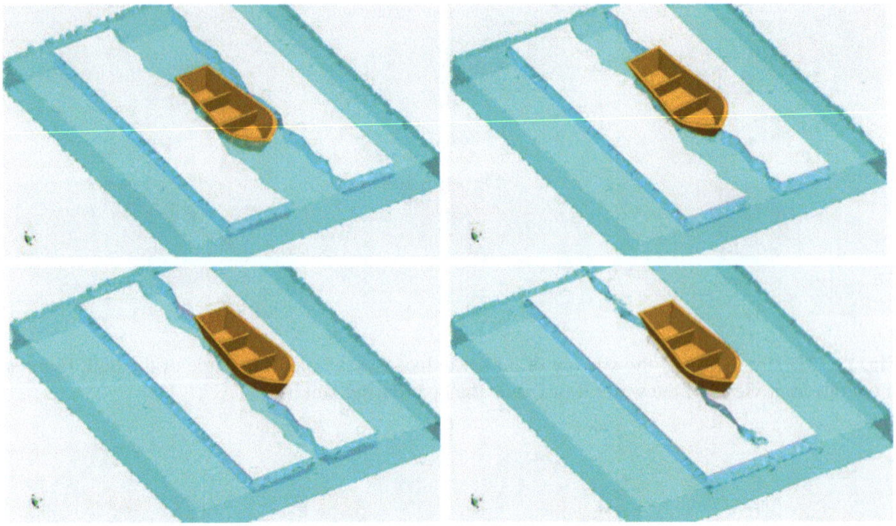

Fig. 13 Simulation of the interaction of two adjacent ice slabs and boat. The ice slabs move towards the boat that ends up out of the water and over the slabs

the variables. Other essential solution ingredients are the efficient regeneration of the finite element mesh, the identification of the boundary nodes using the Alpha-Shape technique and the simple algorithm to treat frictional contact conditions and erosion/wear at fluid-solid and solid-solid interfaces via mesh generation. The examples presented have shown the potential of the PFEM for solving a wide class of practical FSSI problems in naval, marine and harbour engineering.

Acknowledgements This research was partially supported by projects SAFECON and REALTIME of the European Research Council of the European Commission (EC).

References

1. Idelsohn SR, Oñate E, Del Pin F, Calvo N (2002) Lagrangian formulation: the only way to solve some free-surface fluid mechanics problems. In: Mang HA, Rammerstorfer FG, Eberhardsteiner J (eds) 5th World congress on comput mechanics, Vienna, Austria, July 7–12
2. Idelsohn SR, Oñate E, Del Pin F (2003) A Lagrangian meshless finite element method applied to fluid-structure interaction problems. Comput Struct 81:655–671
3. Idelsohn SR, Oñate E, Calvo N, Del Pin F (2003) The meshless finite element method. Int J Numer Methods Eng 58(6):893–912
4. Idelsohn SR, Calvo N, Oñate E (2003) Polyhedrization of an arbitrary point set. Comput Methods Appl Mech Eng 192(22–24):2649–2668
5. Idelsohn SR, Oñate E, Del Pin F (2004) The particle finite element method: a powerful tool to solve incompressible flows with free-surfaces and breaking waves. Int J Numer Methods Eng 61:964–989
6. Oñate E, Idelsohn SR, Del Pin F, Aubry R (2004) The particle finite element method. An overview. Int J Comput Methods 1(2):267–307

7. Aubry R, Idelsohn SR, Oñate E (2005) Particle finite element method in fluid mechanics including thermal convection-diffusion. Comput Struct 83(17–18):1459–1475
8. Idelsohn SR, Oñate E, Del Pin F, Calvo N (2006) Fluid-structure interaction using the particle finite element method. Comput Methods Appl Mech Eng 195:2100–2113
9. Oñate E, García J, Idelsohn SR, Del Pin F (2006) FIC formulations for finite element analysis of incompressible flows. Eulerian, ALE and Lagrangian approaches. Comput Methods Appl Mech Eng 195(23–24):3001–3037
10. Oñate E, Celigueta MA, Idelsohn SR (2006) Modeling bed erosion in free surface flows by the particle finite element method. Acta Geotech 1(4):237–252
11. Del Pin F, Idelsohn SR, Oñate E, Aubry R (2007) The ALE/Lagrangian particle finite element method: a new approach to computation of free-surface flows and fluid-object interactions. Comput Fluids 36:27–38
12. Idelsohn SR, Marti J, Limache A, Oñate E (2008) Unified Lagrangian formulation for elastic solids and incompressible fluids: application to fluid-structure interaction problems via the PFEM. Comput Methods Appl Mech Eng 197:1762–1776
13. Larese A, Rossi R, Oñate E, Idelsohn SR (2008) Validation of the particle finite element method (PFEM) for free surface flows. Eng Comput 25(4):385–425
14. Oñate E, Idelsohn SR, Celigueta MA, Rossi R (2008) Advances in the particle finite element method for the analysis of fluid-multibody interaction and bed erosion in free surface flows. Comput Methods Appl Mech Eng 197(19–20):1777–1800
15. Idelsohn SR, Mier-Torrecilla M, Oñate E (2009) Multi-fluid flows with the particle finite element method. Comput Methods Appl Mech Eng 198:2750–2767
16. Carbonell JM, Oñate E, Suárez B (2010) Modeling of ground excavation with the particle finite element method. J Eng Mech 136(4):455–463
17. Oñate E, Rossi R, Idelsohn SR, Butler K (2010) Melting and spread of polymers in fire with the particle finite element method. Int J Numer Methods Eng 81(8):1046–1072
18. Oñate E, Celigueta MA, Idelsohn SR, Salazar F, Suárez B (2011) Possibilities of the particle finite element method for fluid–soil–structure interaction problems. Comput Mech 48:307–318
19. Edelsbrunner H, Mucke EP (1999) Three dimensional alpha shapes. ACM Trans Graph 13:43–72
20. Oñate E (1998) Derivation of stabilized equations for advective-diffusive transport and fluid flow problems. Comput Methods Appl Mech Eng 151:233–267
21. Oñate E (2004) Possibilities of finite calculus in computational mechanics. Int J Numer Methods Eng 60(1):255–281
22. Oñate E, García J (2001) A finite element method for fluid-structure interaction with surface waves using a finite calculus formulation. Comput Methods Appl Mech Eng 191:635–660
23. Oñate E, Valls A, García J (2006) FIC/FEM formulation with matrix stabilizing terms for incompressible flows at low and high Reynold's numbers. Comput Mech 38(4–5):440–455
24. Zienkiewicz OC, Taylor RL (2005) The finite element method for solid and structural mechanics. Elsevier, Amsterdam
25. Donea J, Huerta A (2003) Finite element method for flow problems. Wiley, New York
26. Zienkiewicz OC, Taylor RL, Zhu JZ (2005) The finite element method. Its basis and fundamentals. Elsevier, Amsterdam
27. Zienkiewicz OC, Taylor RL, Nithiarasu P (2006) The finite element method for fluid dynamics. Elsevier, Amsterdam
28. Archard JF (1953) Contact and rubbing of flat surfaces. J Appl Phys 24(8):981–988

Part III
Isogeometric Methods for Marine Engineering

Part II

Shape-Newton Method for Isogeometric Discretizations of Free-Boundary Problems

K.G. van der Zee, G.J. van Zwieten, C.V. Verhoosel, and E.H. van Brummelen

Abstract We derive Newton-type solution algorithms for a Bernoulli-type free-boundary problem at the continuous level. The Newton schemes are obtained by applying Hadamard shape derivatives to a suitable weak formulation of the free-boundary problem. At each Newton iteration, an updated free boundary position is obtained by solving a boundary-value problem at the current approximate domain. Since the boundary-value problem has a curvature-dependent boundary condition, an ideal discretization is provided by isogeometric analysis. Several numerical examples demonstrate the apparent quadratic convergence of the Newton schemes on isogeometric-analysis discretizations with C^1-continuous discrete free boundaries.

Keywords Newton-type methods · Bernoulli free-boundary problem · Shape derivative · Isogeometric analysis · Smooth discrete boundaries · Shape-linearized free-boundary problem

1 Introduction

In this contribution we consider the classical problem of finding efficient solution algorithms for (steady) free-boundary problems. These algorithms find their application in the simulation of, for example, steady free-surface flow problems in ship hydrodynamics, hydraulics, and free-jet flows [9, 27, 28, 34]. An important model problem for developing algorithms is the Bernoulli free-boundary problem [8, 13]

K.G. van der Zee (✉) · G.J. van Zwieten · C.V. Verhoosel · E.H. van Brummelen
Multiscale Engineering Fluid Dynamics, Eindhoven University of Technology, P.O. Box 513, 5600 MB Eindhoven, The Netherlands
e-mail: k.g.v.d.zee@tue.nl

G.J. van Zwieten
e-mail: g.j.v.zwieten@tue.nl

C.V. Verhoosel
e-mail: c.v.verhoosel@tue.nl

E.H. van Brummelen
e-mail: e.h.v.brummelen@tue.nl

L. Eça et al. (eds.), *MARINE 2011, IV International Conference on Computational Methods in Marine Engineering*, Computational Methods in Applied Sciences 29,
DOI 10.1007/978-94-007-6143-8_5, © Springer Science+Business Media Dordrecht 2013

(or Alt–Caffarelli free-boundary problem [1]). This model problem is also the subject of the current work.

Solution algorithms for the Bernoulli free-boundary problem come in three flavors: trial free-boundary methods, shape-optimization methods and direct Newton-type methods. Trial free-boundary methods alternate between solving a boundary-value problem on a fixed approximate domain, and updating the free boundary using the residual of the free-boundary condition that was not imposed in the boundary-value problem, cf. [3, 13, 18, 20, 25]. These are linear convergence methods, with, generally, parameter-dependent stability and convergence behavior. Shape-optimization methods improve the convergence rate, but are difficult to extend to nonvariational problems since they make use of the particular shape-optimization formulation of the Bernoulli free-boundary problem. For details on these algorithms see, e.g., [12, 14, 15, 22, 25, 26, 30].

Direct Newton-type methods aim at linearizing the system of equations and setting up a Newton–Raphson iteration scheme. Linearization can be done by mapping back to a reference domain [21, 31], or in the current configuration. Methods obtained by linearizing in the current configuration shall be referred to as shape-Newton methods. Their derivation uses either formal asymptotics or rigorous techniques from shape-differential calculus [11, 24]. The resulting schemes are particularly elegant because the effect of domain-perturbations appear only in boundary conditions, and thus have a local effect.[1] Schemes derived based on formal asymptotics may be found in [4, 13, 29] while those based on shape calculus are considered by Karkkainen and Tiihonen [17, 18].

In the current work, we reconsider the derivation of Newton-type schemes using shape-calculus techniques. Our derivation is different from Karkkainen and Tiihonen, in that we do not explicitly identify two sets of equations, but employ a weak form of the complete free-boundary problem.[2] Furthermore, we consider a more general setting with nonconstant Dirichlet data. The resulting scheme yields a boundary-value problem with a curvature-dependent boundary condition (as obtained by Karkkainen and Tiihonen).

Direct discretizations based on the so-derived shape-Newton method require C^1-continuous free-boundaries for a finite curvature. Ordinary interface-fitted finite-element based discretizations have only C^0-continuous free-boundaries, and therefore require a cumbersome implementation with, for example, smoothing or reconstruction [13, 18]. To be able to directly discretize the shape-Newton formulation, we therefore employ isogeometric-analysis discretizations. Isogeometric analysis [6, 16] is a recent extension of finite-element analysis that employs smooth basis functions that are typically used for geometry (such as non-uniform rational B-splines, T-splines, etc.). The relative ease with which such basis functions can be generated with arbitrary smoothness, i.e., up to C^{p-1}-continuity for pth-order

[1] For domain-map linearization (linearization in the reference domain), domain-perturbations have a nonlocal effect, and this nonlocality depends on the particular domain map chosen [31, 32]).

[2] We note that this derivation has been employed previously to obtain the linearized-adjoint operator; see [32].

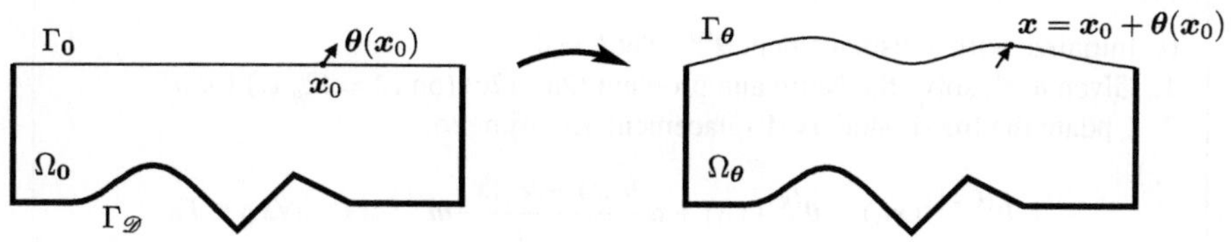

Fig. 1 Illustration of the parametrization of the free boundary $\Gamma_{\boldsymbol{\theta}}$ by its displacement $\boldsymbol{\theta}$ with respect to the reference configuration $\Gamma_{\mathbf{0}}$

splines, is an enormous advantage in the current application. Other applications where isogeometric analysis is advantageous for its smooth geometries are, e.g., shape optimization [5], linearized adjoints for free-boundary problems [33], and direct discretization of shells [19].

The contents of the paper are arranged as follows. In Sect. 2 we introduce the free-boundary problem, as well as a suitable very weak formulation. In Sect. 3 we linearize the free-boundary problem by applying Hadamard shape derivatives on the very weak form. The linearization yields Newton-like schemes which are presented in Sect. 4. Numerical examples with C^1-continuous isogeometric analysis discretizations are presented in Sect. 5. Conclusions are drawn in Sect. 6.

2 Free-Boundary Problem

2.1 Strong Form

The Bernoulli-type free-boundary problem seeks an a priori unknown domain $\Omega_{\boldsymbol{\theta}} \subset \mathbb{R}^N$ (with N the number of space dimensions), for which the boundary $\partial\Omega_{\boldsymbol{\theta}}$ contains a fixed part $\Gamma_{\mathscr{D}}$ and free part $\Gamma_{\boldsymbol{\theta}}$, and a corresponding scalar function $u : \Omega_{\boldsymbol{\theta}} \to \mathbb{R}$ such that

$$-\Delta u = f \quad \text{in } \Omega_{\boldsymbol{\theta}}, \tag{1a}$$

$$\partial_n u = g \quad \text{on } \Gamma_{\boldsymbol{\theta}}, \tag{1b}$$

$$u = h \quad \text{on } \partial\Omega_{\boldsymbol{\theta}} = \Gamma_{\boldsymbol{\theta}} \cup \Gamma_{\mathscr{D}}, \tag{1c}$$

where $\partial_n(\cdot) := \partial(\cdot)/\partial n \equiv \boldsymbol{n} \cdot \nabla(\cdot)$ is the normal derivative. The data f, $g \geq g_0 > 0$ (g_0 is constant) and h are defined on $\mathbb{R}^N$, assumed to be sufficiently smooth, and assumed to be such that there is a nontrivial solution pair $(\Gamma_{\boldsymbol{\theta}}, u)$.

In the above, $\Omega_{\boldsymbol{\theta}}$ and $\Gamma_{\boldsymbol{\theta}}$ have been parametrized by a vectorfield $\boldsymbol{\theta} : \Gamma_{\mathbf{0}} \to \mathbb{R}^N$ which defines the displacement of the current free-boundary $\Gamma_{\boldsymbol{\theta}}$ with respect to the reference free-boundary $\Gamma_{\mathbf{0}}$, i.e.,

$$\Gamma_{\boldsymbol{\theta}} := \{\boldsymbol{x} \in \mathbb{R}^N \mid \boldsymbol{x} = \boldsymbol{x}_0 + \boldsymbol{\theta}(\boldsymbol{x}_0), \forall \boldsymbol{x}_0 \in \Gamma_{\mathbf{0}}\}.$$

An important advantage of introducing this parametrization is that our problem (1a)–(1c) is now posed in terms of the pair $(\boldsymbol{\theta}, u)$. See Fig. 1 for an illustration.

0. Initialize with a displacement $\boldsymbol{\theta}^{(0)}$; Set $k=0$.
1. Given $\boldsymbol{\theta}^{(k)}$, solve the Neumann problem (2a)–(2c) (on $\hat{\Omega} = \Omega_{\boldsymbol{\theta}^{(k)}}$) for u^*.
2. Update the free-boundary displacement according to:

$$\boldsymbol{\theta}^{(k+1)}(\boldsymbol{x}_0) = \boldsymbol{\theta}^{(k)}(\boldsymbol{x}_0) + \alpha \frac{h(\boldsymbol{x}) - u^*(\boldsymbol{x})}{g(\boldsymbol{x})} \boldsymbol{m}^{(k)}(\boldsymbol{x}), \quad \forall \boldsymbol{x}_0 \in \Gamma_0$$

$$\big(\boldsymbol{x} = \boldsymbol{x}_0 + \boldsymbol{\theta}^{(k)}(\boldsymbol{x}_0)\big).$$

3. Set $k \leftarrow k+1$ and repeat from step 1 until convergence.

Fig. 2 The explicit Neumann scheme. In the above, the parameter α is a chosen relaxation parameter, and the vectorfield $\boldsymbol{m}^{(k)} : \Gamma_{\boldsymbol{\theta}^{(k)}} \to \mathbb{R}^N$ is a chosen vectorfield which determines the direction of the update, e.g., $\boldsymbol{m}^{(k)} = \boldsymbol{n}$

Remark 1 (Explicit Neumann scheme) The (continuous) explicit Neumann scheme [13] solves (1a)–(1c) iteratively, by solving the following Neumann problem on a fixed domain $\hat{\Omega}$ with free-boundary $\hat{\Gamma}$:

$$-\Delta u^* = f \quad \text{in } \hat{\Omega}, \tag{2a}$$

$$\partial_n u^* = g \quad \text{on } \hat{\Gamma}, \tag{2b}$$

$$u^* = h \quad \text{on } \Gamma_{\mathcal{D}}, \tag{2c}$$

and subsequently updating the free boundary using the residual of the ignored boundary condition, $h - u^*$ on $\hat{\Gamma}$. An outline of the explicit Neumann scheme is given in Fig. 2.

2.2 Very Weak Form

To be able to linearize the free-boundary problem (1a)–(1c) using shape calculus, we need to introduce a weak formulation of (1a)–(1c) that incorporates all boundary conditions at the free boundary. This can be obtained by multiplying (1a) with a suitable testfunction $v \in V := \{v \in C^\infty(\mathbb{R}^N) \mid v = 0 \text{ on } \Gamma_{\mathcal{D}}\}$, integrating over $\Omega_{\boldsymbol{\theta}}$, and performing an integration-by-parts twice, thereby substituting (1b) and (1c), yielding:

$$\mathcal{N}\big((\boldsymbol{\theta}, u); v\big) = 0 \quad \forall v \in V, \tag{3}$$

where we have introduced the semilinear (nonlinear with respect to its first argument) form $\mathcal{N}((\cdot,\cdot);\cdot)$:

$$\mathcal{N}\big((\boldsymbol{\theta}, u); v\big) := \mathcal{D}(\boldsymbol{\theta}; u, v) - \mathcal{F}(\boldsymbol{\theta}; v) - \mathcal{G}(\boldsymbol{\theta}; v), \tag{4}$$

which is defined in terms of the three semilinear forms:

$$\begin{aligned}
\mathcal{D}\big((\boldsymbol{\theta}, u); v\big) &:= \int_{\Omega_\theta} -u \Delta v \,\mathrm{d}\Omega + \int_{\partial\Omega_\theta} h \partial_n v \,\mathrm{d}\Gamma \\
&= \int_{\Omega_\theta} \big(-(u-h)\Delta v + \nabla h \cdot \nabla v\big) \,\mathrm{d}\Omega, \\
\mathcal{F}(\boldsymbol{\theta}; v) &:= \int_{\Omega_\theta} f v \,\mathrm{d}\Omega, \\
\mathcal{G}(\boldsymbol{\theta}; v) &:= \int_{\Gamma_\theta} g v \,\mathrm{d}\Gamma.
\end{aligned}$$

Remark 2 (Equivalence) Note that a solution of (1a)–(1c) satisfies (3), while standard variational arguments show that if $(\boldsymbol{\theta}, u)$ is a sufficiently smooth solution of (3), then it satisfies (1a)–(1c).

3 Linearizing Free-Boundary Problems

In this section, we consider the linearization of $\mathcal{N}((\boldsymbol{\theta}, u); v)$ at an arbitrary approximation $(\hat{\boldsymbol{\theta}}, \hat{u})$. We assume the approximation-pair to be compatible in the sense that $\hat{u}$ is any approximation that lives on the approximate domain induced by $\hat{\boldsymbol{\theta}}$, i.e., on $\Omega_{\hat{\boldsymbol{\theta}}}$. We furthermore assume that $\hat{u}$ satisfies its boundary condition at the fixed boundary $\Gamma_{\mathcal{D}}$. In the sequel, for notational convenience, let us denote the corresponding approximate domain and free boundary by $\hat{\Omega} \equiv \Omega_{\hat{\boldsymbol{\theta}}}$ and $\hat{\Gamma} \equiv \Gamma_{\hat{\boldsymbol{\theta}}}$, respectively.

Remark 3 (Regularity of approximations) It is important to state the smoothness requirements on $\hat{u}$ and $\hat{\theta}$ that allow the ensuing linearizations to be performed. Natural requirements are that $\hat{u} \in H^1(\hat{\Omega})$ and $\hat{\theta} \in C_0^{1,1}(\Gamma_0)$ such that $\hat{\Omega}$ is of class $C^{1,1}$. Note that the latter requirement implies that the curvature of $\hat{\Omega}$ is bounded [11].

3.1 Linearization with Respect to u

Since the dependence of $\mathcal{N}((\boldsymbol{\theta}, u); v)$ on u is affine, its functional (or Gâteaux) derivative at $\hat{u}$ in the direction δu (for fixed $v \in V$) is simply given by:

$$\begin{aligned}
\big\langle \partial_u \mathcal{N}\big((\hat{\boldsymbol{\theta}}, \hat{u}); v\big), \delta u \big\rangle &:= \lim_{t \to 0} \frac{\mathcal{N}((\hat{\boldsymbol{\theta}}, \hat{u} + \delta u); v) - \mathcal{N}((\hat{\boldsymbol{\theta}}, \hat{u}); v)}{t} \\
&= \int_{\hat{\Omega}} -\delta u \Delta v \,\mathrm{d}\Omega. \qquad (5)
\end{aligned}$$

3.2 Hadamard Shape Derivatives

To linearize $\mathcal{N}((\boldsymbol{\theta}, u); v)$ with respect to $\boldsymbol{\theta}$, we require the Gâteaux derivative of functionals that depend on $\boldsymbol{\theta}$. Inspection of $\mathcal{N}((\boldsymbol{\theta}, u); v)$ reveals that the dependence is either via a domain integral, $\int_{\Omega_\theta}(\cdots)\,\mathrm{d}\Omega$, or a boundary integral $\int_{\Gamma_\theta}(\cdots)\,\mathrm{d}\Omega$. Such derivatives can be computed using techniques of shape calculus and yield so-called Hadamard formulas.

Consider a functional $\boldsymbol{\theta} \mapsto \mathcal{J}(\boldsymbol{\theta})$. The Gâteaux derivative of $\mathcal{J}(\cdot)$ at an approximation $\hat{\boldsymbol{\theta}}$ in the direction $\boldsymbol{\delta\theta}$ is defined as:

$$\langle \partial_\theta \mathcal{J}(\hat{\boldsymbol{\theta}}), \boldsymbol{\delta\theta} \rangle := \lim_{t \to 0} \frac{\mathcal{J}(\hat{\boldsymbol{\theta}} + t\boldsymbol{\delta\theta}) - \mathcal{J}(\hat{\boldsymbol{\theta}})}{t}.$$

In particular, for a domain and boundary integral we have the Hadamard formulas [11, 24]:

$$\mathcal{J}(\boldsymbol{\theta}) = \int_{\Omega_\theta} \phi \,\mathrm{d}\Omega \quad \Rightarrow \quad \langle \partial_\theta \mathcal{J}(\hat{\boldsymbol{\theta}}), \boldsymbol{\delta\theta} \rangle = \int_{\hat{\Gamma}} \phi \boldsymbol{\delta\theta} \cdot \boldsymbol{n} \,\mathrm{d}\Gamma,$$

$$\mathcal{J}(\boldsymbol{\theta}) = \int_{\Gamma_\theta} \varphi \,\mathrm{d}\Gamma \quad \Rightarrow \quad \langle \partial_\theta \mathcal{J}(\hat{\boldsymbol{\theta}}), \boldsymbol{\delta\theta} \rangle = \int_{\hat{\Gamma}} (\partial_n \varphi + \kappa \varphi) \boldsymbol{\delta\theta} \cdot \boldsymbol{n} \,\mathrm{d}\Gamma,$$

respectively, where $\kappa \equiv \kappa(\hat{\Gamma})$ is the additive curvature (sum of $(N-1)$ curvatures) of $\hat{\Gamma}$.

Remark 4 In the above formulas $\boldsymbol{\delta\theta}$ is to be evaluated in the reference domain $\Gamma_{\mathbf{0}}$, i.e. $\boldsymbol{\delta\theta} = \boldsymbol{\delta\theta}(\boldsymbol{x}_0)$ with $\boldsymbol{x}_0 + \hat{\boldsymbol{\theta}}(\boldsymbol{x}_0) = \boldsymbol{x} \in \hat{\Gamma}$.

3.3 Shape Linearization

Applying the Hadamard formulas to the semilinear forms in (4) yields

$$\langle \partial_\theta \mathcal{D}((\hat{\boldsymbol{\theta}}, \hat{u}); v), \boldsymbol{\delta\theta} \rangle = \int_{\hat{\Gamma}} \big(-(\hat{u} - h)\Delta v + \nabla h \cdot \nabla v\big) \boldsymbol{\delta\theta} \cdot \boldsymbol{n} \,\mathrm{d}\Gamma, \tag{6a}$$

$$\langle \partial_\theta \mathcal{F}(\hat{\boldsymbol{\theta}}), \boldsymbol{\delta\theta} \rangle = \int_{\hat{\Gamma}} f v \boldsymbol{\delta\theta} \cdot \boldsymbol{n} \,\mathrm{d}\Gamma, \tag{6b}$$

$$\langle \partial_\theta \mathcal{G}(\hat{\boldsymbol{\theta}}), \boldsymbol{\delta\theta} \rangle = \int_{\hat{\Gamma}} \big((\partial_n g + \kappa g) v + g \partial_n v\big) \boldsymbol{\delta\theta} \cdot \boldsymbol{n} \,\mathrm{d}\Gamma. \tag{6c}$$

It is important to realize that (6a) can be simplified using properties of the exact solution; cf. [17, 18]. Indeed, since $\hat{u}$ is expected to be close to the exact u, and u equals h on the exact free-boundary Γ, we can neglect the $(\hat{u} - h)$-term provided the errors $u - \hat{u}$ and $\boldsymbol{\theta} - \hat{\boldsymbol{\theta}}$ are small. We thus have the approximation:

$$\begin{aligned}\big\langle \partial_\theta \mathcal{D}\big((\hat{\boldsymbol{\theta}}, \hat{u}); v\big), \boldsymbol{\delta\theta}\big\rangle &\approx \int_{\hat{\Gamma}} \nabla h \cdot \nabla v \boldsymbol{\delta\theta} \cdot \boldsymbol{n} \, \mathrm{d}\Gamma \\ &= \int_{\hat{\Gamma}} (\nabla_\Gamma h \cdot \nabla_\Gamma v + \partial_n h \partial_n v) \boldsymbol{\delta\theta} \cdot \boldsymbol{n} \, \mathrm{d}\Gamma \\ &= \int_{\hat{\Gamma}} \big(-\mathrm{div}_\Gamma(\boldsymbol{\delta\theta} \cdot \boldsymbol{n} \nabla_\Gamma h) v + \partial_n h \partial_n v \boldsymbol{\delta\theta} \cdot \boldsymbol{n}\big) \, \mathrm{d}\Gamma, \qquad \text{(6a}\star\text{)}\end{aligned}$$

where in the last two steps we used the tangential gradient splitting, $\nabla_\Gamma(\cdot) = \nabla(\cdot) - \partial_n(\cdot)\boldsymbol{n}$, and the tangential Green's identity (tangential integration-by-parts); see [11, Sect. 8.5]. We furthermore introduced the tangential divergence: $\mathrm{div}_\Gamma(\cdot) = \mathrm{div}(\cdot) - \partial_n(\cdot) \cdot \boldsymbol{n}$.

Combining (6a⋆), (6b) and (6c) yields the approximate shape linearization:

$$\begin{aligned}&\big\langle \partial_\theta \mathcal{N}\big((\hat{\boldsymbol{\theta}}, \hat{u}); v\big), \boldsymbol{\delta\theta}\big\rangle \\ &\quad\approx \int_{\hat{\Gamma}} (\partial_n h - g) \partial_n v \boldsymbol{\delta\theta} \cdot \boldsymbol{n} \, \mathrm{d}\Gamma \\ &\quad\quad - \int_{\hat{\Gamma}} \big((f + \partial_n g + \kappa g) \boldsymbol{\delta\theta} \cdot \boldsymbol{n} + \mathrm{div}_\Gamma(\boldsymbol{\delta\theta} \cdot \boldsymbol{n} \nabla_\Gamma h)\big) v \, \mathrm{d}\Gamma. \qquad (7)\end{aligned}$$

4 Newton-Like Schemes

An exact Newton method applied to (3) would set up the following problem for the update $(\boldsymbol{\delta\theta}, \delta u)$:

$$\big\langle \partial_{(\boldsymbol{\theta},u)} \mathcal{N}\big((\hat{\boldsymbol{\theta}}, \hat{u}); v\big), (\boldsymbol{\delta\theta}, \delta u)\big\rangle = -\mathcal{N}\big((\hat{\theta}, \hat{u}); v\big) \quad \forall v \in V,$$

and subsequently compute the total (updated) approximations:

$$\boldsymbol{\theta}^* = \hat{\boldsymbol{\theta}} + \boldsymbol{\delta\theta} \quad \text{and} \quad u^* = \hat{u} + \delta u$$

(we shall later introduce a mixed total/update form; see Remark 5).

Introducing (5) and our approximation (7) we actually obtain the following Newton-like scheme:

$$\begin{aligned}&\int_{\hat{\Omega}} -\delta u \Delta v \, \mathrm{d}\Omega + \int_{\hat{\Gamma}} (\partial_n h - g) \partial_n v \boldsymbol{\delta\theta} \cdot \boldsymbol{n} \, \mathrm{d}\Gamma \\ &\quad - \int_{\hat{\Gamma}} \big((f + \partial_n g + \kappa g) \boldsymbol{\delta\theta} \cdot \boldsymbol{n} + \mathrm{div}_\Gamma(\boldsymbol{\delta\theta} \cdot \boldsymbol{n} \nabla_\Gamma h)\big) v \, \mathrm{d}\Gamma \\ &= -\mathcal{N}\big((\hat{\boldsymbol{\theta}}, \hat{u}); v\big) \quad \forall v \in V. \qquad (8)\end{aligned}$$

Of course, close to the exact solution, the above scheme is almost an exact Newton scheme. Indeed, at the exact solution the approximation in (7) is exact.

0. Initialize with a displacement $\boldsymbol{\theta}^{(0)}$; Set $k=0$.
1. Given $\boldsymbol{\theta}^{(k)}$, solve the coupled problem (9a)–(9d) (on $\hat{\Omega} = \Omega_{\boldsymbol{\theta}^{(k)}}$) for $(\boldsymbol{\delta\theta} \cdot \boldsymbol{n}, u^*)$.
2. Update the free-boundary displacement according to:

$$\boldsymbol{\theta}^{(k+1)} = \boldsymbol{\theta}^{(k)} + (\boldsymbol{\delta\theta} \cdot \boldsymbol{n})\boldsymbol{m}^{(k)}.$$

3. Set $k \leftarrow k+1$ and repeat from step 1. until convergence.

Fig. 3 The coupled shape-Newton scheme. In the above, the vectorfield $\boldsymbol{m}^{(k)} : \Gamma_{\boldsymbol{\theta}^{(k)}} \to \mathbb{R}^N$ is a chosen vectorfield such that $\boldsymbol{m}^{(k)} \cdot \boldsymbol{n} = 1$ on $\Gamma_{\boldsymbol{\theta}^{(k)}}$

4.1 Coupled Scheme

Using variational arguments, we can extract the boundary-value problem for $(\delta u, \boldsymbol{\delta\theta})$ implied by the Newton-like scheme (8):

$$\begin{aligned}
-\Delta \delta u &= f + \Delta \hat{u} && \text{in } \hat{\Omega},\\
\delta u + (g - \partial_n h)\boldsymbol{\delta\theta} \cdot \boldsymbol{n} + \operatorname{div}_\Gamma(\boldsymbol{\delta\theta} \cdot \boldsymbol{n} \nabla_\Gamma h) &= h - \hat{u} && \text{on } \hat{\Gamma},\\
\partial_n(\delta u) - (f + \partial_n g + \kappa g)\boldsymbol{\delta\theta} \cdot \boldsymbol{n} &= g - \partial_n \hat{u} && \text{on } \hat{\Gamma},\\
\delta u &= 0 && \text{on } \Gamma_{\mathscr{D}}.
\end{aligned}$$

Note that the above system is a coupled linear system, and that the right-hand sides are all residual quantities.

It is natural to write the above system in mixed total/update form, that is, in terms of $u^* = \hat{u} + \delta u$ and $\boldsymbol{\delta\theta} \cdot \boldsymbol{n}$. We shall refer to the resulting system, as the (coupled) shape-Newton scheme:

$$-\Delta u^* = f \qquad \text{in } \hat{\Omega}, \tag{9a}$$

$$u^* + (g - \partial_n h)\boldsymbol{\delta\theta} \cdot \boldsymbol{n} - \operatorname{div}_\Gamma(\boldsymbol{\delta\theta} \cdot \boldsymbol{n} \nabla_\Gamma h) = h \qquad \text{on } \hat{\Gamma}, \tag{9b}$$

$$\partial_n u^* - (f + \partial_n g + \kappa g)\boldsymbol{\delta\theta} \cdot \boldsymbol{n} = g - \partial_n h \qquad \text{on } \hat{\Gamma}, \tag{9c}$$

$$u^* = h \qquad \text{on } \Gamma_{\mathscr{D}}. \tag{9d}$$

An outline of the algorithm implied by (9a)–(9d) is given in Fig. 3.

Remark 5 (Mixed total/update form) The importance of writing the scheme in mixed total/update form (9a)–(9d) should not be underestimated. During iterations, it is not required to update functions that live on different domains. Indeed, the total u^* is directly computed on the last domain.

Remark 6 (Invariance) It is clear from (9a)–(9d) that the shape-Newton scheme is, in essence, independent of the approximation for $\hat{u}$. It is only driven by the approximation for the free-boundary, $\hat{\Gamma}$.

Remark 7 (Exact Newton scheme) Because of the above-mentioned invariance, one may interpret the shape-Newton scheme as being obtained at the approximation $(\hat{\Gamma}, \hat{u})$ with $\hat{u}$ such that $\hat{u} = h$ at $\hat{\Gamma}$. The approximation in (7) is then exact, so that we have an exact Newton scheme.

4.2 Decoupled Scheme

The particular case of constant Dirichlet data at the free boundary is often considered in literature. Assuming that $h = h_0$ (= constant) at the free boundary, $\nabla_\Gamma h = \mathbf{0}$ and $\partial_n h = 0$, so that we can simplify (9a)–(9d) by eliminating $\boldsymbol{\delta\theta} \cdot \boldsymbol{n}$ using (9b). We shall refer to the resulting system as the decoupled shape-Newton scheme:

$$-\Delta u^* = f \qquad \text{in } \hat{\Omega}, \tag{10a}$$

$$\partial_n u^* + \big((f + \partial_n g)/g + \kappa\big)\big(u^* - h_0\big) = g \qquad \text{on } \hat{\Gamma}, \tag{10b}$$

$$u^* = h \qquad \text{on } \Gamma_{\mathscr{D}}, \tag{10c}$$

$$\boldsymbol{\delta\theta} \cdot \boldsymbol{n} = \big(h_0 - u^*\big)/g \qquad \text{on } \hat{\Gamma}. \tag{10d}$$

An outline of the algorithm implied by (10a)–(10d) is given in Fig. 4.

Remark 8 (Robin boundary condition) The boundary condition for u^* in (10b) is the well-known curvature-dependent Robin boundary condition; see, e.g., [4, 13, 17, 32].

5 Numerical Experiments

We will consider two 2D free-surface flow problems, both of constant Dirichlet type, i.e. $h = h_0$, such that we can employ the decoupled scheme (10a)–(10d) outlined in Fig. 4.

5.1 Isogeometric-Analysis Based Discretizations

We consider Galerkin approximations for (10a)–(10d): Introducing a discrete test function v and integrating by parts leads to the weak formulation:

$$\text{Find } u^* \in V(\Omega_{\hat{\theta}}): \quad \int_{\Omega_{\hat{\theta}}} \nabla u^* \cdot \nabla v \,\mathrm{d}\Omega + \int_{\Gamma_{\hat{\theta}}} k_{\hat{\theta}} u^* v \,\mathrm{d}\Gamma$$

$$= \int_{\Omega_{\hat{\theta}}} f v \,\mathrm{d}\Omega + \int_{\Gamma_{\hat{\theta}}} (k_{\hat{\theta}} h_0 + g) v \,\mathrm{d}\Gamma \quad \forall v \in V(\Omega_{\hat{\theta}}), \tag{11}$$

0. Initialize with a displacement $\boldsymbol{\theta}^{(0)}$; Set $k=0$.
1. Given $\boldsymbol{\theta}^{(k)}$, solve the Robin problem (10a)–(10c) (on $\hat{\Omega} = \Omega_{\boldsymbol{\theta}^{(k)}}$) for u^*.
2. Update the free-boundary displacement according to:

$$\boldsymbol{\theta}^{(k+1)} = \boldsymbol{\theta}^{(k)} + \frac{h_0 - u^*}{g}\boldsymbol{m}^{(k)}.$$

3. Set $k \leftarrow k+1$ and repeat from step 1. until convergence.

Fig. 4 The decoupled shape-Newton scheme. In the above, the vectorfield $\boldsymbol{m}^{(k)} : \Gamma_{\boldsymbol{\theta}^{(k)}} \to \mathbb{R}^N$ is a chosen vectorfield such that $\boldsymbol{m}^{(k)} \cdot \boldsymbol{n} = 1$ on $\Gamma_{\boldsymbol{\theta}^{(k)}}$

where $V(\Omega_{\hat{\boldsymbol{\theta}}})$ is a discrete space, to be explained below, and where we adopted, for notational convenience,

$$k_{\hat{\boldsymbol{\theta}}} := (f + \partial_n g)/g + \kappa(\Gamma_{\hat{\boldsymbol{\theta}}}).$$

In accordance with Remark 3, and directly apparent from the curvature term in $k_{\hat{\boldsymbol{\theta}}}$, the discrete domain $\Omega_{\hat{\boldsymbol{\theta}}}$ needs to be of class C^1. This is most conveniently achieved via a tensorial reference domain $\mathcal{D} = [0, L_x] \times [0, L_y]$, which is mapped onto the physical domain by means of a smoothness preserving map

$$\mathbf{T}_{\hat{\boldsymbol{\theta}}} : \mathcal{D} \to \Omega_{\hat{\boldsymbol{\theta}}}. \tag{12}$$

The tensorial structure allows for function spaces of the following kind

$$V(\mathcal{D}) := \left\{ v_x \otimes v_y : v_x \in V\big([0, L_x]\big), v_y \in V\big([0, L_y]\big) \right\}, \tag{13}$$

where $V([0, L])$ are function spaces of the following type. A finite, piecewise polynomial spline basis of degree p is obtained by partitioning the domain into disjoint intervals $[0, L] = [\xi_0, \xi_1] \cup [\xi_1, \xi_2] \cup \cdots \cup [\xi_{N+2p}, \xi_{N+2p+1}]$ and applying Cox–de Boor recursion [7, 10]:

$$\begin{cases} N_{i,0}(\xi) = \mathbf{1}_{(\xi_i, \xi_{i+1})}(\xi) \\ N_{i,p}(\xi) = \dfrac{\xi - \xi_i}{\xi_{i+p} - \xi_i} N_{i,p-1}(\xi) + \dfrac{\xi_{i+p+1} - \xi}{\xi_{i+p+1} - \xi_{i+1}} N_{i+1,p-1}(\xi). \end{cases} \tag{14}$$

Choosing in particular

$$\xi_i = \begin{cases} 0 & i \le p \\ (i-p)L/N & p < i \le p+N \\ L & N+p < i, \end{cases} \tag{15}$$

we obtain a uniform N-element mesh supporting $N+p$ piecewise polynomial basis functions, with $p-1$ order smoothness over element boundaries. Figure 5 shows two realisations of this construction.

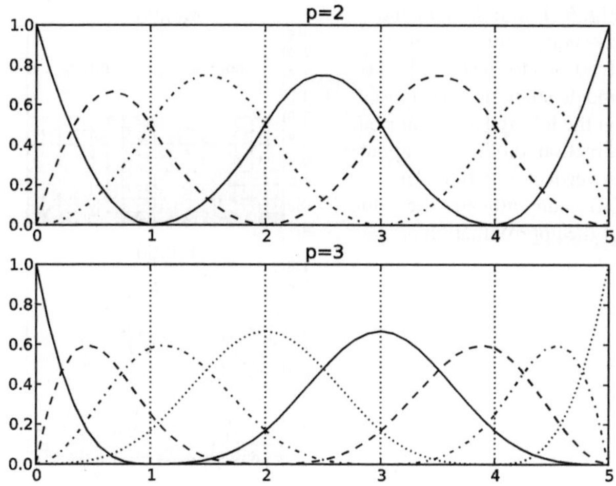

Fig. 5 Spline bases of polynomial order $p = 2$ and $p = 3$ on a $N = 5$ element mesh. Note the interpolation property at both domain ends. Further note that (like is the case for Lagrange finite elements) each element supports exactly $p + 1$ basis functions

Per standard finite element practice, the thus constructed function space $V(\mathcal{D})$ is used for two purposes; for constructing a geometric map $\mathbf{T}_{\hat{\theta}} \in [V(\mathcal{D})]^2$, mapping $\mathcal{D}$ onto the intended geometry, and, through inverse mapping, as test and trial space $V(\Omega_{\hat{\theta}}) = V(\mathcal{D}) \circ \mathbf{T}_{\hat{\theta}}^{-1}$. With emphasis on the former, the resulting set-up is commonly referred to as *isogeometric analysis*.

Remark 9 (Implementation) The above construction via Cox–de Boor relations is mainly instructive. A faster construction that is commonly used in practice transforms from a standard polynomial basis to a spline basis of equal order via extraction operators. This process is referred to as Bézier extraction [2, 23]. An additional advantage of the Bézier representation is that it can provide a unified interface to the myriad of spline technologies available, in particular T-splines.

5.2 Testcase I: Parabolic Free-Boundary

The first test case is derived from a manufactured solution U, designed such that $U|_{\Gamma_\theta} = 1$ on a parabolic surface $\Gamma_\theta = \{(x, y) : y = (3 - x^2)/2, 0 < x < 1\}$; see [31, Sect. 6.2] for more details. The data defining the free-boundary problem is then obtained as follows:

$$\begin{cases} f = -\Delta U, \\ g = \nabla U \cdot (-x, 1)/\sqrt{1 + x^2}, \\ h = 1. \end{cases} \tag{16}$$

On the fixed boundary $\partial\Omega_\theta \setminus \Gamma_\theta$ we have the Dirichlet data $h = y$.

Starting from an initial domain $\Omega_0 = [-1, 1] \times [0, 1]$, Fig. 6 shows the first three iterations (not counting the initial condition) of the algorithm outlined in

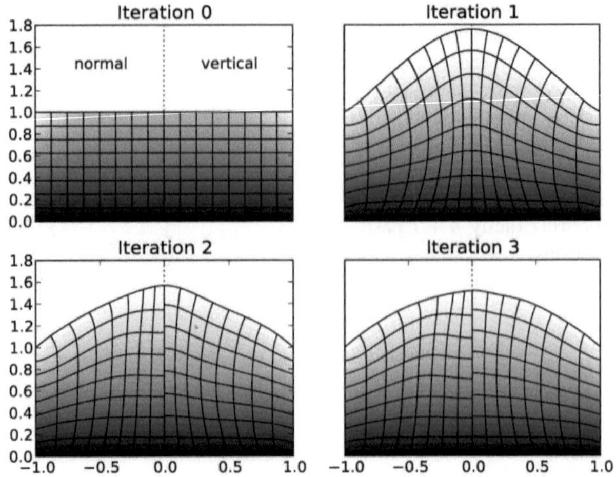

Fig. 6 Initial domain and subsequent three Newton iterations for testcase I, two update strategies juxtaposed; on the left updates in normal direction, on the right updates in vertical direction. The *grayscale* indicates the value of the approximation u^*

Fig. 4, solving the decoupled system weakly by (11). After each surface update the finite element mesh is warped to fit the new contour by solving a Dirichlet-constrained elasticity problem, with constraints resulting from L^2 projection of the surface update. By the third iteration the anticipated parabolic shape begins to manifest.

Recall that the surface updates require definition of a field $\boldsymbol{m}$ to determine the direction of surface updates after each Newton iteration, satisfying the constraint $\boldsymbol{n} \cdot \boldsymbol{m} = 1$. The iterations shown in the left half of Fig. 6 result from the trivial choice $\boldsymbol{m} = \boldsymbol{n}$, performing surface updates strictly in normal direction. This results in visible skewing of the center most elements. An alternative choice for $\boldsymbol{m}$ that satisfies the constraint is $(0, 1/n_y)$, thus performing updates in strictly vertical direction. Iterates for this scheme are shown in the right half of Fig. 6. The main qualitative difference is a visibly enhanced mesh regularity.

Figure 7 shows a side by side comparison of the first four Newton and Neumann iterations, with the latter formed according to the algorithm of Fig. 2 with relaxation parameter $\alpha = \frac{1}{2}$. The juxtaposed update strategies show no significant difference for Neumann iterations, offset against the clear qualitative difference observed earlier in Newton iterations. Inter-algorithm comparison further shows marked differences in convergence trajectories.

A quantitative comparison of the four permutations of algorithm and update strategy is presented in Fig. 8, which shows the convergence in terms of the Dirichlet error $\|u^* - h\|_{L^2(\hat{\Gamma})}$, and the surface error (which compares the vertical position of the exact and approximate free-boundary in L^2-sense). Directly apparent is the superlinear convergence of the Dirichlet error for Newton iterations, versus the much slower linear convergence of Neumann. Figure 8 furthermore shows that the update strategies are mutually interchangeable from a convergence perspective. The plateau in the convergence of the surface error is due to the coarse discretization used.

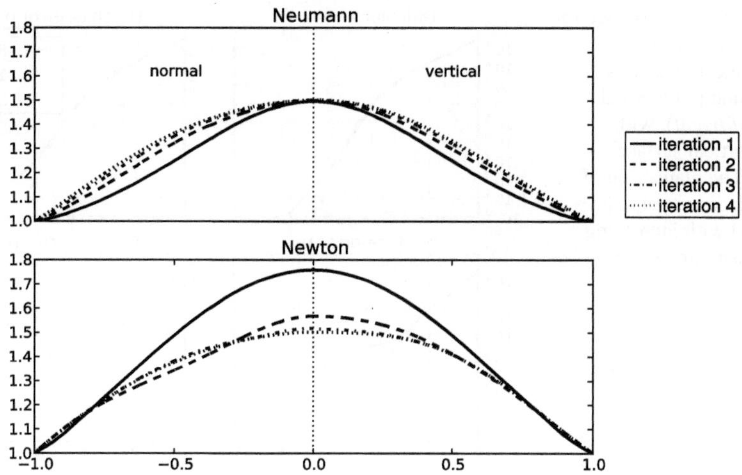

Fig. 7 Side by side comparison of the first four Neumann (relaxation 0.5) and Newton iterations for testcase I, superposing surface updates. Two update strategies juxtaposed as in Fig. 6

Fig. 8 Error convergence for Neumann and Newton iterations, comparing Dirichlet error $u^* - h|_{\hat{\Gamma}}$ (*above*) and surface error $\Gamma_\theta - \hat{\Gamma}$ (*below*), both measured in L^2 norm. The latter error can be evaluated by construction of the testcase, to which the exact solution is a parabola

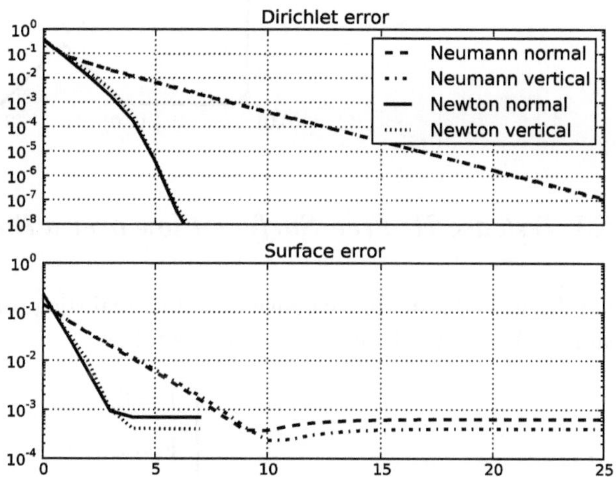

A more detailed analysis of Newton convergence is found in Fig. 9, which compares convergence for different mesh densities and polynomials orders. The surface error plateau that is reached after 4–6 iterations can again be explained as running into the limits of the current parametrization. This is in agreement with the observation that enriching the parametrization, by increasing either mesh density or polynomial order, results in lowering of the plateau. Another interesting observation to be made is that Dirichlet error convergence turns linear when this plateau is reached. This can be explained from error domination of the geometric component in the Newton update, which obstructs superlinear convergence: The scheme is a Newton scheme up to the level of discretization errors.

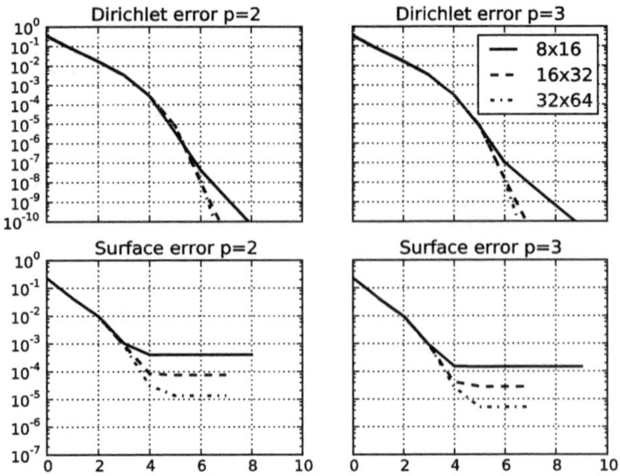

Fig. 9 Error convergence for Newton iterations for different mesh densities (graphs) and polynomial orders (*left/right*), with strictly vertical surface updates. The plateauing of surface error is seen to be postponed with increasing mesh density and polynomial order

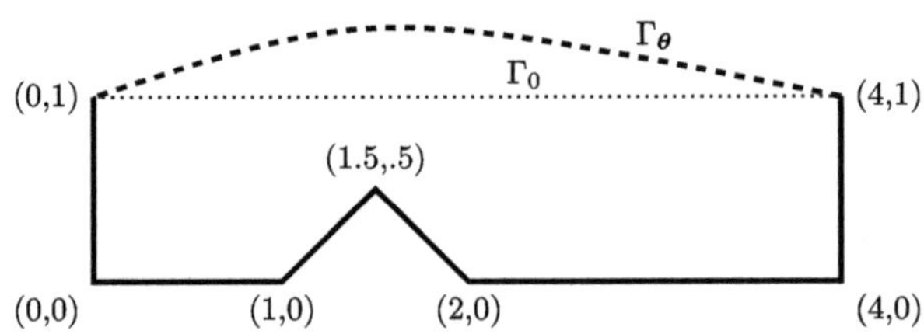

Fig. 10 Geometrical layout of the domain in the notched channel test-case

5.3 Testcase II: Free-Surface Flow over a Bump

For the second testcase we consider a more realistic problem corresponding to free-surface flow over a bump. The data is as follows:

$$\begin{cases} f = 0, \\ g = 1, \\ h = 1. \end{cases} \tag{17}$$

The domain is rectangular with a horizontally obstructing triangular notch, as depicted in Fig. 10. On the fixed boundary $\partial\Omega_\theta \setminus \Gamma_\theta$ the solution is constrained at $u = 0$ at the notch, and $u = y$ everywhere else.

A spline-discretization issue arises at the notch. Having gone through lengths to create a smooth geometric map, the three sharp corners cannot be represented on the resulting parametrization. Although solutions to this problem do exist in the form of T-splines [33] or degenerate meshes, we choose to approximate the notch by straightforward spline interpolation, resulting in the domain as shown in Fig. 11. This procedure is the direct isogeometric-analysis analogue to the usual treatment of Dirichlet boundary conditions in finite element analysis.

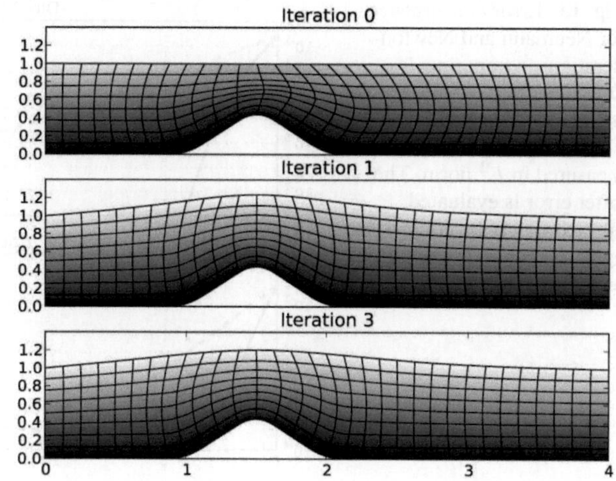

Fig. 11 Initial domain and subsequent two Newton iterations for testcase II, updating in normal direction. The *grayscale* indicates the value of the approximation u^*

Figure 11 shows the first three iterations of the Newton algorithm. As the normal and vertical update schemes are virtually indistinguishable, we present results for vertical updates only. The shape-Newton algorithm converges very rapidly. This is especially clear from Fig. 12 showing the first four iterations superposed. Like in testcase I, the Neumann algorithm shows a markedly slower convergence.

Convergence for both Neumann and Newton is quantified in Fig. 13. As the exact solution is unknown, the surface error cannot be evaluated directly. Instead we compare against a converged solution on a sufficiently refined mesh. The qualitative pattern that emerges is identical to that of testcase I.

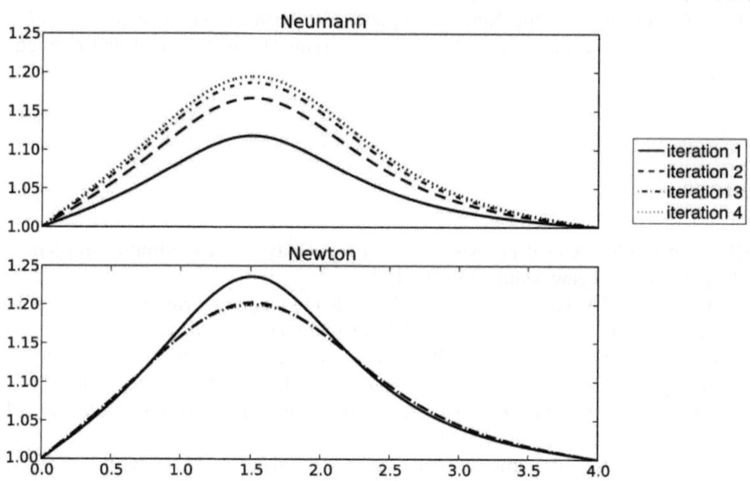

Fig. 12 Side by side comparison of the first four Neumann (relaxation 0.5) and Newton iterations for testcase II, superposing surface updates

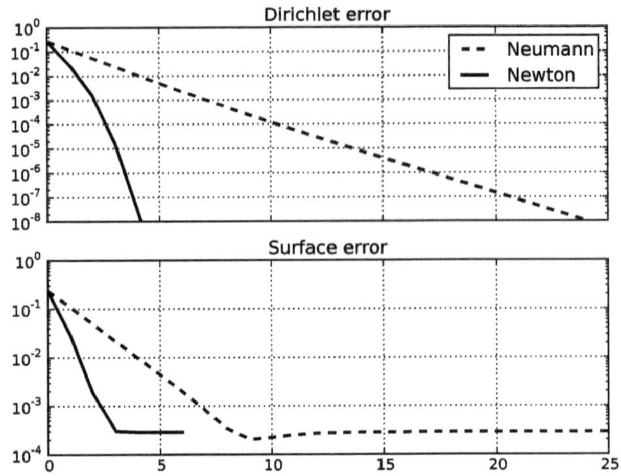

Fig. 13 Error convergence for Neumann and Newton iterations, comparing Dirichlet error $u - h|_{\hat{\Gamma}}$ (*above*) and surface error $\Gamma_\theta - \hat{\Gamma}$ (*below*), both measured in L^2 norm. The latter error is evaluated approximately by comparison against a converged solution on a four times denser mesh

6 Concluding Remarks

We derived a shape-Newton scheme for Bernoulli-type free-boundary problems. To obtain the linearized problem, we applied Hadamard shape derivatives to a suitable weak formulation of the free-boundary problem. The shape-Newton method corresponds to a boundary-value problem with a curvature-dependent boundary condition. Because of the curvature requirement, we employed isogeometric-analysis based discretizations with C^1-continuous discrete free boundaries. Numerical experiments showed rapid convergence behavior which seemed to point to quadratic convergence.

Acknowledgements This work is supported by NanoNextNL, a micro and nanotechnology consortium of the Government of the Netherlands and 130 partners. The research of K.G. van der Zee and C.V. Verhoosel is funded by the Netherlands Organisation for Scientific Research (NWO), VENI scheme.

References

1. Alt HW, Caffarelli LA (1981) Existence and regularity for a minimum problem with free boundary. J Reine Angew Math 325:105–144
2. Borden MJ, Scott MA, Evans JA, Hughes TJR (2011) Isogeometric finite element data structures based on Bézier extraction of NURBS. Int J Numer Methods Eng 87:15–47
3. Bouchon F, Clain S, Touzani R (2005) Numerical solution of the free boundary Bernoulli problem using a level set formulation. Comput Methods Appl Mech Eng 194:3934–3948
4. Bouchon F, Clain S, Touzani R (2008) A perturbation method for the numerical solution of the Bernoulli problem. J Comput Math 26:23–36
5. Cho S, Ha S-H (2009) Isogeometric shape design optimization: exact geometry and enhanced sensitivity. Struct Multidiscip Optim 38:53–70
6. Cottrell JA, Hughes TJR, Bazilevs Y (2009) Isogeometric analysis: toward integration of CAD and FEA. Wiley, Chichester

7. Cox MG (1972) The numerical evaluation of B-splines. IMA J Appl Math 10:134–149
8. Crank J (1987) Free and moving boundary problems. Oxford University Press, London
9. Cuvelier C, Schulkes RMSM (1990) Some numerical methods for the computation of capillary free boundaries governed by the Navier–Stokes equations. SIAM Rev 32:355–423
10. de Boor C (1972) On calculating with B-splines. J Approx Theory 6:50–62
11. Delfour MC, Zolésio J-P (2001) Shapes and geometries: analysis, differential calculus, and optimization. SIAM series on advances in design and control, vol 4. Society for Industrial and Applied Mathematics, Philadelphia
12. Eppler K, Harbrecht H (2006) Efficient treatment of stationary free boundary problems. Appl Numer Math 56:1326–1339
13. Flucher M, Rumpf M (1997) Bernoulli's free-boundary problem, qualitative theory and numerical approximation. J Reine Angew Math 486:165–204
14. Harbrecht H (2008) A Newton method for Bernoulli's free boundary problem in three dimensions. Computing 82:11–30
15. Haslinger J, Kozubek T, Kunisch K, Peichl G (2003) Shape optimization and fictitious domain approach for solving free boundary problems of Bernoulli type. Comput Optim Appl 26:231–251
16. Hughes TJR, Cottrell JA, Bazilevs Y (2005) Isogeometric analysis: CAD, finite elements, NURBS, exact geometry and mesh refinement. Comput Methods Appl Mech Eng 194:4135–4195
17. Kärkkäinen KT, Tiihonen T (1999) Free surfaces: shape sensitivity analysis and numerical methods. Int J Numer Methods Eng 44:1079–1098
18. Kärkkäinen KT, Tiihonen T (2004) Shape calculus and free boundary problems. In: Neittaanmäki P, Rossi T, Korotov S, Oñate E, Périaux J, Knörzer D (eds) European Congress on computational methods in applied sciences and engineering, ECCOMAS 2004, Jyväskylä, 24–28 July 2004
19. Kiendl J, Bletzinger K-U, Linhard J, Wüchner R (2009) Isogeometric shell analysis with Kirchhoff–Love elements. Comput Methods Appl Mech Eng 198:3902–3914
20. Kuster CM, Gremaud PA, Touzani R (2007) Fast numerical methods for Bernoulli free boundary problems. SIAM J Sci Comput 29:622–634
21. Mejak G (1994) Numerical solution of Bernoulli-type free boundary value problems by variable domain method. Int J Numer Methods Eng 37:4219–4245
22. Mejak G (1997) Finite element solution of a model free surface problem by the optimal shape design approach. Comput Methods Appl Mech Eng 40:1525–1550
23. Scott MA, Borden MJ, Verhoosel CV, Sederberg TW, Hughes TJR (2011) Isogeometric finite element data structures based on Bézier extraction of T-splines. Int J Numer Meth Eng 88
24. Sokołowski J, Zolésio J-P (1992) Introduction to shape optimization: shape sensitivity analysis. Springer series in computational mathematics, vol 16. Springer, Berlin
25. Tiihonen T (1998) Fixed point methods for internal free boundary problems. Numer Funct Anal Optim 19:399–413
26. Toivanen JI, Haslinger J, Mäkinen RAE (2008) Shape optimization of systems governed by Bernoulli free boundary problems. Comput Methods Appl Mech Eng 197:3802–3815
27. Tsai W-T, Yue DKP (1996) Computation of nonlinear free-surface flows. Annu Rev Fluid Mech 28:249–278
28. van Brummelen EH (2002) Numerical methods for steady viscous free-surface flows. PhD thesis, Universiteit van Amsterdam, Amsterdam, The Netherlands
29. van Brummelen EH, Raven HC, Koren B (2001) Efficient numerical solution of steady free-surface Navier–Stokes flow. J Comput Phys 174:120–137
30. van Brummelen EH, Segal A (2003) Numerical solution of steady free-surface flows by the adjoint optimal shape design method. Int J Numer Methods Fluids 41:3–27
31. van der Zee KG, van Brummelen EH, de Borst R (2010) Goal-oriented error estimation and adaptivity for free-boundary problems: the domain-map linearization approach. SIAM J Sci Comput 32:1064–1092

32. van der Zee KG, van Brummelen EH, de Borst R (2010) Goal-oriented error estimation and adaptivity for free-boundary problems: the shape linearization approach. SIAM J Sci Comput 32:1093–1118
33. van der Zee KG, Verhoosel CV (2011) Isogeometric analysis-based goal-oriented error estimation for free-boundary problems. Finite Elem Anal Des 47:600–609
34. Wackers J, Koren B, Raven H, van der Ploeg A, Starke A, Deng G, Queutey P, Visonneau M, Hino T, Ohashi K (2011) Free-surface viscous flow solution methods for ship hydrodynamics. Arch Comput Methods Eng 18:1–42

Part IV
Marine/Offshore Renewable Energy

Enabling Computational Methods for Offshore Wind Turbines

Y. Bazilevs, M.-C. Hsu, I. Akkerman, and D.J. Benson

Abstract In this book chapter we give an overview of our recent work in computational methods for geometry modeling, fluid mechanics, structural mechanics, and fluid–structure interaction that enable high-fidelity simulations of offshore wind turbines. Computational examples involving free-surface flow and full-scale wind turbine simulations are presented to illustrate the capabilities of the computational techniques developed.

Keywords Fluid-structure interaction · ALE-VMS · Offshore wind turbines · Isogeometric analysis · Kirchhoff–Love shells · Free-surface flow

1 Introduction

In this work, the developments are focused on the formulation and application of high-fidelity fluid–structure interaction (FSI) methods to the simulation of offshore wind turbines. This class of applications is characterized by the presence of free-surface turbulent flow, time-dependent, topological changes of the free surface (e.g., wave breaking), 3D complex geometry, moving spatial domains, and fluid–structure coupling. The main challenges in this work are:

- The spatial scales involved in the applications of interest are large. Besides the obvious mesh resolution requirements, this presents a major challenge for the validation of the proposed numerical technology, as high-quality experimental results are typically reported for lab-scale models.

Y. Bazilevs (✉) · D.J. Benson
Department of Structural Engineering, University of California, San Diego, 9500 Gilman Drive, La Jolla, CA 92093, USA
e-mail: yuri@ucsd.edu

M.-C. Hsu
Institute for Computational Engineering and Sciences, University of Texas, Austin, 201 E. 24th Street, Austin, TX 78712, USA

I. Akkerman
School of Engineering and Computing Sciences, Durham University, South Road, Durham, UK

L. Eça et al. (eds.), *MARINE 2011, IV International Conference on Computational Methods in Marine Engineering*, Computational Methods in Applied Sciences 29,
DOI 10.1007/978-94-007-6143-8_6, © Springer Science+Business Media Dordrecht 2013

- The flow Reynolds number is very high, which gives rise to fully developed and physically complex turbulent flows. The presence of the free surface and boundary layers near solid surfaces exacerbate the situation further. In order to simulate the relevant turbulent phenomena for this range of Reynolds numbers and in complex geometry configurations, a mathematically sound numerical formulation, together with increased grid resolution, is required.
- The applications of interest involve moving and stationary components. This requires advanced mesh management and exchange of the kinematic and traction data between the nonmatching interfaces. The interfaces may or may not be in relative motion.
- FSI coupling needs to be sufficiently robust in order to preclude divergence of the computational procedures.
- Implementation of the proposed framework in the HPC environment needs to be such that good parallel scalability is attained for the fully coupled FSI computations, not just for the individual fluid and structural mechanics computations.

To address the above challenges we developed a computational framework, which consists of the following items and features: (a) Multiscale modeling of free-surface turbulent flows in geometrically complex configurations; (b) Modeling of geometrically complex structures based on the concept of Isogeometric Analysis (IGA); (c) Accurate and efficient treatment of fluid–structure and fluid-fluid interfaces present in the modeling and simulation of the coupled FSI phenomena; (d) Algorithmic implementation suitable for modern parallel architectures. In this book chapter we present our aerodynamics, free-surface flow, and FSI computational techniques, as well as example simulations that illustrate the potential of the methods developed.

2 Enabling Computational Technology

2.1 Isogeometric Analysis

The concept of Isogeometric Analysis (IGA) was introduced in [1]. The motivation for introducing IGA comes from the need for a tighter integration between engineering design, which is primarily done using Computer-Aided Design (CAD), and engineering simulation, which is primarily based on the Finite Element Method (FEM). The main idea behind the IGA is to focus on one, and only one, geometric model, which can be utilized directly as an analysis model, or from which geometrically precise analysis models can be automatically built. To instantiate such an idea requires a change from the classical FEM to an analysis procedure based on CAD representations. There are several candidate technologies from computational geometry that may be used in IGA. The most widely used in engineering design are NURBS (non-uniform rational B-splines), the industry standard (see [2]). The major strengths of NURBS are that they are convenient for free-form surface modeling, can exactly represent all conic sections, and therefore circles, cylinders, spheres, ellipsoids, and other special geometries, and that there exist many efficient and numerically stable algorithms to generate NURBS objects. They also possess useful mathematical properties, such as good approximation, and the ability to be refined through knot insertion. Because NURBS are a CAD standard, representing many years in development, they were the natural starting point for IGA.

T-splines [3] are a recently developed generalization of NURBS technology. They extend NURBS to permit local refinement (and coarsening). They are backward- and forward-compatible with NURBS, which makes them an attractive CAD technology. Preliminary investigations of T-splines as IGA technology may be found in [4]. Recent results on linear independence and improved local refinement algorithms may be found in [5, 6]. A recent attempt to construct solid T-splines from existing hexahedral meshes may be found in [7].

The recent book [8], and the references therein, present in detail the mathematical developments, basis function research, geometry modeling, model quality assessment, and early applications in IGA. In this work, we use IGA primarily for the modeling of structural mechanics. Very good results using turbulent and free-surface flows with IGA were reported in [9–16].

2.2 *Variational Multiscale Methods for Fluids and Turbulence*

Due to their multiscale nature, advection-dominated and turbulent phenomena to this day remain a significant computational challenge, requiring sophisticated discrete formulations and many grid points for improved resolution. Standard Galerkin methods are not a sufficiently robust technology for advection-dominated flows. For this reason, stabilized methods [17–25] were designed to circumvent this shortcoming of the Galerkin technique. Stabilized methods, which are essentially residual-based modifications of the Galerkin method, exhibit uniform stability and convergence behavior across the full range of advective and diffusive phenomena.

The basic theory of variational multiscale (VMS) methods was developed in [26], wherein stabilized methods were first identified as multiscale methods. Relationship between stabilized methods and subgrid scale modeling was also identified in [26], and now presents an important research direction [27]. Recently, in [9], the authors proposed a residual-based turbulence modeling and computational framework that is based on the VMS theory, named RBVMS. The main feature of this approach is that turbulence models, viewed as subgrid-scale models, are derived directly from the Navier–Stokes equations. The fine scales are introduced in the coarse-scale equations in a consistent manner and are modeled in a residual-based fashion. The underlying methodology performs well on both laminar and turbulent flows, for a wide range of Reynolds numbers. The extension of the RBVMS framework to the moving-domain case, where the motion of the fluid mechanics domain is handled using the Arbitrary Lagrangian–Eulerian (ALE) formulation [28], was named ALE-VMS in [29, 30].

An important additional feature of the ALE-VMS methodology is weak enforcement of essential boundary conditions. Weakly enforced essential boundary conditions were introduced in [31]. Weak boundary conditions produce significantly more accurate solutions than strongly enforced boundary conditions on meshes with insufficient boundary layer resolution [10, 12, 14, 32], which is almost always the case in practice. ALE-VMS with weakly enforced boundary conditions presents the main computational technology behind aerodynamics and free-surface flow computations shown in this book chapter.

2.3 Free-Surface Flow

Accurate prediction of wave loading on structures, and their motion in response to such loads, necessitates the development of a free-surface flow simulation framework, which involves the interaction between two fluids: water and air. There are two classes of techniques that are applied to free-surface problems: interface-tracking and interface-capturing (see [22] for the terminology and detailed discussion of the relative merits of the methods). In the interface-tracking techniques, element boundaries exactly coincide with the boundary of the two-fluid interface. They work well when the free-surface motion is mild and enhance the fidelity of the computational results. However, in the case of violent free-surface motions that involve wave breaking and other topological changes, interface-capturing techniques are significantly more convenient and practical to use.

In the case of an interface-capturing formulation based on the level set method, the boundary between the two fluids is described implicitly as a zero level set of a scalar function defined in the problem domain (see, e.g., [33–35]). The subdomains corresponding to negative and positive values of the level set function are those occupied by water and air, respectively. The level set function is simultaneously a signed-distance function, meaning its magnitude at a point in 3D space is the distance from that point to the air-water interface, and its sign determines if the point is in the water or air domain. The signed distance property of the scalar function is not necessary in general, however, it has several accuracy benefits.

The level set function is convected with the flow speed. This is accomplished by solving a time-dependent pure advection equation for the level set function together with fluid flow equations. In order to preserve the signed-distance property of the level set function, the Eikonal partial differential equation is employed. For stability, a smoothed Heaviside function is used to distribute the fluid properties in the epsilon-layer adjacent to the interface. Mass conservation, in its global or local form (see [16, 36]), is enforced, which is critical for long-time integration of the equations and accurate load prediction on structures. This interface-capturing methodology was tested in the context of FEM in [36] and IGA in [16]. It was shown in the latter reference that the per-degree-of-freedom accuracy is better for IGA than for the FEM. It was also shown that besides the clear advantages for violent free-surface motion, the interface-capturing methodology gives reasonably accurate results also for the cases of mild free-surface deformations. As a result, the interface-capturing methodology is applicable to the cases that involve both types of free surface behavior, as in the case of offshore wind turbines.

2.4 IGA Modeling of Thin Structures

Offshore wind turbine structures are predominantly curved thin shells reinforced with structural stiffeners. As a result, to simulate such structures, discretization of thin shell theories are employed for computational efficiency, and are key to structural modeling.

Low-order, bi-linear quadrilateral finite elements, which are widely used and are considered standard shell element technology, exhibit several shortcomings: These elements require the use of displacement and rotation degrees of freedom to describe shell kinematics; One needs a fine mesh to represent shell geometries with high local curvature and to simultaneously achieve the desired solution accuracy; Ad-hoc element technology (e.g., hourglass stabilization) is necessary to overcome locking; In the case of explicit time stepping, the rotational inertias must be artificially increased to alleviate severe time step size stability restrictions; In the case implicit time stepping is employed, the presence of rotational degrees of freedom doubles the size of the solution and right-hand-side residual arrays, quadruples the size of the left-hand-side matrix, and results in an order-of-magnitude increase in linear solver time. Higher-order Lagrange finite elements present an alternative to the low-order approaches. However, their use in thin shell analysis is not common due to the observed lack of robustness relative to low-order elements. Furthermore, the fact that Lagrange elements are C^1-continuous at the inter-element boundaries requires one to use rotational degrees of freedom.

Isogeometric shell analysis was recently proposed in [37] to address the shortcomings of standard finite element technology for thin shells listed above. It was found that the higher-order continuity (C^1 and above) of the IGA basis functions significantly improved the per-degree-of-freedom accuracy and robustness of thin shell discretizations, as compared to the FEM. Furthermore, the increased continuity of the IGA discretizations enabled the use of shell kinematics without rotational degrees of freedom [38–40], leading to further computational cost savings. The isogeometric rotation-free Kirchhoff–Love shell formulation for structures composed of multiple structural patches, called the bending strip method, was developed in [41], which enabled the application of the rotation-free IGA technology to real-life structures, such as wind turbine rotors (see [42–44]). Besides significant savings in computational time, the rotation-free shell discretization makes FSI coupling simpler than the discretization with rotational degrees-of-freedom.

Although NURBS are employed in this work to discretize the structural mechanics equations, T-splines [4, 45] or subdivision surfaces [46–48] are also well suited for the proposed structural modeling approach. For other rotation-free shell formulations the reader is referred to [46–51].

2.5 Free-Surface Modeling on Moving Spatial Domains

Simulating the interaction of free-surface flow with rigid or flexible structures requires additional computational technology that is able to track the interface between the structure and the surrounding air-water medium. We do this as follows. The fluid mesh is conforming to the boundaries of the floating structure and is constructed with appropriate boundary layer discretization. However, the free surface is described with a level set function, which does not conform to the mesh boundaries (see section on free-surface flow above). During the course of the simulation, the structure undergoes large motions under the action of forces coming from wind, surface waves, and underwater currents. This motion changes the geometry of the fluid mechanics domain, and must be reflected in the discretization. This may be accomplished by moving the problem mesh, or remeshing in the case the mesh quality is no longer acceptable, such that the structural surface discretization remains conforming to the fluid–structure interface.

The proposed methodology is an instantiation of the more general Mixed Interface-Tracking/Interface-Capturing Technique (MITICT) where the fluid-structure interface is tracked with the mesh, while the air-water interface is captured on that mesh. While boundary-layer meshing is employed near solid surfaces, which is important for boundary-layer accuracy, resolution of the air-water interface depends on the mesh size in the neighborhood of where the interface is. The MITICT [52] was introduced primarily for fluid-object interactions with multiple fluids (see, for example, [53]). The MITICT was successfully tested in [54], where the interface-tracking technique used was a space–time formulation [55, 56], and the interface-capturing method was the Edge-Tracked Interface Locator Technique (ETILT) [52]. It was also tested in [57] by using a moving Lagrangian interface technique [58] for interface tracking and the ETILT for interface capturing.

The interface-tracking technique used in the MITICT can also be the ALE method, employed in this work. The interface-capturing technique rides on this, and is based on solving, over a moving mesh, in addition to the Navier–Stokes equations, the advection equation governing the time-evolution of the scalar interface function. The ALE-based MITCT technique was formulated in [59]. The current implementation assumes that the structures are approximated as rigid objects. The method was tested on several problems, which involve moving and nonmoving domains. Reference [60] shows validation of the proposed MITCT framework on the planing Fridsma hull. Good agreement with experimental results is achieved. In the case of planning hulls, unlike in the majority of free-surface flow applications, modeling the air domain plays an important role. This is also the case for offshore wind turbines.

2.6 FSI Methods and Coupling

Wind-turbine simulations incorporate FSI coupling. Recently, a fully coupled FSI simulation methodology for wind turbine rotors was developed in [42]. The full machine was simulated with FSI coupling in [44]. The simulations showed the importance of FSI modeling for full-scale wind turbines.

We assume strong coupling between the fluid and structure, and employ Newton–Raphson linearization at each time step. At the level of the Newton–Raphson iteration, the implementation allows one to choose the degree of coupling between the fluid and structure in the left-hand-side matrix. This leads to an hierarchy of coupling techniques, which may be explored for the best efficiency for a given application. However, the right-hand-side vector of the coupled equation system is unchanged, which automatically guarantees convergence to a correctly coupled solution independent of the degree of coupling in the left-hand-side matrix. Given that the applications in this work involve relatively heavy structures and light fluids, we are able to remove the non-standard coupling terms from the left-hand-side matrix without degrading nonlinear convergence. This strategy was successfully employed in simulating wind-turbine FSI in [30, 44, 61].

The FSI coupling methodology assumes a nonmatching fluid–structure interface discretization. As a result, the mesh resolution of the structure and fluid may be tailored to the analysis requirements of the individual subsystems, leading to better computational efficiency. In order to take advantage of the superior accuracy of IGA for structural mechanics applications, and to leverage the existing advanced volumetric mesh generation tools for the FEM, we choose to couple low-order FEM for fluid and IGA for structural mechanics. Although IGA discretizations were shown to produce results that are of superior per-degree-of-freedom quality to standard FEM for fluid mechanics and turbulence applications, good-quality fluid mechanics results for similar applications may also be achieved with standard low-order FEM with a manageable number of degrees-of-freedom (see [32, 62–64]). It is felt that the proposed combination of FEM and IGA has the highest potential for adoption of IGA by industry and research labs.

Nonmatching interface discretizations in the FSI and sliding-interface problems necessitate the use of interpolation or projection of kinematic and traction data between the nonmatching surface meshes (see, e.g., [61, 65–67], where [67] is more comprehensive than [66]). A computational procedure, which can simultaneously handle the data transfer for IGA and FEM discretizations, was proposed in [61]. The procedure also includes a robust approach in identifying "closest points" for arbitrary shaped surfaces. While such interface projections are rather straightforward for weakly-coupled FSI algorithms, they require special techniques [24, 67, 68] for strongly-coupled, "direct" and "quasi-direct" methods [24, 67–70] that are monolithic-like (i.e. become monolithic for matching discretizations).

Using nonmatching discretizations at the fluid–structure interface can serve purposes beyond an implementation convenience typically associated with weakly-coupled FSI methods. What we do here is an example of taking advantage of the flexibility associated with using nonmatching discretizations, combining the most appropriate discretization for each part of an FSI problem. Exploiting this kind of flexibility, in conjunction with FSI homogenization methods [71–73], enabled computation of some of the most challenging parachute FSI problems [73–75], including parachute clusters [74, 75], where the contact model [68, 75] needed to deal with the interaction between the parachutes took advantage of this flexibility.

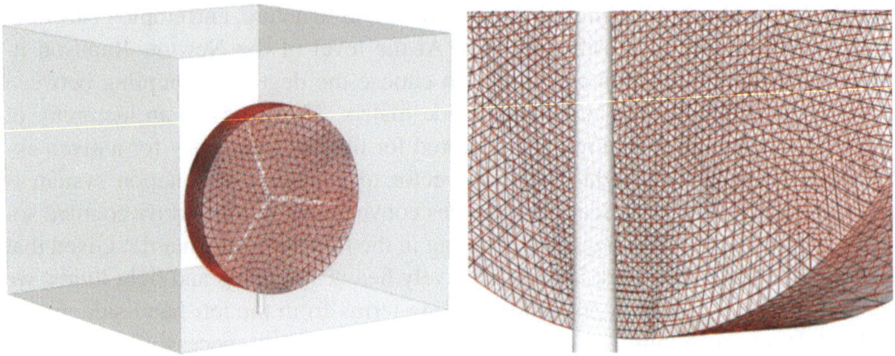

Fig. 1 Nonmatching meshes at the sliding interface between the stationary and moving subdomains

2.7 Sliding-Interface Formulation for Objects in Relative Motion

In order to simulate the full wind turbine configuration, rotor–tower interaction should be included in the modeling. In [76] a sliding-interface coupling technique was formulated in the context of the ALE-VMS method to handle flows about mechanical components in relative motion. In [44, 64], the technique was applied to the simulation of full-scale wind turbines. The sliding-interface method makes use of a moving subdomain, which encloses the entire wind turbine rotor, and a stationary subdomain that contains the rest of the wind turbine (see Fig. 1). The moving subdomain rotates with the rotor and accommodates deflection of the turbine blades. The two domains are in relative motion and share a sliding interface. The meshes on each side of the interface are nonmatching because of the relative motion (see Fig. 1). As a result, a numerical procedure is needed to impose the continuity of the fluid velocities and tractions at the stationary and rotating subdomain interface despite the fact that the interface discretizations are incompatible. Such a procedure was developed in [76] in the context of IGA for computing flows about rotating components. The advantage of IGA for rotating-component flows is that the sliding interfaces are represented exactly, and no geometry errors are incurred. In the case of standard FEM employed here, the geometric compatibility is only approximate. We note that in application of the FEM to flows with moving mechanical components, the Shear–Slip Mesh Update Method [77–79] and its more general versions [52, 80] may also be employed to handle objects in relative motion.

3 Computational Results

In this section we show two computational examples, one from [44] and another from [59], that illustrate the performance of the computational techniques described

Table 1 Mesh statistics for Cases I, II and III

	Number of nodes	Number of elements
Case I	1,319,427	4,218,459
Case II	1,326,129	4,205,519
Case III	1,440,425	4,828,692

in the previous section. In all cases, we employ the Generalized-α time integration [81–83], which is a fully-implicit second-order accurate method with control over the dissipation of high-frequency modes.

3.1 FSI Simulations of a 5 MW Offshore Wind Turbine

In this section, we present our computations of the NREL 5 MW offshore baseline wind turbine from [44]. The NREL 5 MW wind turbine is a conventional three-blade upwind turbine proposed in [84] to support concept studies aimed at assessing offshore wind technology [85–91]. The rotor-only configuration of this wind turbine was simulated earlier in [13, 15, 32, 42, 61–63, 92] using both NURBS-based IGA and standard FEM. The detailed geometry description and construction for the blade surface was documented in [13]. The 63 m blade is composed of a series of DU airfoils and the NACA64 profile. In this work, the proposed framework is applied to the simulation of the full wind turbine configuration, including the rotor (blades and hub), nacelle, and tower. The land-based tower is assumed to be rigid, and has a base diameter of 6 m and a top diameter of 3.87 m. The tower height is 87.6 m and the hub height is 90 m [84].

The cases considered in this work are as follows. Case I is a rotor-only setup, where the rotor is enclosed in a cylindrical domain, and the rotation is applied to the entire computational domain to simulate the spinning rotor. Case II is also a rotor-only setup, however, the rotor is housed in a cylindrical rotating subdomain that is enclosed by a stationary exterior flow subdomain. The sliding-interface formulation is applied at the interface between the subdomains to weakly impose the continuity of the kinematics and tractions. We compare the simulation results from Cases I and II to study the effect of the sliding-interface formulation. Case III has a similar setup as Case II, but this time the stationary subdomain contains the nacelle and tower, and thus enables us to study the rotor–tower interaction effects.

All simulations are performed at the rated wind speed of 11.4 m/s. This setup corresponds to one of the cases reported in [84]. The wind speed is prescribed at the inflow boundary, the traction vector is set to zero at the outflow boundary, and the slip condition is set on the top, bottom, and lateral boundaries. The air density and viscosity are 1.2 kg/m^3 and 2.0×10^{-5} kg/(m s), respectively.

The aerodynamics volume mesh statistics for all three cases are summarized in Table 1. Case III has more elements and nodes compared with Cases I and II due to the presence of the nacelle and tower. A 2D cut of the mesh at $x = 0$ for Case III

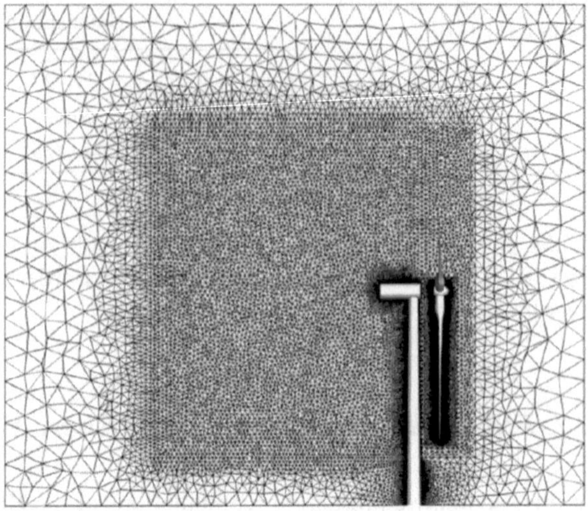

Fig. 2 A 2D cut at $x = 0$. The mesh is refined in the inner region for better flow resolution near the wind turbine

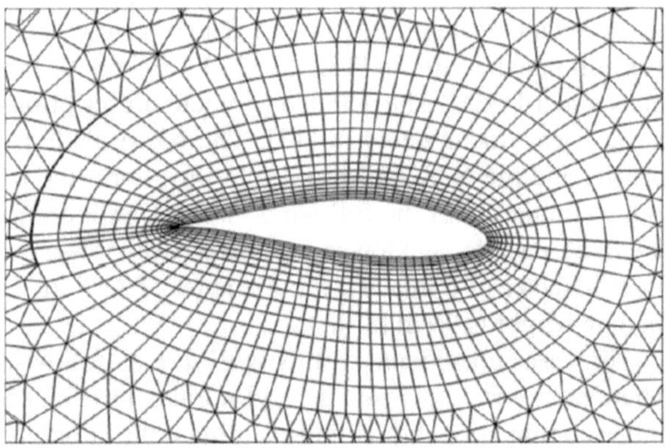

Fig. 3 A 2D cut at 75 % spanwise station to illustrate the boundary-layer mesh. The size of the first element in the wall-normal direction is 0.02 m, and a growth ratio of 1.2 is used to generate 15 layers of prismatic elements

is shown in Fig. 2 to show the mesh quality used in our computations. The mesh is refined in the inner region for better flow resolution. Figure 3 shows a 2D cut at 75 % spanwise station to illustrate the boundary-layer mesh used in our computations. Near the blade surface, the size of the first element in the wall-normal direction is 0.02 m, and 15 layers of prismatic elements were generated with a growth ration of 1.2. The Reynolds number based on the chord length and relative speed at this location is $O(10^7)$. The same boundary-layer mesh is used in all three cases. The time-step size is 2.5×10^{-4} s for all cases.

The rotor blades are discretized using NURBS-based IGA. The structural mechanics problem is governed by the isogeometric Kirchhoff–Love composite shell formulation with the aid of the bending strip method [38, 41, 42]. Only displacement degrees of freedom are employed. A symmetric fiberglass/epoxy composite with $[\pm 45/0/90_2/0_3]_s$ lay-up was proposed in [42] for the blade material. The de-

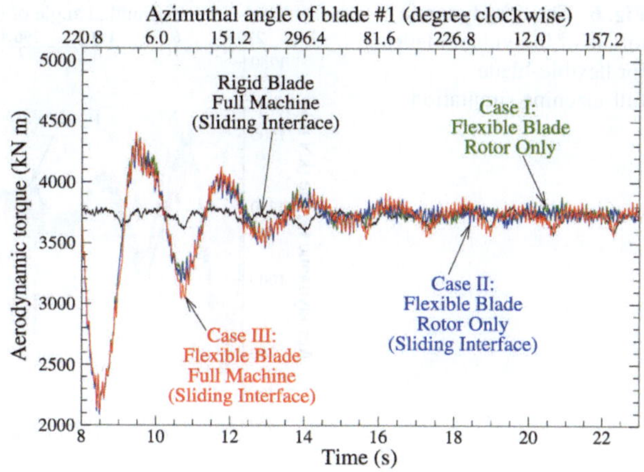

Fig. 4 The aerodynamic torque for flexible-blade FSI computations

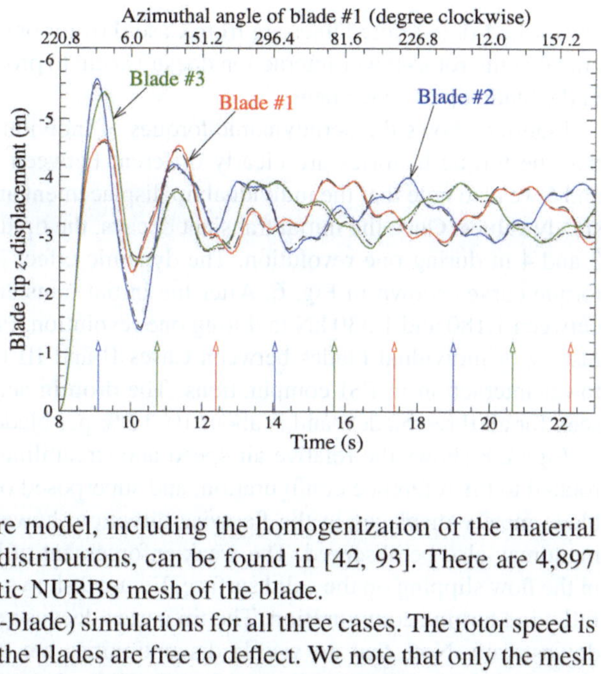

Fig. 5 The blade tip z-displacement for rotor-only (*dashed*, *dashed-dotted*, and *dashed-dotted-dotted lines*) and full-machine (*solid lines*) simulations. The *arrows* near the bottom of the figure show the instants where the blade passes in front of the tower

tails of the NURBS structure model, including the homogenization of the material and the laminate thickness distributions, can be found in [42, 93]. There are 4,897 control points in the quadratic NURBS mesh of the blade.

We perform FSI (flexible-blade) simulations for all three cases. The rotor speed is fixed at 12.1 rpm, however, the blades are free to deflect. We note that only the mesh motion part associated with blade deflections is computed using the mesh moving method that we typically use, which is an elastic mesh moving technique with Jacobian-based stiffening [94–96]. The mesh motion part associated with global blade rotation is computed exactly (see [42] for details).

The time histories of the aerodynamic torque are shown in Fig. 4. The comparison between Cases I and II shows that the presence of the sliding interface has no effect on the prediction of aerodynamic loads on the rotor.

Figure 5 shows the tip z-displacement history of the individual blades. The solid lines are the results from Case III (full machine), while the dashed, dashed-doted,

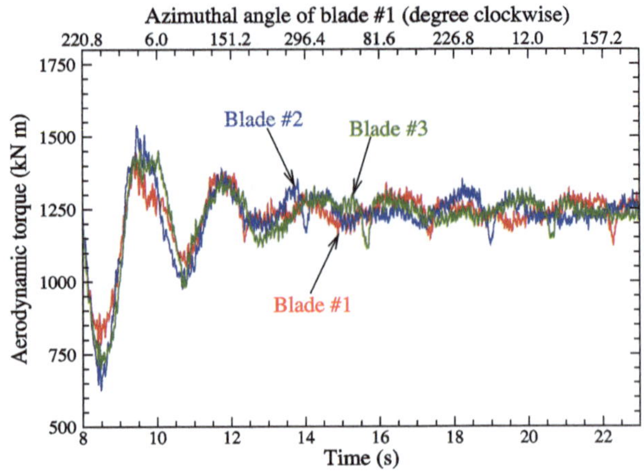

Fig. 6 The aerodynamic torque of individual blades for flexible-blade full-machine simulation

and dashed-doted-doted lines are from Case II (rotor only). From the comparison we note that the rotor–tower interaction doesn't seem to produce a pronounced response in the blade tip displacement.

Figure 6 shows the aerodynamic torques of individual blades for Case III. Note that the torque histories are clearly different between three blades. From Figs. 5 and 6 we also note that the individual tip displacement and torque do not settle at the steady values. Once the initial transient decays, the tip displacement varies between 3 and 4 m during one revolution. The dynamic effect is also present in individual torque curves shown in Fig. 6. After the initial transient decays, the torque varies between 1,180 and 1,330 kN m during one revolution. Figure 7 compares the torque history of individual blades between Cases II and III to show the effect of rotor–tower interaction in FSI computations. The drop in aerodynamic torque is clearly seen for all three blades, and is about 10–12 % per blade.

Figure 8 shows the relative air speed and streamlines at 50 % spanwise station, rotated to the reference configuration, and superposed on the moving blade #2. The blade mostly displaces in the flapwise direction, however, the edgewise displacement may also be observed. The weak enforcement of boundary conditions results in the flow slipping on the solid surface. Figure 9 shows the flow visualization of the full wind turbine configuration. The tip vortex decays very slowly as it is convected downstream. Note that no visible discontinuities are seen in the flow field at the sliding interface, which indicates that the method correctly handles the kinematic compatibility at this location.

The time-averaged aerodynamic torque values for all cases are summarized in Table 2. From the FSI computation of Case III, and without considering any power loss, the energy generated by this wind turbine design is about 4.73 MW. According to the Betz' law (see, e.g. [97]), the maximum power that this wind turbine design is able to extract for the wind and rotor speeds considered is 6.57 MW, leading to the aerodynamic efficiency of 72 %, which is quite good for modern wind turbine designs.

Fig. 7 The aerodynamic torque of individual blades for rotor-only and full-machine simulations (flexible blade). The drop in the aerodynamic torque is clearly visible

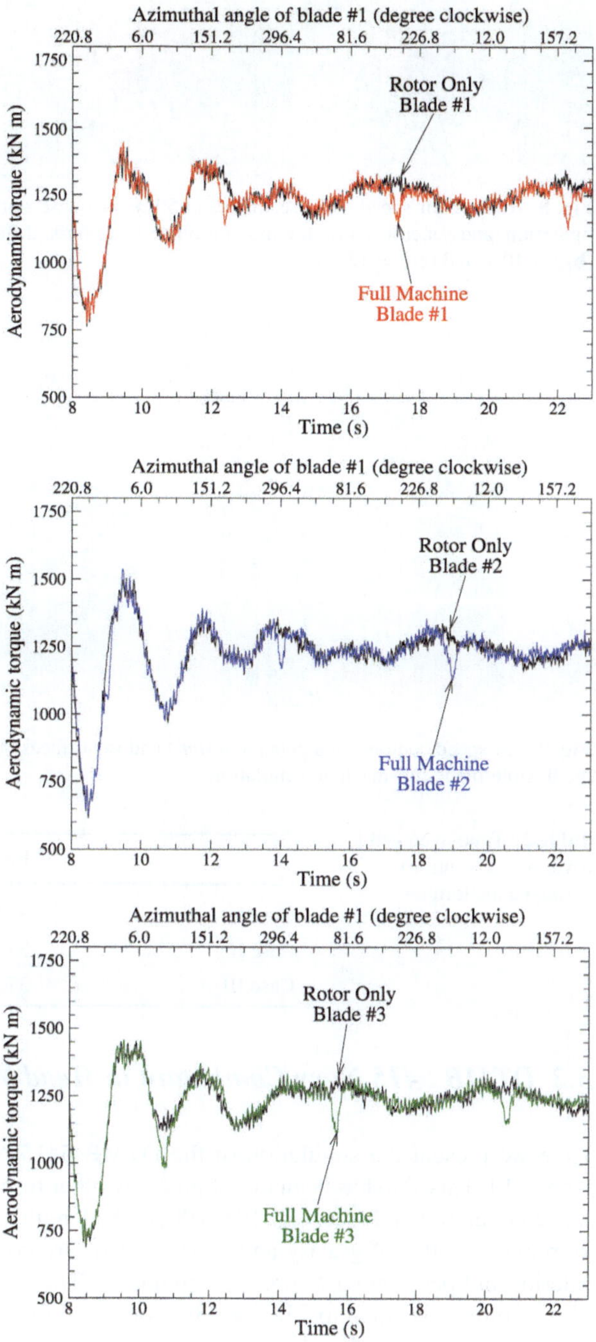

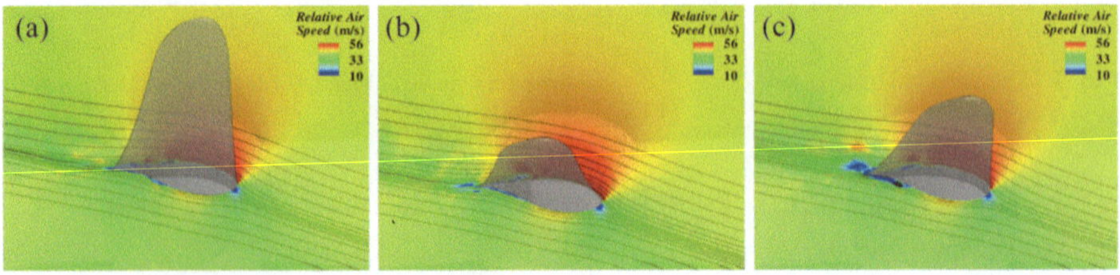

Fig. 8 Relative air speed and streamlines at 50 % spanwise station, rotated to the reference configuration, and superposed on the moving blade #2 at three different time instants: (**a**) $t = 9.0$ s, (**b**) $t = 10$ s, and (**c**) $t = 12.5$ s

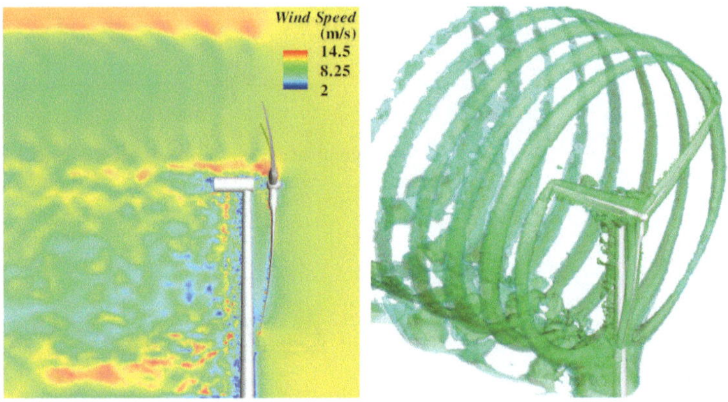

Fig. 9 Air speed contours at a planar cut (*left*) and isosurfaces of air speed (*right*) at an instant for the flexible-blade full-machine simulation

Table 2 Time-averaged (over one revolution) aerodynamic torques

	Flexible-Blade Aero. Torque (kN m)
Case I	3749
Case II	3751
Case III	3734

3.2 DTMB 5415 Navy Combatant in Head Waves

Here we present the simulation of the DTMB 5415 Navy combatant at lab scale from [59]. This ship has been investigated by other researchers, both experimentally and computationally (see, e.g., [98–100]). The length of the ship hull is 5.72 m. The ship mass, center of gravity and inertia tensor are computed by meshing the ship interior and performing a direct computation. The total ship volume is 1,366 m^3. The ship mass is equal to 532.3 kg. It is obtained by multiplying the volume of the ship below the water line by the constant water density. The center of gravity and the inertia tensor are computed assuming the ship's effective density (i.e., the ship

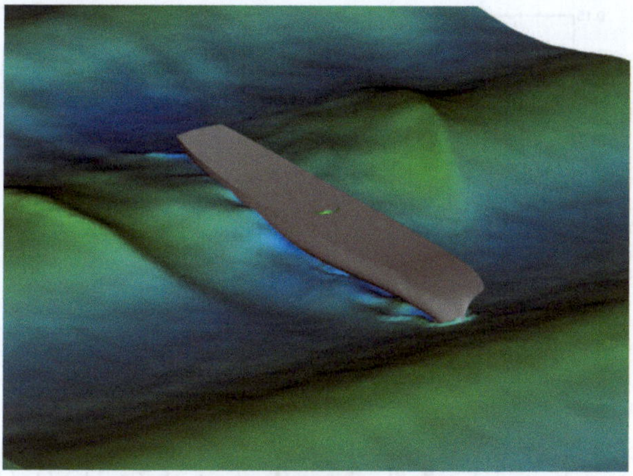

(a) t =9.00 s

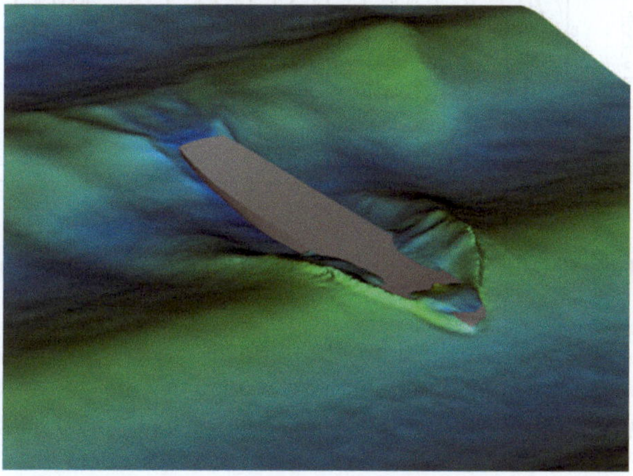

(b) t =9.50 s

Fig. 10 DTMB 5415 in head waves. Snapshots of the ship negotiating high-amplitude waves. The water surface is colored by the fluid speed

mass divided by its total volume), which results in

$$\boldsymbol{X}_0 = \begin{bmatrix} 2.761 \\ 0 \\ 0.280 \end{bmatrix} \text{m} \tag{1}$$

and

$$\boldsymbol{I} = \begin{bmatrix} 7.256\text{E}{-2} & 2.69\text{E}{-7} & 5.35\text{E}{-2} \\ 2.69\text{E}{-7} & 2.89 & -2.44\text{E}{-8} \\ 5.35\text{E}{-2} & -2.44\text{E}{-8} & 2.91 \end{bmatrix} \text{kg}\,\text{m}^2, \tag{2}$$

respectively.

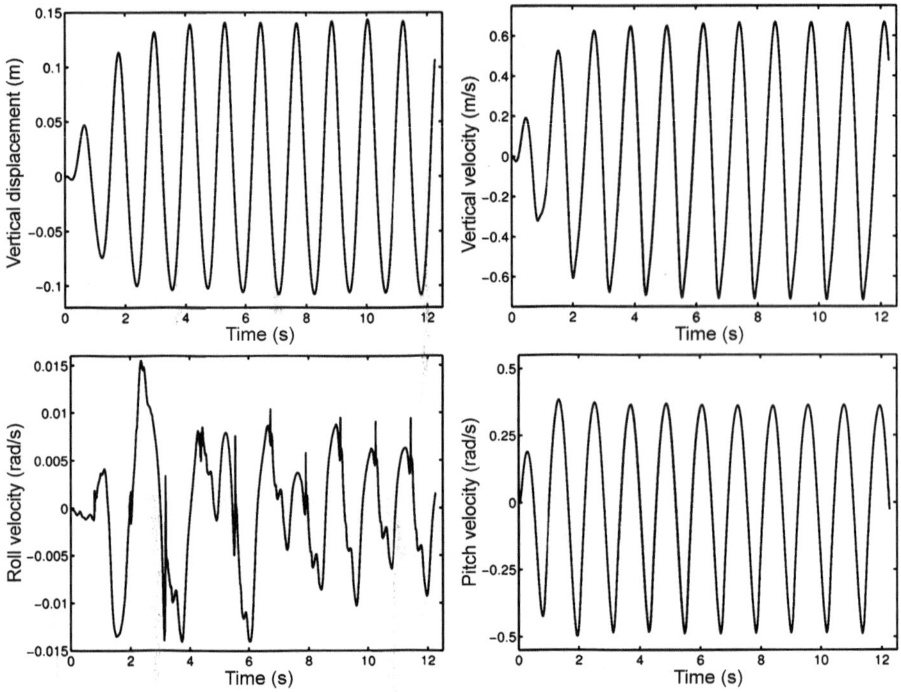

Fig. 11 DTMB 5415 in head waves. Time history of ship motion

We compute the ship in head waves, meaning the waves that travel in the direction opposite to that of the ship. We assume that the ship speed is $U_{in} = 1.873$ m/s, which gives the Froude number of 0.25 based on the ship length.

We make use of the linear Airy waves [101] to prescribe inlet boundary conditions. The Airy waves may be derived using potential theory, and are specified as follows: Given, the wave amplitude, wave length and water depth, $A_w = 0.2$ m, $L_w = 5.72$ m and $h = 3.49$ m, respectively, we compute $k = 2\pi/L_w$, the angular wavenumber, $\omega = \sqrt{gk\tanh(kh)}$, the wave phase speed, and $A_v = \frac{\omega A_w}{\sinh(kh)}$, the velocity amplitude. With these definitions, the Airy waves are given by

$$u = A_v \cosh(kz)\cos(kx - \omega t) + U_{in} \tag{3}$$

$$v = 0 \tag{4}$$

$$w = A_v \sinh(kz)\sin(kx - \omega t) \tag{5}$$

$$\phi = A_w \cos(kx - \omega t) + h - z, \tag{6}$$

where $(u, v, w)^T$ is the fluid velocity vector and the air-water interface in the hydrostatic configuration is assumed to be located at $z = 0$.

The simulation was performed on a mesh consisting of 6,285,445 linear tetrahedral elements and 1,059,174 nodes. The simulation took 5,000 time steps at a fixed

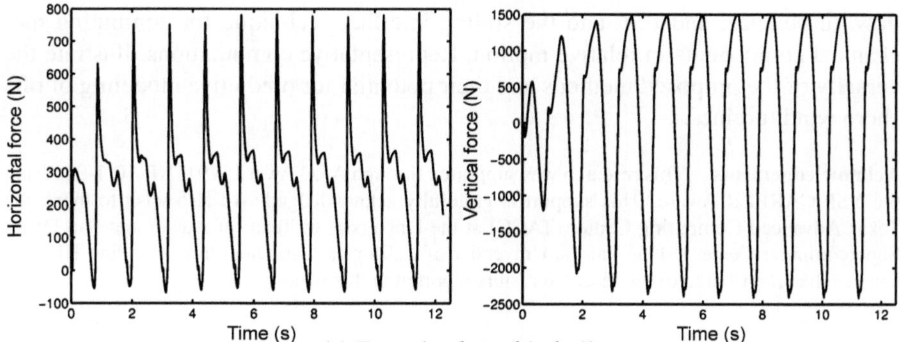

(a) Force load on ship hull

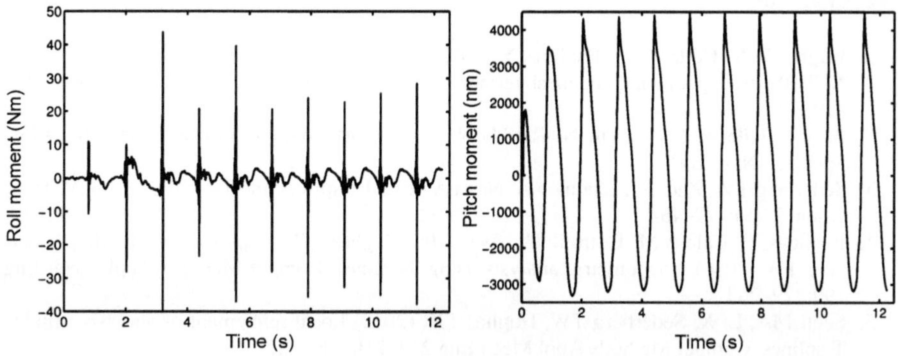

(b) Moment load on ship hull

Fig. 12 DTMB 5415 in head waves. Time history of forces and moments acting on the ship

time step size of $\Delta t = 0.0025$ s. The ship was allowed to move vertically, to pitch and to roll, while the rest of the rigid body degrees-of-freedom were constrained.

Figure 10 shows the snapshots of the ship negotiating high-amplitude waves. The bottom part of Fig. 10 shows the ship partially submerged in water, which is a result of the oncoming wave hitting the bow of the ship. In this case, near the bow, the free surface experiences topological changes, which necessitates the use of an interface-capturing method for this class of problems.

The time history of the ship motion is given in Fig. 11. Although the prescribed Airy waves only have a single frequency, multiple frequencies are present in the ship's response. Note that the ship develops a low-amplitude, chaotic rolling motion.

Figure 12a shows the time history of the thrust force necessary to maintain the ship moving forward at constant speed. The time history of the forces and moments in the unconstrained directions are shown in Figs. 12b–d.

4 Conclusions

We presented a collection of numerical techniques for modeling fluids, structures, and FSI, which include isogeometric analysis, ALE-VMS methods for free-surface flow, turbulence, and FSI, and the sliding interface technique for computing mechanical components in relative motion. Representative computations illustrate the veracity of the proposed methods and their potential for predictive modeling of offshore wind turbines.

Acknowledgements This research was supported through ARO Award W911NF-10-1-0247 and the NSF CAREER Award. This support is gratefully acknowledged. We also wish to thank the Texas Advanced Computing Center (TACC) at the University of Texas at Austin and San Diego Supercomputer Center (SDSC) at the University of California, San Diego for providing HPC resources that contributed to the research results reported in this paper.

References

1. Hughes TJR, Cottrell JA, Bazilevs Y (2005) Isogeometric analysis: CAD, finite elements, NURBS, exact geometry, and mesh refinement. Comput Methods Appl Mech Eng 194:4135–4195
2. Piegl L, Tiller W (1997) The NURBS book, 2nd edn. Monographs in visual communication. Springer, New York
3. Sederberg TW, Zheng J, Bakenov A, Nasri A (2003) T-splines and T-NURCCS. ACM Trans Graph 22(3):477–484
4. Bazilevs Y, Calo VM, Cottrell JA, Evans JA, Hughes TJR, Lipton S, Scott MA, Sederberg TW (2010) Isogeometric analysis using T-splines. Comput Methods Appl Mech Eng 199:229–263
5. Scott MA, Li X, Sederberg TW, Hughes TJR (2012) Local refinement of analysis-suitable T-splines. Comput Methods Appl Mech Eng 213–216:206–222
6. Li X, Zheng J, Sederberg TW, Hughes TJR, Scott MA (2012) On linear independence of T-spline blending functions. Comput-Aided Geom Des 29:63–76
7. Wang W, Zhang Y, Xu G, Hughes TJR (2012) Converting an unstructured quadrilateral/hexahedral mesh to a rational T-spline. Comput Mech. doi:10.1007/s00466-011-0674-6
8. Cottrell JA, Hughes TJR, Bazilevs Y (2009) Isogeometric analysis: toward integration of CAD and FEA. Wiley, Chichester
9. Bazilevs Y, Calo VM, Cottrel JA, Hughes TJR, Reali A, Scovazzi G (2007) Variational multiscale residual-based turbulence modeling for large eddy simulation of incompressible flows. Comput Methods Appl Mech Eng 197:173–201
10. Bazilevs Y, Michler C, Calo VM, Hughes TJR (2007) Weak Dirichlet boundary conditions for wall-bounded turbulent flows. Comput Methods Appl Mech Eng 196:4853–4862
11. Akkerman I, Bazilevs Y, Calo VM, Hughes TJR, Hulshoff S (2008) The role of continuity in residual-based variational multiscale modeling of turbulence. Comput Mech 41:371–378
12. Bazilevs Y, Michler C, Calo VM, Hughes TJR (2010) Isogeometric variational multiscale modeling of wall-bounded turbulent flows with weakly enforced boundary conditions on unstretched meshes. Comput Methods Appl Mech Eng 199:780–790
13. Bazilevs Y, Hsu M-C, Akkerman I, Wright S, Takizawa K, Henicke B, Spielman T, Tezduyar TE (2011) 3D simulation of wind turbine rotors at full scale. Part I: Geometry modeling and aerodynamics. Int J Numer Methods Fluids 65:207–235
14. Bazilevs Y, Akkerman I (2010) Large eddy simulation of turbulent Taylor–Couette flow using isogeometric analysis and the residual–based variational multiscale method. J Comput Phys 229:3402–3414
15. Hsu M-C, Akkerman I, Bazilevs Y (2011) High-performance computing of wind turbine aerodynamics using isogeometric analysis. Comput Fluids 49:93–100
16. Akkerman I, Bazilevs Y, Kees CE, Farthing MW (2011) Isogeometric analysis of free-surface flow. J Comput Phys 230:4137–4152

17. Brooks AN, Hughes TJR (1982) Streamline upwind/Petrov–Galerkin formulations for convection dominated flows with particular emphasis on the incompressible Navier–Stokes equations. Comput Methods Appl Mech Eng 32:199–259
18. Hughes TJR, Tezduyar TE (1984) Finite element methods for first-order hyperbolic systems with particular emphasis on the compressible Euler equations. Comput Methods Appl Mech Eng 45:217–284
19. Tezduyar TE, Park YJ (1986) Discontinuity capturing finite element formulations for nonlinear convection-diffusion-reaction equations. Comput Methods Appl Mech Eng 59:307–325
20. Hughes TJR, Franca LP, Balestra M (1986) A new finite element formulation for computational fluid dynamics: V. Circumventing the Babuška–Brezzi condition: a stable Petrov–Galerkin formulation of the Stokes problem accommodating equal-order interpolations. Comput Methods Appl Mech Eng 59:85–99
21. Tezduyar TE, Osawa Y (2000) Finite element stabilization parameters computed from element matrices and vectors. Comput Methods Appl Mech Eng 190:411–430
22. Tezduyar TE (2003) Computation of moving boundaries and interfaces and stabilization parameters. Int J Numer Methods Fluids 43:555–575
23. Hughes TJR, Scovazzi G, Franca LP (2004) Multiscale and stabilized methods. In: Stein E, de Borst R, Hughes TJR (eds) Encyclopedia of computational mechanics, vol. 3, fluids. Wiley, New York, Chapter 2
24. Tezduyar TE, Sathe S (2007) Modeling of fluid–structure interactions with the space–time finite elements: Solution techniques. Int J Numer Methods Fluids 54:855–900
25. Hsu M-C, Bazilevs Y, Calo VM, Tezduyar TE, Hughes TJR (2010) Improving stability of stabilized and multiscale formulations in flow simulations at small time steps. Comput Methods Appl Mech Eng 199:828–840
26. Hughes TJR (1995) Multiscale phenomena: Green's functions, the Dirichlet-to-Neumann formulation, subgrid scale models, bubbles, and the origins of stabilized methods. Comput Methods Appl Mech Eng 127:387–401
27. Hughes TJR, Sangalli G (2007) Variational multiscale analysis: the fine-scale Green's function, projection, optimization, localization, and stabilized methods. SIAM J Numer Anal 45:539–557
28. Hughes TJR, Liu WK, Zimmermann TK (1981) Lagrangian–Eulerian finite element formulation for incompressible viscous flows. Comput Methods Appl Mech Eng 29:329–349
29. Takizawa K, Bazilevs Y, Tezduyar TE (2012) Space–time and ALE-VMS techniques for patient-specific cardiovascular fluid–structure interaction modeling. Arch Comput Methods Eng 19:171–225
30. Bazilevs Y, Hsu M-C, Takizawa K, Tezduyar TE (2012) ALE-VMS and ST-VMS methods for computer modeling of wind-turbine rotor aerodynamics and fluid–structure interaction. Math Models Methods Appl Sci 22(suppl2):1230002
31. Bazilevs Y, Hughes TJR (2007) Weak imposition of Dirichlet boundary conditions in fluid mechanics. Comput Fluids 36:12–26
32. Hsu M-C, Akkerman I, Bazilevs Y (2012) Wind turbine aerodynamics using ALE–VMS: Validation and the role of weakly enforced boundary conditions. Comput Mech. doi:10.1007/s00466-012-0686-x
33. Osher S, Fedkiw R (2003) Level set methods and dynamic implicit surfaces. Applied mathematical sciences, vol 153. Springer, New York
34. Sethian JA (1999) Level set methods and fast marching methods. Cambridge University Press, Cambridge
35. Sussman M, Smereka P, Osher SJ (1994) A level set approach for computing solutions to incompressible two-phase flows. J Comput Phys 114:146–159
36. Kees CE, Akkerman I, Farthing MW, Bazilevs Y (2011) A conservative level set method suitable for variable-order approximations and unstructured meshes. J Comput Phys 230:4536–4558
37. Benson DJ, Bazilevs Y, Hsu M-C, Hughes TJR (2010) Isogeometric shell analysis: the Reissner–Mindlin shell. Comput Methods Appl Mech Eng 199:276–289

38. Kiendl J, Bletzinger K-U, Linhard J, Wüchner R (2009) Isogeometric shell analysis with Kirchhoff–Love elements. Comput Methods Appl Mech Eng 198:3902–3914
39. Benson DJ, Bazilevs Y, De Luycker E, Hsu M-C, Scott M, Hughes TJR, Belytschko T (2010) A generalized finite element formulation for arbitrary basis functions: from isogeometric analysis to XFEM. Int J Numer Methods Eng 83:765–785
40. Benson DJ, Bazilevs Y, Hsu M-C, Hughes TJR (2011) A large deformation, rotation-free, isogeometric shell. Comput Methods Appl Mech Eng 200:1367–1378
41. Kiendl J, Bazilevs Y, Hsu M-C, Wüchner R, Bletzinger K-U (2010) The bending strip method for isogeometric analysis of Kirchhoff–Love shell structures comprised of multiple patches. Comput Methods Appl Mech Eng 199:2403–2416
42. Bazilevs Y, Hsu M-C, Kiendl J, Wüchner R, Bletzinger K-U (2011) 3D simulation of wind turbine rotors at full scale. Part II: Fluid–structure interaction modeling with composite blades. Int J Numer Methods Fluids 65:236–253
43. Bazilevs Y, Hsu M-C, Kiendl J, Benson DJ (2012) A computational procedure for prebending of wind turbine blades. Int J Numer Methods Eng 89:323–336
44. Hsu M-C, Bazilevs Y (2012) Fluid–structure interaction modeling of wind turbines: simulating the full machine. Comput Mech. doi:10.1007/s00466-012-0772-0
45. Dörfel MR, Jüttler B, Simeon B (2010) Adaptive isogeometric analysis by local h-refinement with T-splines. Comput Methods Appl Mech Eng 199:264–275
46. Cirak F, Ortiz M, Schröder P (2000) Subdivision surfaces: a new paradigm for thin shell analysis. Int J Numer Methods Eng 47:2039–2072
47. Cirak F, Ortiz M (2001) Fully C^1-conforming subdivision elements for finite deformation thin shell analysis. Int J Numer Methods Eng 51:813–833
48. Cirak F, Scott MJ, Antonsson EK, Ortiz M, Schröder P (2002) Integrated modeling, finite-element analysis, and engineering design for thin-shell structures using subdivision. Comput Aided Des 34:137–148
49. Oñate E, Zarate F (2000) Rotation-free triangular plate and shell elements. Int J Numer Methods Eng 47:557–603
50. Oñate E, Flores FG (2005) Advances in the formulation of the rotation-free basic shell triangle. Comput Methods Appl Mech Eng 194:2406–2443
51. Nguyen-Thanh N, Kiendl J, Nguyen-Xuan H, Wüchner R, Bletzinger KU, Bazilevs Y, Rabczuk T (2011) Rotation-free isogeometric thin shell analysis using PHT-splines. Comput Methods Appl Mech Eng 200:3410–3424
52. Tezduyar TE (2001) Finite element methods for flow problems with moving boundaries and interfaces. Arch Comput Methods Eng 8:83–130
53. Takizawa K, Tanizawa K, Yabe T, Tezduyar TE (2007) Ship hydrodynamics computations with the CIP method based on adaptive Soroban grids. Int J Numer Methods Fluids 54:1011–1019
54. Akin JE, Tezduyar TE, Ungor M (2007) Computation of flow problems with the mixed interface-tracking/interface-capturing technique (MITICT). Comput Fluids 36:2–11
55. Tezduyar TE, Behr M, Liou J (1992) A new strategy for finite element computations involving moving boundaries and interfaces—the deforming-spatial-domain/space–time procedure: I. The concept and the preliminary numerical tests. Comput Methods Appl Mech Eng 94(3):339–351
56. Tezduyar TE, Behr M, Mittal S, Liou J (1992) A new strategy for finite element computations involving moving boundaries and interfaces—the deforming-spatial-domain/space–time procedure: II. Computation of free-surface flows, two-liquid flows, and flows with drifting cylinders. Comput Methods Appl Mech Eng 94(3):353–371
57. Cruchaga MA, Celentano DJ, Tezduyar TE (2007) Collapse of a liquid column: numerical simulation and experimental validation. Comput Mech 39:453–476
58. Cruchaga M, Celentano D, Tezduyar T (2001) A moving Lagrangian interface technique for flow computations over fixed meshes. Comput Methods Appl Mech Eng 191:525–543
59. Akkerman I, Bazilevs Y, Benson DJ, Farthing MW, Kees CE (2012) Free-surface flow and fluid–object interaction modeling with emphasis on ship hydrodynamics. J Appl Mech

79:010905
60. Akkerman I, Dunaway J, Kvandal J, Spinks J, Bazilevs Y (2012) Toward free-surface modeling of planing vessels: simulation of the Fridsma hull using ALE-VMS. Comput Mech. doi:10.1007/s00466-012-0770-2
61. Bazilevs Y, Hsu M-C, Scott MA (2012) Isogeometric fluid–structure interaction analysis with emphasis on non-matching discretizations, and with application to wind turbines. Comput Methods Appl Mech Eng. doi:10.1016/j.cma.2012.03.028
62. Takizawa K, Henicke B, Montes D, Tezduyar TE, Hsu M-C, Bazilevs Y (2011) Numerical-performance studies for the stabilized space–time computation of wind-turbine rotor aerodynamics. Comput Mech 48:647–657
63. Takizawa K, Henicke B, Tezduyar TE, Hsu M-C, Bazilevs Y (2011) Stabilized space–time computation of wind-turbine rotor aerodynamics. Comput Mech 48:333–344
64. Hsu M-C, Akkerman I, Bazilevs Y (2012) Finite element simulation of wind turbine aerodynamics: validation study using NREL Phase VI experiment. Wind Energy, accepted
65. Farhat C, Lesoinne M, Le Tallec P (1998) Load and motion transfer algorithms for fluid/structure interaction problems with non-matching discrete interfaces: Momentum and energy conservation, optimal discretization and application to aeroelasticity. Comput Methods Appl Mech Eng 157:95–114
66. Takizawa K, Tezduyar TE (2011) Multiscale space–time fluid–structure interaction techniques. Comput Mech 48:247–267
67. Takizawa K, Tezduyar TE (2012) Space–time fluid–structure interaction methods. Math Models Methods Appl Sci. doi:10.1142/S0218202512300013
68. Takizawa K, Tezduyar TE (2012) Computational methods for parachute fluid–structure interactions. Arch Comput Methods Eng 19:125–169
69. Tezduyar TE, Sathe S, Stein K (2006) Solution techniques for the fully-discretized equations in computation of fluid–structure interactions with the space–time formulations. Comput Methods Appl Mech Eng 195:5743–5753
70. Tezduyar TE, Sathe S, Keedy R, Stein K (2006) Space–time finite element techniques for computation of fluid–structure interactions. Comput Methods Appl Mech Eng 195:2002–2027
71. Tezduyar TE, Sathe S, Pausewang J, Schwaab M, Christopher J, Crabtree J (2008) Interface projection techniques for fluid–structure interaction modeling with moving-mesh methods. Comput Mech 43:39–49
72. Tezduyar TE, Sathe S, Pausewang J, Schwaab M, Christopher J, Crabtree J (2008) Fluid–structure interaction modeling of ringsail parachutes. Comput Mech 43:133–142
73. Takizawa K, Moorman C, Wright S, Spielman T, Tezduyar TE (2011) Fluid–structure interaction modeling and performance analysis of the Orion spacecraft parachutes. Int J Numer Methods Fluids 65:271–285
74. Takizawa K, Wright S, Moorman C, Tezduyar TE (2011) Fluid–structure interaction modeling of parachute clusters. Int J Numer Methods Fluids 65:286–307
75. Takizawa K, Spielman T, Tezduyar TE (2011) Space–time FSI modeling and dynamical analysis of spacecraft parachutes and parachute clusters. Comput Mech 48:345–364
76. Bazilevs Y, Hughes TJR (2008) NURBS-based isogeometric analysis for the computation of flows about rotating components. Comput Mech 43:143–150
77. Tezduyar T, Aliabadi S, Behr M, Johnson A, Kalro V, Litke M (1996) Flow simulation and high performance computing. Comput Mech 18:397–412
78. Behr M, Tezduyar T (1999) The shear-slip mesh update method. Comput Methods Appl Mech Eng 174:261–274
79. Behr M, Tezduyar T (2001) Shear-slip mesh update in 3D computation of complex flow problems with rotating mechanical components. Comput Methods Appl Mech Eng 190:3189–3200
80. Tezduyar TE (2007) Finite elements in fluids: special methods and enhanced solution techniques. Comput Fluids 36:207–223

81. Chung J, Hulbert GM (1993) A time integration algorithm for structural dynamics withimproved numerical dissipation: the generalized-α method. J Appl Mech 60:371–375
82. Jansen KE, Whiting CH, Hulbert GM (2000) A generalized-α method for integrating the filtered Navier–Stokes equations with a stabilized finite element method. Comput Methods Appl Mech Eng 190:305–319
83. Bazilevs Y, Calo VM, Hughes TJR, Zhang Y (2008) Isogeometric fluid–structure interaction: theory, algorithms, and computations. Comput Mech 43:3–37
84. Jonkman J, Butterfield S, Musial W, Scott G (2009) Definition of a 5-MW reference wind turbine for offshore system development. Technical report NREL/TP-500-38060, National Renewable Energy Laboratory, Golden, CO
85. Jonkman JM (2009) Dynamics of offshore floating wind turbines–model development and verification. Wind Energy 12:459–492
86. Jonkman JM, Matha D (2011) Dynamics of offshore floating wind turbines–analysis of three concepts. Wind Energy 14:557–569
87. Lackner MA, Rotea MA (2011) Passive structural control of offshore wind turbines. Wind Energy 14:373–388
88. Nicklasson PJ, Homola MC, Virk MS, Sundsbø PA (2012) Performance losses due to ice accretion for a 5 MW wind turbine. Wind Energy 15:379–389
89. Chow R, van Dam CP (2011) Verincation of computational simulations of the NREL 5 MW rotor with a focus on inboard now separation. Wind Energy. doi:10.1002/we.529
90. Lackner MA (2012) An investigation of variable power collective pitch control for load mitigation of floating offshore wind turbines. Wind Energy. doi:10.1002/we.1500
91. Sebastian T, Lackner MA (2012) Characterization of the unsteady aerodynamics of offshore floating wind turbines. Wind Energy. doi:10.1002/we.545
92. Bazilevs Y, Hsu M-C, Takizawa K, Tezduyar TE (2012) ALE–VMS and ST–VMS methods for computer modeling of wind-turbine rotor aerodynamics and fluid–structure interaction. Math Models Methods Appl Sci. doi:10.1142/S0218202512300025
93. Hsu M-C (2012) Fluid–structure interaction analysis of wind turbines. PhD thesis, University of California, San Diego
94. Tezduyar TE, Behr M, Mittal S, Johnson AA (1992) Computation of unsteady incompressible flows with the finite element methods—space–time formulations, iterative strategies and massively parallel implementations. In: New methods in transient analysis. ASME, New York, pp 7–24. PVP-Vol. 246/AMD-Vol.143
95. Tezduyar T, Aliabadi S, Behr M, Johnson A, Mittal S (1993) Parallel finite-element computation of 3D flows. Computer 26(10):27–36
96. Johnson AA, Tezduyar TE (1994) Mesh update strategies in parallel finite element computations of flow problems with moving boundaries and interfaces. Comput Methods Appl Mech Eng 119:73–94
97. Hau E (2006) Wind turbines: fundamentals, technologies, application, economics, 2nd edn. Springer, Berlin
98. Longo J, Stern F (2005) Uncertainty assessment for towing tank tests with example for surface combatant DTMB model 5415. J Ship Res 49:55–68
99. Garciá J, Oñate E (2003) An unstructured finite element solver for ship hydrodynamics problems. J Appl Mech 70. doi:10.1115/1.1530631
100. Longo J, Shao J, Irvine M, Stern F (2007) Phase-averaged piv for the nominal wake of a surface ship in regular head waves. J Fluids Eng 129:524–541
101. McCormick ME (2010) Ocean engineering mechanics. With applications. Cambridge University Press, Cambridge

Advances in the Development of a Time-Domain Unstructured Finite Element Method for the Analysis of Waves and Floating Structures Interaction

Borja Servan-Camas and Julio Garcia-Espinosa

Abstract Being capable of predicting wave-structure interaction in the time domain is of great interest for the offshore industry. However, most computer programs used in the industry work in the frequency domain. Therefore, the main objective of this work is the development a time domain solver based on the finite element method capable of solving wave-structure interaction problems using unstructured meshes. We found good agreement between the numerical results we obtained and analytical solutions as well as numerical solutions obtained by other numerical method.

Keywords Wave structure interaction · Ocean energy, finite element · Time domain · Unstructured mesh, seakeeping

1 Introduction

Wave structure interaction is a topic of great interest in naval and offshore engineering. This interest is growing in the last years due to the boost given by the development of the marine renewable energy field. In this context, the development of time-domain wave structure interaction programs is a main request from the industry.

Up to date the numerical simulation of wave structure interaction has been mostly carried out using the frequency domain. The reason might be that the computational cost of time domain simulations were too high and computational time was too large. Moreover, assumptions like linear waves and the harmonic nature of water waves made the frequency domain to be the right choice. However, nowadays computing capabilities make possible to carry out numerical simulations in the time

B. Servan-Camas (✉) · J. Garcia-Espinosa
Centre Internacional de Mètodes Numèrics en Enginyeria (CIMNE), Campus Norte UPC, 08034 Barcelona, Spain
e-mail: bservan@cimne.upc.edu
url: http://www.cimne.com

J. Garcia-Espinosa
Universitat Politècnica de Catalunya – BarcelonaTech (UPC), Campus Nàutica, Edif. NT3, C. Escar 6-8, 08039 Barcelona, Spain

L. Eça et al. (eds.), *MARINE 2011, IV International Conference on Computational Methods in Marine Engineering*, Computational Methods in Applied Sciences 29,
DOI 10.1007/978-94-007-6143-8_7, © Springer Science+Business Media Dordrecht 2013

domain in a reasonable time, with the advantage of making easier the coupling with other phenomena.

Regarding the numerical method usually adopted, the boundary element method (BEM) has dominated over others like finite element method (FEM). The main advantage of BEM over FEM resides in the fact that only boundary meshes are required, while FEM demands meshing the whole volume, with the corresponding increase in the number of variables of the discrete problem. However, despite of the higher number of variables required by FEM, it is not clear that BEM has to be more efficient. Mostly due to the sparse pattern in FEM and the large availability of iterative solver preconditioners that can improve the resolution of the corresponding linear system of equations. In [1] Cai et al. a heuristic comparison between both methods is given and demonstrate that a solution to the boundary value problem can be obtained more efficiently by the FEM.

In the last decade, there have been extensive applications of the finite element method (FEM) to free surface problems. For example, Oñate and García [2] presented a stabilized FEM for fluid structure interaction in the presence of free surface where the latter was modelled by solving a fictitious elastic problem on the moving mesh. In [3, 4] Löhner et al. developed a FEM capable of tracking violent free surface flows interacting with objects. Also García et al. [5] developed a new technique to track complex free surface shapes. However, many works like the previous ones are based on solving the Navier-Stokes equations, too expensive computationally speaking when it comes to simulating many real problems regarding ocean waves interacting with floating structures, which can be more cheaply simulated using potential flow theory along with Stokes perturbation approximation.

Despite of the great effort invested in the last years to the development of FEM algorithms, to the authors' knowledge, yet there has not been developed a FEM for solving first order waves, based on Stokes perturbation, interacting with structures in the time domain using unstructured meshes. The use of structure or semi-structures meshes is a big limitation since it limits the complexity of the geometry to be used. In this study we present a FEM for wave-structure interaction that can be used with unstructured meshes. Besides, since it is based on Stokes wave theory, no re-meshing or moving mesh technique are needed, which keeps computational costs and times low. The algorithm has been adapted to include non-linear external forces, like those used to define mooring systems, and variations on the pressure over the free surface.

2 Problem Statement

2.1 Governing Equations

We consider the first order diffraction-radiation problem of a floating body.

$$\nabla^2 \varphi = 0 \quad \text{in } \Omega \tag{1}$$

$$\partial_t \varphi + g\eta = -P_a/\rho + C \quad \text{in } z = 0$$

$$\text{(dynamic free surface boundary condition)} \tag{2}$$

$$\partial_t \eta - \partial_z \varphi = 0 \quad \text{in } z = 0 \quad \text{(kinematic free surface boundary condition)} \tag{3}$$

$$\partial_z \varphi = 0 \quad \text{in } z = -H \tag{4}$$

$$\nabla\varphi \cdot \mathbf{n}_B = \mathbf{v}_B \cdot \mathbf{n}_B \quad \text{in } \Gamma_B \tag{5}$$

where φ and η are the first order potential and free surface elevation respectively; Ω is the fluid domain bounded by $z = 0$; P_a is the atmospheric pressure; ρ is the water density; C is a constant value; Γ_B represents the wetted surface of a floating body; and H is the water depth. The domain is assumed to be infinite in the horizontal directions.

2.2 Velocity Potential Decomposition

The aim of this work is to simulate the dynamics of a floating body subjected to the action of waves. To do so, we will first model the environment as the sum of a number of airy waves. This can be expressed in terms of a velocity potential given by:

$$\psi = \sum_m \frac{A_m g}{\omega_m} \frac{\cosh(|\mathbf{k}_m|(H+z))}{\cosh(|\mathbf{k}_m|H)} \cos\big(|\mathbf{k}_m|(x\cos\alpha_m + y\sin\alpha_m - \omega_m t + \delta_m)\big) \tag{6}$$

where A_m are the wave amplitudes; ω_m are the wave frequencies; $\mathbf{k}_m$ are the wave numbers; α_m are the wave directions; and δ_m are the wave phases. From this point on, we will refer to ψ as the incident potential. This potential, along with the dispersion relation $\omega_m^2 = g|\mathbf{k}_m|\tanh(|\mathbf{k}_m|H)$, fulfils Eqs. (1)–(4), and therefore is solution of the mathematical model in the absence of bodies.

Let $\varphi = \psi + \phi$ be the solution to the governing equations. The equations to be fulfilled by ϕ are

$$\nabla^2 \phi = 0 \quad \text{in } \Omega \tag{7}$$

$$\partial_t \phi + g\eta = -P_a/\rho + C' \quad \text{in } z = 0$$

$$\text{(dynamic free surface boundary condition)} \tag{8}$$

$$\partial_t \eta - \partial_z \phi = 0 \quad \text{in } z = 0 \quad \text{(kinematic free surface boundary condition)} \tag{9}$$

$$\partial_z \phi = 0 \quad \text{in } z = -H \tag{10}$$

$$\nabla\phi \cdot \mathbf{n}_B = (\mathbf{v}_B - \nabla\psi) \cdot \mathbf{n}_B \quad \text{in } \Gamma_B \tag{11}$$

2.3 Radiation Condition and Wave Absorption

We will make use of a Sommerfeld radiation condition at the edge of the computational domain: $\partial_t \phi + c\nabla\phi \cdot \mathbf{n}_R = 0$ in the surface limiting of the domain in the horizontal directions, and c is a prescribed wave velocity. Wave dissipation is also introduced into the dynamic free surface boundary condition by varying the pressure such that $P_a/\rho = P_0 + \kappa(\mathbf{x})\partial_z\phi$ where $\kappa(\mathbf{x})$ is a damping coefficient. Combining the dynamic and kinematic boundary condition, introducing the wave absorption and choosing $C' = P_0$, the governing equations for ϕ become:

$$\nabla^2\phi = 0 \quad \text{in } \Omega \tag{12}$$

$$\partial_{tt}\phi = -g\partial_z\phi - \kappa(\mathbf{x})\partial_t\partial_z\phi \quad \text{in } z = 0 \tag{13}$$

$$\partial_z\phi = 0 \quad \text{in } z = -H \tag{14}$$

$$\nabla\phi \cdot \mathbf{n}_B = (\mathbf{v}_B - \nabla\psi) \cdot \mathbf{n}_B \quad \text{in } \Gamma_B \tag{15}$$

$$\partial_t\phi + c\nabla\phi \cdot \mathbf{n}_R = 0 \quad \text{in } \Gamma_R \tag{16}$$

$$\eta = -\frac{1}{g}\partial_t\phi - \frac{P_a}{\rho g} + \frac{C'}{g} \quad \text{in } z = 0$$
$$\text{(kinematic free surface boundary condition)} \tag{17}$$

3 Finite Element Formulation

The discrete variational problem can be written as:

$$\int_\Omega \nabla v_h \cdot \nabla\phi_h \, d\Omega = \int_{\Gamma^B} v_h \cdot \widehat{\phi}_n^B \, d\Gamma + \int_{\Gamma^R} v_h \cdot \widehat{\phi}_n^R \, d\Gamma + \int_{\Gamma^{Z_0}} v_h \cdot \widehat{\phi}_n^{Z_0} \, d\Gamma$$
$$+ \int_{\Gamma^{Z_{-H}}} v_h \cdot \widehat{\phi}_n^{Z_{-H}} \, d\Gamma \quad \forall v_h \in Q_h \tag{18}$$

where $\widehat{\phi}_n^B$, $\widehat{\phi}_n^R$, $\widehat{\phi}_n^{Z_0}$ and $\widehat{\phi}_n^{Z_{-H}}$ are the potential normal gradients corresponding to the Neumann boundary conditions on body, radiation boundary, free surface and bottom, respectively. The associated matrix form is:

$$\overline{\overline{\mathbf{L}}}\boldsymbol{\phi} = \mathbf{b}^B + \mathbf{b}^R + \mathbf{b}^{Z_0} + \mathbf{b}^{Z_{-H}} \tag{19}$$

3.1 Boundary Conditions

The right hand side of Eq. (19) is implemented as follows:

$$\left(\mathbf{b}^{Z_0}\right)^{n+1} = -\overline{\overline{\mathbf{M}}}_{\Gamma^{Z_0}}\left(\frac{12}{g\Delta t^2 + 6\kappa(\mathbf{x})\Delta t}\left(\boldsymbol{\phi}^{n+1} - 2\boldsymbol{\phi}^n + \boldsymbol{\phi}^{n-1}\right)\right)$$
$$-\overline{\overline{\mathbf{M}}}_{\Gamma^{Z_0}}\left(\frac{10g\Delta t}{g\Delta t + 6\kappa(\mathbf{x})}\left(\boldsymbol{\phi}_z^{Z_0}\right)^n + \left(\frac{g\Delta t - 6\kappa(\mathbf{x})}{g\Delta t + 6\kappa(\mathbf{x})}\right)\left(\boldsymbol{\phi}_z^{Z_0}\right)^{n-1}\right) \tag{20}$$

$$(\mathbf{b}^R)^{n+1} = \overline{\overline{\mathbf{M}}}_{\Gamma^R}(\boldsymbol{\phi}_n^R)^{n+1} = \frac{1}{\Delta t}\overline{\overline{\mathbf{M}}}_{\Gamma^R}(\boldsymbol{\phi}^n - \boldsymbol{\phi}^{n-1}) \tag{21}$$

$$(\mathbf{b}^B)^{n+1} = \overline{\overline{\mathbf{M}}}_{\Gamma^B}(\boldsymbol{\phi}_n^B)^{n+1} \tag{22}$$

The free surface and pressure are computed by the following fourth order finite difference scheme:

$$\eta^{n+1} = -\frac{1}{g\Delta t}\left(\frac{25}{12}\varphi^{n+1} - 4\varphi^n + 3\varphi^{n-1} - \frac{4}{3}\varphi^{n-2} + \frac{1}{4}\varphi^{n-3}\right) \tag{23}$$

$$p^{n+1} = -\rho g z - \frac{\rho}{\Delta t}\left(\frac{25}{12}\varphi^{n+1} - 4\varphi^n + 3\varphi^{n-1} - \frac{4}{3}\varphi^{n-2} + \frac{1}{4}\varphi^{n-3}\right) \tag{24}$$

3.2 Body Dynamics

Integrating the pressure over the body surface, the resulting forces and moments are obtained. On the other hand, the body dynamics is given by the equation of motion:

$$\overline{\overline{M}}\mathbf{X}_{tt} + \overline{\overline{K}}\mathbf{X} = \mathbf{F} \tag{25}$$

where $\overline{\overline{M}}$ is the mass matrix; $\overline{\overline{K}}$ is the hydrostatic restoring coefficient matrix; $\mathbf{F}$ are the hydrodynamic forces induced over the body plus any other external forces; and $\mathbf{X}$ represent the movement of the six degrees of freedom. We use an implicit Newmark's average acceleration method to carry out the temporal integration of Eq. (25).

3.3 Free Surface Boundary Condition for OWC Calculations

In order to be able to simulate OWC devices, a non-linear free surface boundary condition has been developed based on the characteristic curves of the Wells-turbines type used in these devices. For instance, based on working ideal conditions, an actual 18.5 Kw Wells turbine get a power output and pressure that relates to the flux as:

$$p(q) = 0.0779|q|^3 - 0.065q^2 + 0.1933|q| \tag{26}$$

$$P(q) = 164.07 \cdot q \cdot |q| \tag{27}$$

where p is the output power in kilowatts, P is the pressure drop across the turbine in Pascals, and q is the instantaneous air flux flowing through the turbine in cubic

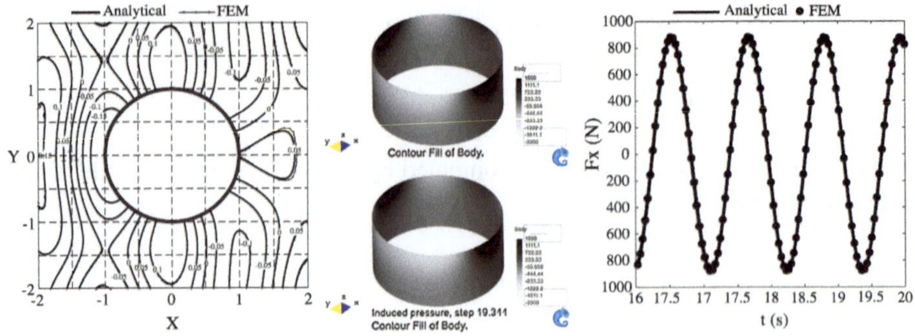

Fig. 1 *Left*: Contour lines of free surface elevation at $t = nT$. Comparison between analytical (*solid line*) and FEM (*dot line*) results. *Center*: Pressure induced on the cylinder by the velocity potential at time $t = nT$. Comparison between analytical (*up*) and FEM (*down*) results. *Right*: Horizontal force induced over the cylinder. Comparison between analytical (*solid line*) and FEM (*dots*) results

meters per second. The pressure obtained by Eq. (27) is introduced into the dynamic condition of the free surface as

$$\frac{\phi^{n+1} - 2\phi^n + \phi^{n-1}}{\Delta t^2} = -\frac{1}{12} g\big(\partial_z \phi^{n+1} + 10\partial_z \phi^n + \partial_z \phi^{n-1}\big) - \frac{1}{12\rho}\big(\partial_t P^{n+1} + 10\partial_t P^n + \partial_t P^{n-1}\big) \tag{28}$$

$$\partial_t P^{n+1} = \big(3P^{n+1} - 4P^n + P^{n-1}\big)/(2\Delta t) \tag{29}$$

4 Numerical Results

4.1 Waves Refracted by a Vertical Circular Cylinder

In this section we solve the problem of a monochromatic wave interacting with a fix bottom mounted circular cylinder. The analytical solution for the incident and scattered waves can be found in [6]. Next we compare numerical results obtained by the analytical solution with numerical results obtained by our FEM schemes for the specific case of $R = 1$, $H = 1$, $A = 0.1$, $L = 2$. Using $g = 9.81$ and by mean of the dispersion relation for first order waves, we obtain the frequency value $\omega = \sqrt{g\pi \tanh(\pi)} = 5.5411$ rad/s, and the wave period $T = 1.1339$ s. Figure 1 left compares the contour lines of the free surface elevation at any time $t = nT$. It can be observed that the FEM solution mostly lie over the analytical solution. Figure 1 (center) compares the pressure distribution over the cylinder obtained by the analytical solution and the FEM solution. Both pressure distributions are obtained using the same colour scale, with a maximum value of 1500 and a minimum of 2000. Figure 1 right compares the force induced over the cylinder obtained by the analytical

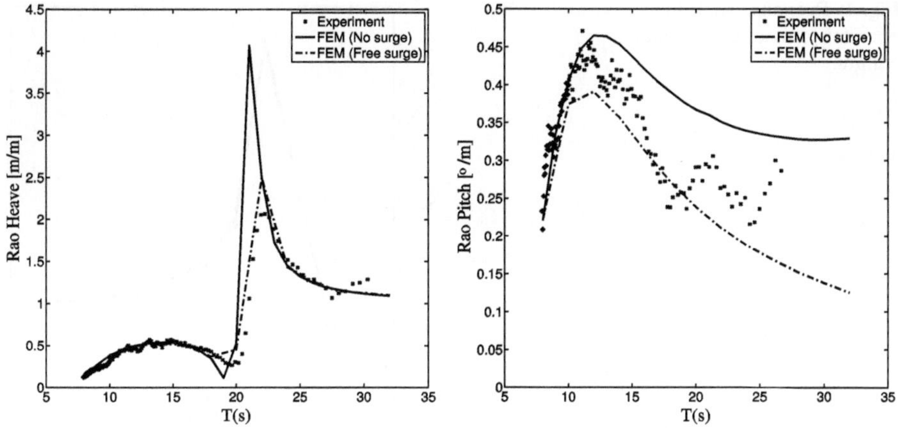

Fig. 2 Heave and pitch response amplitude operators for the GVA 4000. *Dots*: experimental data. *Solid line*: FEM

solution and FEM. It can be seen that the forces obtained in both ways are quite close to each other.

4.2 Seakeeping of a GVA 4000 Semisubmersible Platform

Next we carry out comparison, between experimental data [7] and numerical results obtained via the method presented in this work, regarding the seakeeping behaviour of the GVA 4000 semisubmersible platform. The results to be compared are the heave and pitch response amplitude operators of the GVA 400 in heading seas, with a range of wave periods between 6 and 32 seconds.

The platform displacement is 25940 tonnes. The center of gravity is located 21.35 m above the keel, and the horizontal position corresponds to the geometric center of the platform. The radii of inertia are $rxx = 30.40$ m, $ryy = 31.06$ m, and $rzz = 37.54$ m. The geometry of the GVA 4000 can be found in [7].

Based on [7] the model tests were carried out with the surge movement constraint by the action of a pre-stressed spring whose mission is to keep in place the structure during the testing. Besides, this spring creates also a pitching moment. Therefore, the pitch movement will be influenced not just by the waves, but also by the surge movement and the spring.

Since we have no data regarding the mechanism used by the model basin to keep in place the platform during the tests, we cannot include it in the simulation. Instead, simulations have been carried out in two cases: no surge and free surge. Figure 2 compares the experimental results and numerical FEM results. The experimental results lay within the numerical results obtained for free surge and no surge cases as expected.

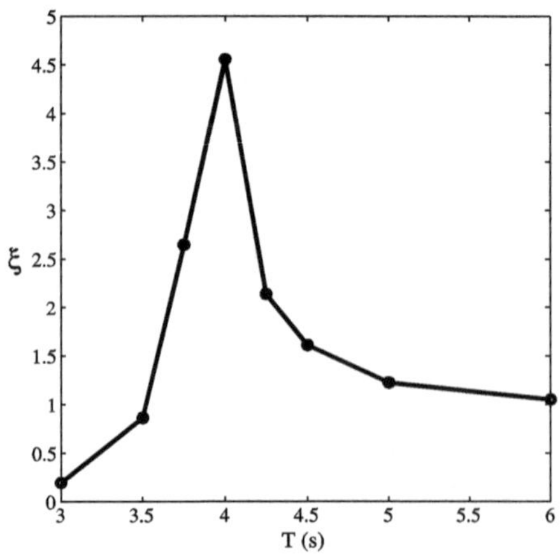

Fig. 3 Response amplitude operator for oscillating water column

4.3 Oscillating Water Column Test

In this section, the FEM model was used to analyze the performance of an OWC device to absorb energy from waves. The OWC consist of a circular bell of 5 meters in diameter (inner), 0.5 meters in thickness, and 2.5 meter in draft. The water depth is 20 meters and the OWC device will be subject to the action of monochromatic waves with periods ranging between 3 and 5 seconds. The free surface elevation was analyzed within the chamber in the absence of the turbine ($P = 0$). Figure 3 shows response amplitude operators obtained as $\xi = Q_{\max}/(A\pi R^2\omega)$, where $Q_{\max}$ is the instantaneous flux amplitude; A is the wave amplitude; R is the inner radius of the OWC device; ω is the wave frequency; and ξ represents the average amplitude of the free surface elevation within the device.

The performance of the same OWC device was also analyzed after installing a Wells-type turbine, the Wavegen 18.5 Kw model. The imposition of Eq. (26) implies that the problem is no linear anymore. Therefore, the performance of the system was analyzed for three wave amplitudes and a range of periods.

Table 1 summarizes the results obtained from the simulation, where the columns named power provide the average power supplied by the device, and the efficiency is obtained from dividing the average power by the power transported by a wave front of 5 meters.

5 Conclusions

A FEM solver for wave structure interaction in the time domain has been presented. The wave modelling is based on Stokes perturbation theory, which allows keeping the same computational domain along the simulation. The FEM has been developed so unstructured meshes can be used, no matter the complexity of the structure.

Table 1 Oscillating water column results

T (s)	$A = 0.1$ m		$A = 0.5$ m		$A = 1$ m	
	Power (kw)	Efficiency	Power (kw)	Efficiency	Power (kw)	Efficiency
3	0.079	13.38 %	0.379	2.58 %	0.923	1.57 %
3.5	0.150	21.85 %	0.762	4.44 %	1.936	2.82 %
4	0.203	25.87 %	1.189	6.06 %	3.135	3.99 %
4.5	0.234	26.44 %	1.587	7.17 %	4.310	4.87 %
5	0.254	25.65 %	2.002	8.09 %	5.590	5.64 %
6	0.259	21.00 %	2.654	8.62 %	7.790	6.32 %
7	0.242	15.97 %	3.082	8.14 %	9.483	6.26 %

Both, the free surface and outlet boundary conditions are based on implicit schemes. They have been introduced within the system matrix so that no iterations are required within the time step to reach convergence among the Laplace and boundary conditions.

FEM results have been compared to analytical results available for a circular vertical cylinder. The agreement between both solutions shows that the algorithms develop in this work perform well. Furthermore a comparison with experimental data for a offshore structure has been carried out, obtaining a good agreement.

Moreover, since the present study is based on a time domain formulation, nonlinear external forces and moments acting over structures can be easily brought into the dynamics of the structure interacting with waves. Nonlinear pressure boundary conditions have been implemented to show how the algorithms can be used to evaluate wave energy converters based on oscillating water columns.

Acknowledgements This study was partially supported by the Ministry for Science and Innovation of Spain in the AIDMAR project CTM2008-06565-C03-01.

The authors also want to acknowledge Prof. Clauss and Dr. Schmittner, for providing the main characteristics of the GVA4000 model used in the experimental tests.

References

1. Cai X, Langtangen HP, Nielsen BF, Tveito A (1998) A finite element method for fully nonlinear water waves. J Comput Phys 143:544–568
2. Oñate E, García J (2001) A finite element method for fluid-structure interaction with surface waves using a finite calculus formulation. Comput Methods Appl Mech Eng 191:635–660
3. Löhner R, Yang C, Oñate E (2006) On the simulation of flows with violent free surface motion. Comput Methods Appl Mech Eng 195:5597–5620
4. Löhner R, Yang C, Oñate E (2007) On the simulation of flows with violent free surface motion and moving objects using unstructured meshes. Comput Methods Appl Mech Eng 53:1315–1338
5. García J, Valls A, Oñate E (2008) ODDLS: a new unstructured mesh finite element method for the analysis of free surface flow problems. Int J Numer Methods Fluids 76(9):1297–1327

6. McCamy R, Fuchs R (1954) Wave forces on piles: a diffraction theory. Tech Memo No 69, US Army Corps of Engrs
7. Clauss GF, Schmittner C, Stutz K (2002) Time-domain investigation of a semisubmersible in rouge waves. In: 21st international conference on offshore mechanics and arctic engineering, Oslo, Norway, June 23–28, 2002. OMAE2002-28450

The Variation in Wake Structure of a Tidal Stream Turbine with Flow Velocity

R. Malki, I. Masters, A.J. Williams, and T.N. Croft

Abstract A combined Blade Element Momentum—Computational Fluid Dynamics (BEM-CFD) model is applied to a 10 m diameter tidal stream turbine blade and the supporting nacelle and tower structure in a 700 m long rectangular channel. The modelling approach is computationally efficient and is suitable for capturing the time-averaged influence of the turbine on the flow. A range of simulations are conducted for the purpose of undertaking a comparative study of the influence of the turbine on mean flow characteristics. Variations in flow structure around the turbine for different flow conditions were evaluated.

Simulations are conducted for a range free-stream velocities typical of potential tidal stream deployment sites, typically up to 3.0 m s^{-1}. Velocity deficit profiles and wake dimensions are evaluated for each flow condition implemented. Downstream flow recovery is strongly linked to the flow velocity, and occurs over a longer distance with increasing velocity. For the range of velocities considered, some properties, such as wake length and the maximum wake length location increase linearly, or nearly-linearly with velocity. Other properties, such as the maximum wake width, and the recovery distance downstream demonstrate a tendency to converge towards a constant value. The key findings of this study highlight the significance of the free-stream velocity as an influence on the flow structure around and downstream of a tidal stream turbine.

Keywords BEM · CFD · Tidal stream turbines · Wakes

1 Introduction

The extraction of tidal stream energy is a relatively new concept although it draws upon the wide knowledge base developed for the wind power industry. Nevertheless, there is much uncertainty associated with tidal power due to the significant contrasts

R. Malki (✉) · I. Masters · A.J. Williams · T.N. Croft
Marine Energy Research Group, Civil and Computational Research Centre, Swansea University, Swansea, UK
e-mail: ramimalki@hotmail.com
url: http://www.swan.ac.uk/engineering/marineenergy/

L. Eça et al. (eds.), *MARINE 2011, IV International Conference on Computational Methods in Marine Engineering*, Computational Methods in Applied Sciences 29,
DOI 10.1007/978-94-007-6143-8_8, © Springer Science+Business Media Dordrecht 2013

between atmospheric and aquatic environments, for example the differences in fluid density and viscosity. These properties have implications on the loads experienced by the devices as well as flow Reynolds numbers.

A number of deployment initiatives are currently underway to investigate the performance of tidal stream turbines in offshore environments [1–3]. Although the outcomes from such initiatives will be very valuable, these deployments are very costly and there can be significant risks to project development investments due to the highly dynamic flows which are likely to occur at the most suitable sites for exploiting tidal stream energy. Understanding how turbine designs are likely to perform in various conditions can be useful at the planning stage to maximise the return on the investment without subjecting the devices to avoidable damage. This is particularly important at the early test trial stages on which the fate of future and more significant scaled-up deployments for large scale power generation may depend.

Future deployments are likely to consist of numerous tidal stream turbines being deployed in predetermined configurations. To maximise the power generation potential per device, blade diameters will be significant to increase their projected area of influence across the flow, and hence, turbine wakes will be large. Ideally, the spacing between the devices should be minimised to maximise the number of devices deployed across a given site without compromising the performance of the devices. This could occur if the devices are too close to each other and hence, turbine blades fall within the wakes generated by upstream or neighbouring turbines. Ultimately, the economics will be the deciding factor when determining the number of devices to deploy. A loss in efficiency is acceptable if more devices increase profits. However, an understanding of wake structures around the turbines is necessary to aid the planning process.

To numerically simulate turbine blades, a moving mesh approach would be suitable for capturing the transient nature of the flow as it interacts with the blades [4]. However, this approach is very computationally expensive. An alternative method, which is presented and used here, is a combined blade element momentum—computational fluid dynamics (BEM-CFD) model where the time-averaged forces acting on the flow due to the blades are introduced into the flow domain in the form of momentum sink terms.

The majority of tidal stream turbine studies to date have focused on the performance of the devices and power outputs in hypothetical flow conditions [5]. An investigation into the wake structure downstream of a turbine and how it varies with flow velocity is presented in this paper to assist the planning of tidal stream turbine array layouts.

2 The BEM-CFD Model

2.1 Governing Equations

The steady-state CFD model solves the Navier-Stokes Equations, which consist of the equations for mass continuity (1) and the conservation of momentum (2) whereby the fluid is treated as incompressible and turbulent.

$$\nabla.u = 0 \tag{1}$$

$$\nabla.(\rho u u_i) = -\frac{\partial p}{\partial x_i} + \nabla.(\mu_{lam} + \mu_t)\nabla u_i + S_i \tag{2}$$

k In these equations, ρ is the fluid density, u_i is the ith component of the velocity vector, p is the pressure, μ_{lam} and μ_t are the dynamic laminar and turbulent viscosities respectively and S_i represents additional source terms into the momentum equations. Turbulence in the flow is resolved using the $k-\varepsilon$ model [6] in which (3) and (4) are solved:

$$\nabla.(\rho u k) = \nabla.\left(\mu_{lam} + \frac{\mu_t}{\sigma_k}\right)\nabla k + \mu_t G - \rho\varepsilon \tag{3}$$

$$\nabla.(\rho u \varepsilon) = \nabla.\left(\mu_{lam} + \frac{\mu_t}{\sigma_\varepsilon}\right)\nabla\varepsilon + \frac{\varepsilon}{k}(C_{1\varepsilon}\mu_t G - C_{2\varepsilon}\rho\varepsilon) \tag{4}$$

Equation (3) is the conservation equation for turbulent kinetic energy, k, and (4) is the conservation equation for the turbulent dissipation rate, ε. The turbulent viscosity, μ_t, is calculated using (3) and (4) as follows:

$$\mu_t = \frac{\rho C_\mu k^2}{\varepsilon} \tag{5}$$

In (3), (4) and (5), σ_k, σ_ε, $C_{1\varepsilon}$, $C_{2\varepsilon}$ and C_μ are constants for which standard values are used, and G is the turbulent generation rate.

2.2 Blade Element Momentum Method

For long time scales, the time-averaged influence of turbine blades on the flow is considered to act over all parts of a circular area with a diameter equal to that of the blade. Forces acting on the flow due to the blade are assumed to be equal across the circular area for equal radial distances from the centre. This time-averaged approach allows sources that represent the force on the fluid due to the blades of the turbines to be applied to each of the momentum equations. The advantage of this approach is that the physical characteristics of the blade are built into the source rather than the mesh consequently allowing the use of better quality meshes. The disadvantage is that because of the time average principle of the approach it fails to resolve any transient flow features due to blade position.

To define the characteristics of the rotor according to axial and radial position the blade element momentum (BEM) [5] method is employed. Figure 1a shows diagrammatically how a three bladed rotor is discretised using the BEM approach. The blade properties for a blade element at a certain radius, r, are determined and

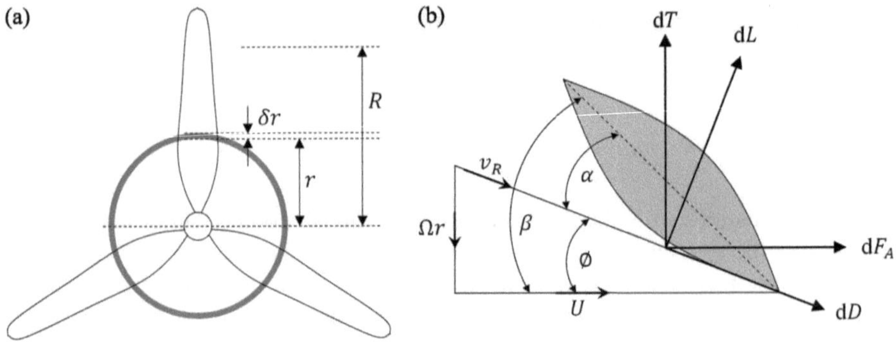

Fig. 1 (**a**) Schematic diagram of the discretisation of the rotor; (**b**) resolution of the lift and drag forces

are then averaged throughout the whole of the shaded region. This is performed for each blade element throughout the radius of the rotor, R.

Each blade element has a chord length, c_{Fb}, and radial width, δr. Each element experiences forces acting on it due to the fluid. These are shown in Fig. 1b. $\mathrm{d}T$ and $\mathrm{d}F_A$ represent the torque and axial forces respectively. The lift, $\mathrm{d}L$, and drag, $\mathrm{d}D$, are dependent on the angle of attack, α, between the blade element and the resultant velocity, v_R. Ω is the angular velocity and U is the upstream longitudinal flow velocity. The chord inclination angle and flow inclination angle are denoted by β and ϕ respectively.

Following the approach described by Masters et al. [5], an axial force on a blade element is defined by:

$$\mathrm{d}F_A = \mathrm{d}L \sin\phi + \mathrm{d}D \cos\phi \tag{6}$$

and the tangential force on a blade element, which is equal to the torque/radius, i.e. dT/r, is defined by:

$$\mathrm{d}F_T = \mathrm{d}L \cos\phi - \mathrm{d}D \sin\phi \tag{7}$$

The flow inclination angle, ϕ, is defined by:

$$\phi = \tan^{-1}\left(\frac{\Omega r - u_\theta}{u_z}\right) \tag{8}$$

where u_θ, u_z are the local fluid tangential and axial velocities respectively, and Ω is the angular velocity of the blades in rad/s. The lift force, $\mathrm{d}L$, and drag force, $\mathrm{d}D$, are given by:

$$\mathrm{d}L = \frac{1}{2}\rho |v_R|^2 C_L c_{Fb} \delta r \tag{9}$$

$$\mathrm{d}D = \frac{1}{2}\rho |v_R|^2 C_D c_{Fb} \delta r \tag{10}$$

Here, C_L and C_D are the lift and drag coefficients respectively, and:

$$|v_R|^2 = u_z^2 + (\Omega r - u_\theta)^2 \tag{11}$$

Substituting (9) and (10) into (6) and (7) gives the following equations:

$$S_z = \mathrm{d}F_A = \frac{1}{2}\rho|v_R|^2 c_{Fb}\delta r(C_L \sin\phi + C_D \cos\phi) \tag{12}$$

$$S_\theta = \mathrm{d}F_T = \frac{1}{2}\rho|v_R|^2 c_{Fb}\delta r(C_L \cos\phi - C_D \sin\phi) \tag{13}$$

which, when resolved into Cartesian components and converted into force per volume, are substituted into the momentum equations (2) through S_i.

2.3 Computational Solution

The solution to the governing equations (1)–(4) is achieved using a collocated cell-centred finite volume scheme [7] within the software PHYSICA [8]. In the finite volume scheme, the physical domain is represented by a mesh consisting of non-overlapping polyhedral elements. In the cell-centred scheme, a solution node is placed at the centre of each of these elements. The governing equations are integrated over each mesh element, and the discretised equations lead to a series of relationships between the unknown values in an element and the values in neighbouring elements. The solution procedure is based on a variant of the SIMPLE [9] algorithm, and since a collocated scheme is used, the Rhie-Chow approximation is used to prevent checker-boarding of the pressure field [10]. The CFD equations are typically solved for using a diagonally preconditioned conjugate gradient algorithm. Further details of the solution procedure can be found in [8, 11].

The model domain used for the simulations was 700 m long, 100 m wide and 30 m deep. A 10 m diameter blade was centred laterally and vertically across the channel, and positioned 300 m from the upstream boundary. This is to ensure the full development of the boundary layer and the achievement of a uniform flow upstream of the turbine. Also, a considerable distance is required downstream of the turbine to evaluate the development of the flow.

An element size of 0.19 m by 0.19 m by 0.1 m was implemented within the blade box. This is the region within which the momentum sink terms associated with the blade are introduced, and hence, a relatively fine mesh is required to capture the variation in lift and drag forces across the blade. The element sizes were gradually increased away from the blade box to reduce the overall mesh size, however, to a lesser extent within the wake region downstream of the blade where the flow structure is to be studied.

Each simulation was carried out for 15,000 iterations, which required 44 hours of computation on a single processor. However, the simulations were conducted in parallel on a multi-node cluster and hence, actual wall-times were significantly reduced.

2.4 Definitions

The tip speed ratio (TSR), the ratio between the rotational speed of the tip of the blades and the free-stream longitudinal flow velocity, is defined by:

$$\mathrm{TSR} = \frac{D\Omega}{2U} \tag{14}$$

where U is the upstream free velocity and D is the turbine diameter.

The 'velocity deficit', U_{def}, is used in this paper in accordance with the definition suggested by Myers and Bahaj [12] as follows:

$$U_{def} = 1 - \frac{u}{U} \tag{15}$$

such that a velocity deficit of 0.25 for example, is equivalent to a local velocity equivalent to 75 % of the free-stream velocity.

3 Results and Discussion

3.1 Simulations

The simulations presented in this paper were conducted for a tip speed ratio of 3.0, which is approximately the optimum for the blade used. A range of velocities typical of offshore environments where tidal stream turbines are likely to be deployed were implemented, typically ranging up to 3.0 $\mathrm{m\,s^{-1}}$. Since power is a function of the cubic power of velocity, there is a significant amplification of power output with increasing velocity (see Fig. 2b).

To fully appreciate the significance of this curve, consider a typical diurnal tidal cycle consisting of two periods of high tide annexed together by an intermediate period of low tide. Maximum velocities may be expected during a 'turning' tide which occurs between low and high tides. However, during a slack tide, which occurs around low and high tides, there is relatively little flow movement. When conducting feasibility studies, tidal energy developers may be tempted to estimate potential power outputs using a rated velocity factor of 70 % [13], based on the maximum flow velocities observed during a mean spring tide across a potential deployment site. However, to gain a more accurate appreciation of the true return of the investment, a more detailed assessment of the power outputs over the range of velocities, and range of tidal cycles would be required. Velocities will not only vary across each tidal cycle, but also between different tidal cycles due to a natural variation in low and high tide sizes, in addition to the influence of meteorological factors, particularly wind-generated waves.

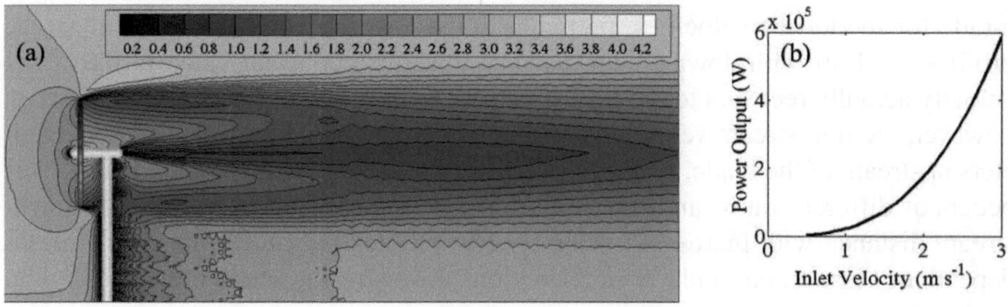

Fig. 2 (**a**) A lateral view of a turbine in a channel showing a velocity contour plot for an inlet velocity of 3.0 m s^{-1}; and (**b**) variation in power output with inlet velocity. Tip speed ratio is 3.0

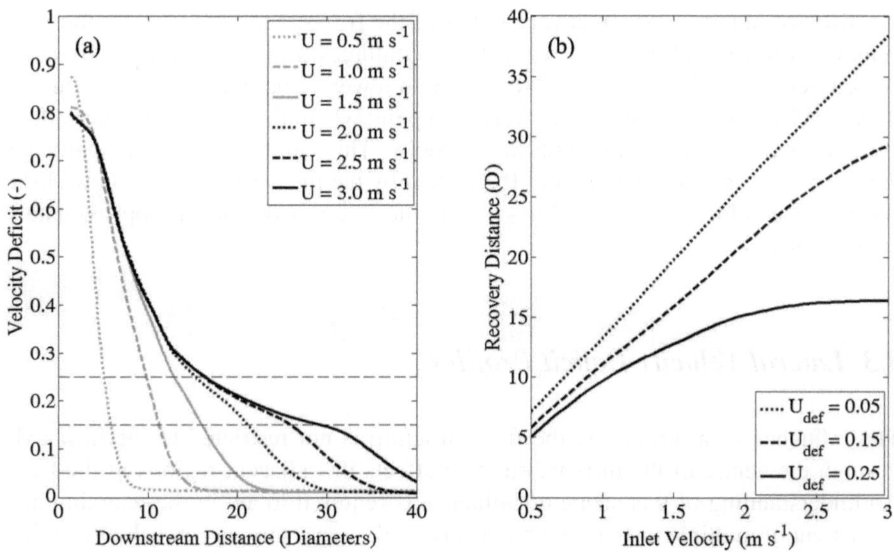

Fig. 3 (**a**) Velocity deficit profiles downstream of a 10 m diameter turbine for inlet velocities between 0.5 and 3.0 m s^{-1}. (**b**) Downstream distance required for recovery to a velocity deficit of 0.05, 0.15 and 0.25 for the range of velocities considered. Tip speed ratio is 3.0

3.2 Downstream Velocity Deficit

The reduction in velocity deficit with downstream distance from a turbine along the flow axis through the centre of the blade is a measure of flow recovery following the momentum extraction and velocity reduction arising due to the interaction of the blade with the flow. This can be used to determine suitable longitudinal distances between turbines within an array to minimise wake interference between the devices, and hence, avoid compromising the performance of downstream devices.

Velocity deficit profiles downstream of the blade along its centreline are presented in Fig. 3a for the range of inlet velocities considered. Initially, the velocity deficit profiles are similar in magnitude, although they are slightly greater in magnitude for the lowest velocities, particularly the lowest velocity of 0.5 $\mathrm{m\,s^{-1}}$. The profiles level off with downstream distance towards a constant value of 0.01. The velocity actually recovers to the free-stream velocity (i.e. a velocity deficit of zero), however, the free-stream velocity is estimated here based on the velocity two diameters upstream of the blade, hence the slight surplus in the final velocity deficit. This occurs at different rates, and the minimum value is reached over a longer downstream distance with increasing velocity. The profiles indicate some minor mesh-dependency issues, particularly in far regions downstream, where the mesh is relatively coarser than near the turbine.

To evaluate flow recovery downstream of the turbine, the distances required for the velocity along centreline through the middle of the blade to return to velocity deficits of 0.05, 0.15 and 0.25 are considered in Fig. 3b. The relationship between the distance required for a recovery to 0.1 of the free-stream velocity and the inlet velocity is linear and the recovery distance continues to increase at the maximum velocity considered of 3.0 $\mathrm{m\,s^{-1}}$. Recovery to lower velocity deficits, for example 0.15 or 0.25, departs from a linear trend to a point where the recovery distance does not appear to increase with increasing velocity. This can be seen most clearly for the recovery to a velocity deficit of 0.25 whereby the recovery distance shows little variation for velocities above 2.5 $\mathrm{m\,s^{-1}}$, and the recovery distance is approximately 16.4 diameters.

3.3 Lateral Velocity Deficit Profiles

The influence of a turbine on the flow structure is not restricted to the flow axis through the centre of the turbine, but also affects the adjacent regions of the flow. An understanding of this scope of influence is required to aid in the planning and placement of multiple turbines within an array. To assess this aspect, velocity deficit profiles are considered for lateral sections perpendicular to the flow direction.

Lateral profiles are presented at a range of downstream distances from the turbine blade for flow velocities of 1.0 $\mathrm{m\,s^{-1}}$ and 3.0 $\mathrm{m\,s^{-1}}$ in Fig. 4a and b respectively. These represent velocities around the lower and upper end of the operating range of velocities in a typical offshore environment suitable for tidal stream turbine deployment. In both flows, there is initially little difference in the velocity deficit profiles with downstream distance up to 4.0 diameters downstream from the blade. Within this range, the velocity deficit is significant.

Beyond 4.0 diameters downstream, there is a sharp decrease in velocity deficit, particularly for a lower flow velocity, where the velocity deficit is negligible beyond 8.0 diameters downstream. For the higher flow velocity, there is a steady decrease in velocity deficit with distance downstream beyond 4.0 diameters, although the influence of the turbine is still evident beyond 28.0 diameters downstream.

Now lateral profiles of velocity deficit are compared for different inflow velocities at two different sections: 2.0 and 10.0 diameters downstream from the blade (Fig. 5a and b respectively). Immediately downstream of the blade, there is little

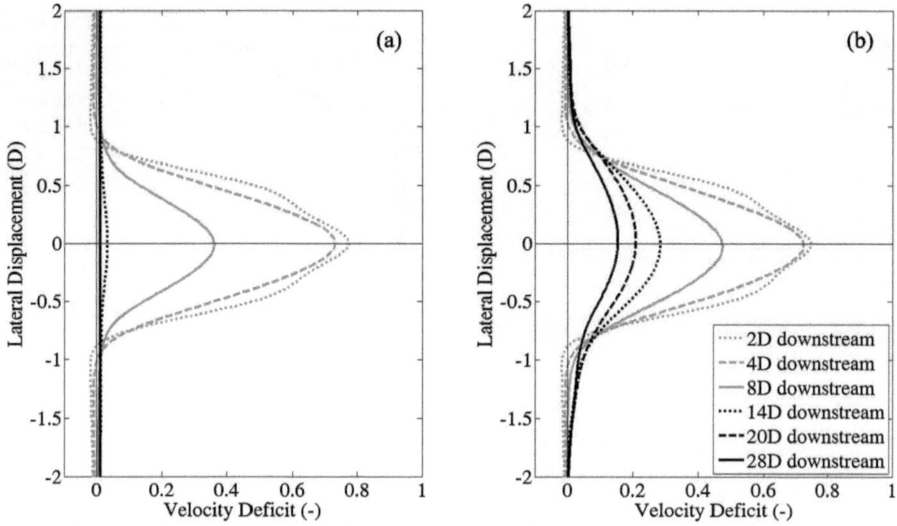

Fig. 4 Lateral profiles of velocity deficit downstream from a 10.0 m diameter turbine for inlet velocities of (**a**) 1.0 m s^{-1}; and (**b**) 3.0 m s^{-1}. Tip speed ratio is 3.0

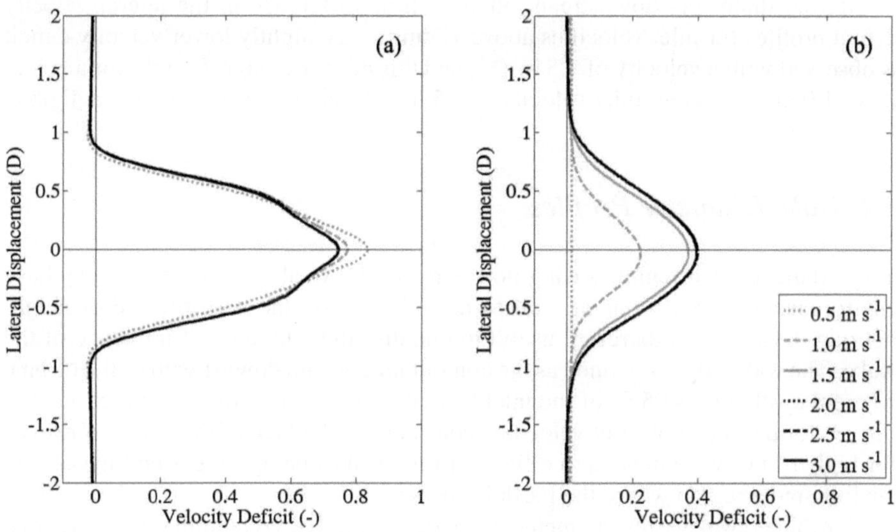

Fig. 5 Lateral profiles of velocity deficit at a downstream distance of (**a**) 2.0 diameters; and (**b**) 10 diameters from a 10.0 m diameter turbine for inlet velocities between 0.5 and 3.0 m s^{-1}. Tip speed ratio is 3.0

difference in velocity deficit profiles for different inflow velocities. Velocity deficits are slightly higher for the lowest velocities considered, particularly 0.5 m s^{-1}, and to a lesser extent, 1.0 m s^{-1}.

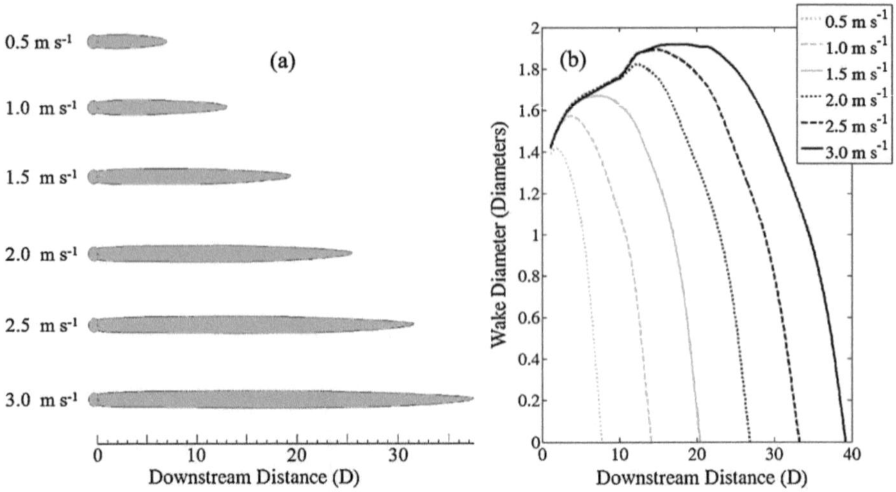

Fig. 6 (**a**) Wake outlines based on a wake edge velocity of 95 % of the inlet velocity for inlet velocities between 0.5 and 3.0 $m\,s^{-1}$; and (**b**) wake diameter profiles downstream of a 10 m diameter turbine for inlet velocities between 0.5 and 3.0 $m\,s^{-1}$. Tip speed ratio is 3.0

At 10.0 diameters downstream, there is little difference in the lateral velocity deficit profiles for inlet velocities above 1.5 $m\,s^{-1}$. A slightly lower velocity deficit is observed with a velocity of 1.5 $m\,s^{-1}$, and this decreases significantly for a velocity of 1.0 $m\,s^{-1}$. For an inlet velocity of 0.5 $m\,s^{-1}$, the velocity deficit is negligible.

3.4 Wake Diameter Profiles

The turbulent wake region arising downstream of a tidal stream turbine may have environmental implications as well as an influence on the performance of neighbouring devices. It is therefore useful to quantify the total area of influence of the wake. The wake edge is defined as the contour around and downstream of the turbine where the velocity is 95 % of the inlet velocity. The wake outlines are presented in Fig. 6a for the range of inlet velocities considered. The lack of symmetry observed for higher inlet velocities can be linked to the wakes being longer and hence, extending into a region where the mesh is coarse.

To evaluate wake sizes, diameters were determined by calculating the lateral distance between the two outer edges of the wake along the horizontal plane at hub-level, and this was performed at 1.0 m increments over the length of the wake. Wake diameter profiles are presented in Fig. 6b for the range of inlet velocities considered. The wake size, both in terms of total length and maximum diameter, are linked to the Reynolds number of the flow, as dictated by the inlet velocity. Wake expansion is initially evident downstream of the turbine in all cases, followed by a rapid reduction due to wake dissipation. Wake expansion continues over a considerably greater distance downstream of the blade with an increasing free-stream velocity.

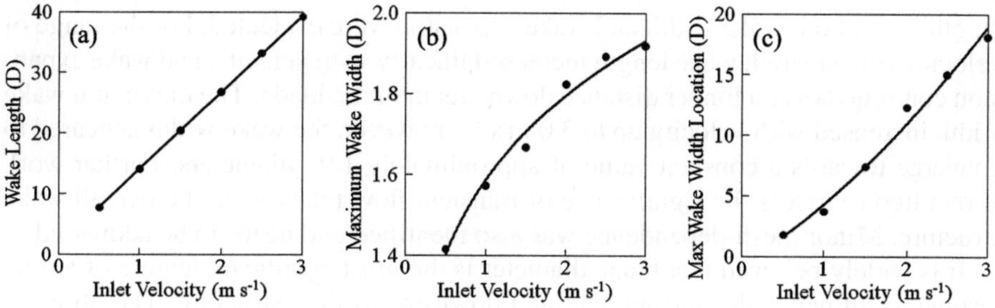

Fig. 7 Variation in (**a**) wake length; (**b**) maximum wake width; and (**c**) maximum wake width location with inlet velocity for a tip speed ratio of 3.0

The increases in wake length and diameter with inlet velocity are summarised in Fig. 7a and b respectively. A significant increase in wake length occurs. The relationship is linear and equates to approximately 80.0 % of the increase in velocity. Increases in the maximum wake diameter were less significant ranging between 1.42 and 1.92 diameters for velocities between 0.5 $\mathrm{m\,s^{-1}}$ and 3.0 $\mathrm{m\,s^{-1}}$. There was less of an increase in wake diameter with velocity at the higher velocities considered indicating that there is a limit to the increase in the diameter of the wake.

Wake expansion is observed in the simulation results. The expansion of the wake continues over a longer downstream distance with increasing velocity as shown in Fig. 7c. Despite the maximum wake width showing less variation with velocity at higher inlet velocities as shown in Fig. 7b, its location continues to move further downstream from the blade.

4 Conclusion

A combined Blade Element Momentum - Computational Fluid Mechanics model was used for modelling a tidal stream turbine in a range of flow conditions typical of offshore aquatic environments. Simulation results were used to assess the influence of the turbine on flow structure and the size of the wake arising downstream.

The velocity downstream along the turbine centreline recovers to the free-stream velocity, and the distance required for flow recovery increases linearly with inlet velocity initially. However, above a certain free-stream velocity, the recovery distance length converges to a certain value. For instance, a recovery to a velocity deficit of 0.25 occurs over a distance equivalent to approximately 16.4 blade diameters for inlet velocities greater than 2.5 $\mathrm{m\,s^{-1}}$.

The influence of the turbines on the flow extends across the flow direction away from the centreline through the middle of the blade, and this was reflected in lateral velocity profiles. Immediately downstream of the blade, velocity deficits were slightly higher for lower velocities. However, for lower flow velocities, there was also a rapid reduction in velocity deficits with downstream distance.

These trends were evaluated by characterising the wake edge as having a velocity equivalent to 95 % of the inlet velocity. Using this definition, variation in wake length, maximum wake width and wake expansion were evaluated. For the range of velocities considered, wake length increased linearly with velocity, and wake expansion continued over a longer distance downstream of the blade. The maximum wake width increased with velocity up to 3.0 m s^{-1}, however, the wake width appeared to converge towards a constant value of approximately 1.92 diameters. Further work is required to assess the significance of transient flow features on the overall wake structure. Minor mesh-dependence was also identified and needs to be addressed.

It is widely believed that blade diameter is the most significant feature of a tidal turbine to influence the surrounding and downstream flow structure. The observations identified in this paper highlight that some wake characteristics are linearly related to the flow velocity, while others are not, or even converge towards a constant value. The key point is that the flow velocity is a significant parameter in determining wake characteristics. The relationships identified here can serve as a predictor of likely levels of wake interference between devices for different tidal turbine array arrangements to aid in layout planning.

Acknowledgements This work was undertaken as part of the Low Carbon Research Institute Marine Consortium (www.lcrimarine.org).

The Authors wish to acknowledge the financial support of the Welsh Assembly Government, the Higher Education Funding Council for Wales, the Welsh European Funding Office, and the European Regional Development Fund Convergence Programme.

References

1. Gilson P (2010) Alstom's first tidal in stream energy converter demonstration project. In: Proc ICOE
2. Fraenkel PL (2007) Marine current turbines: pioneering the development of kinetic energy converters. J Power Energy 221(2):159–169. Proc IMechE
3. Paish M, Giles J, Panahandeh B (2010) The pulse stream concept, and the development of the pulse stream demonstrator. In: Proc ICOE
4. O'Doherty T, Egarr DA, Mason-Jones A, O'Doherty DM (2009) An assessment of axial loading on a five-turbine array. In: Proc ICE (energy), vol 162 (EN2), pp 57–65
5. Masters I, Chapman JC, Willis MR, Orme JAC (2011) A robust blade element momentum theory model for tidal stream turbines including tip and hub loss corrections. J Marine Eng Technol 10(1):25–35
6. Launder BE, Spalding DB (1972) Mathematical models of turbulence. Academic Press, London
7. Versteeg HK, Malalasekera W (2007) An introduction to computational fluid dynamics: the finite volume method. Pearson Prentice Hall, London
8. Croft TN, Pericleous K, Cross M (1995) PHYSICA: a multiphysics environment for complex flow processes. In: Taylor C et al (eds) Numerical methods in laminar and turbulent flows, vol 9. Pineridge Press, Swansea
9. Patankar SV (1980) Numerical heat transfer and fluid flow. McGraw-Hill, New York
10. Rhie CM, Chow WL (1983) Numerical study of the turbulent flow past an aerofoil with trailing edge separation. AIAA J 22:1525–1532
11. Chow P, Cross M, Periclious K (1995) A natural extension of standard control volume CFD procedures to polygonal unstructured meshes. Appl Math Model 20:170–183
12. Myers L, Bahaj A (2010) Design of 1st-generation marine current energy converter arrays. In: Proc WREC
13. Phase II (2005) UK Tidal Stream Energy Resource Assessment. Black & Veatch, London

Part V
Maneuvering and Seakeeping

Viscous-Flow Calculations for KVLCC2 in Deep and Shallow Water

Serge L. Toxopeus

Abstract In the SIMMAN 2008 workshop, the capability of CFD tools to predict the flow around manoeuvring ships has been investigated. It was decided to continue this effort but to extend the work to the flow around ships in shallow water. In this paper, CFD calculations for the KLVCC2 are presented. The aim of the study is to verify and validate the prediction of the influence of the water depth on the flow field and the forces and moments on the ship for a full-block hull form.

An extensive numerical investigation has been conducted. For each water depth, several grid densities were used to investigate the discretisation error in the results. In general, the uncertainties were found to increase with increased flow complexity, i.e. for larger drift angles or yaw rates. A dependency of the uncertainty on the water depth was not found. The predicted resistance values were used to derive water-depth dependent form factors. Comparisons with resistance measurements and with an empirical formula given by Millward show good agreement for deep as well as for shallow water depths. The CFD results give insight into the forces and moments acting on the ship as a function of the drift angle, yaw rate and water depth. A clear dependence of the forces and moments on the water depth is found for steady drift conditions. For pure rotation, this dependence is much more complex and only develops fully for larger non-dimensional rotation rates. The paper shows that CFD is a useful tool when studying the flow around ships in restricted water depths.

Keywords KVLCC2 · Manoeuvring · Shallow water · CFD · RANS

1 Introduction

In the SIMMAN 2008 workshop [1], the capability of CFD tools to predict the flow around manoeuvring ships has been investigated. Within the NATO RTO Applied Vehicle Technology group on Assessment of Stability and Control Prediction

S.L. Toxopeus (✉)
Maritime Research Institute Netherlands (MARIN)/Delft University of Technology, P.O. Box 28, 6700 AA Wageningen, The Netherlands
e-mail: s.l.toxopeus@marin.nl
url: http://www.marin.nl

L. Eça et al. (eds.), *MARINE 2011, IV International Conference on Computational Methods in Marine Engineering*, Computational Methods in Applied Sciences 29,
DOI 10.1007/978-94-007-6143-8_9,

Methods for NATO Air & Sea Vehicles (AVT-161) it was decided to continue this effort but to extend the work to the flow around ships in shallow water, see e.g. Toxopeus [2]. Accordingly, the KVLCC2 hull form was selected, since for this ship captive model tests in various water depths are available. In this paper, CFD calculations performed by MARIN for the KLVCC2 are presented. The aim of the study is to investigate the uncertainty in predicting the forces and moments on a full-block hull form in various water depths; validate the predictions using model test results; and to obtain information about the influence of the water depth on the flow around the ship and the forces and moments on the hull.

2 Coordinate System

The origin of the right-handed system of axes used in this study is located at the intersection of the water plane, midship and centre-plane, with x directed forward, y to starboard and z vertically downward. The forces and moments presented in this paper are given according to this coordinate system.

In the present calculations, a positive drift angle β corresponds to the flow coming from port side (i.e. $\beta = \arctan -v/u$). The non-dimensional yaw rate γ is calculated with $\gamma = r \cdot L_{\text{pp}}/V$ and is positive for a turn to starboard when sailing at positive forward speed.

3 KVLCC2

The KVLCC2 (KRISO Very Large Crude Carrier) hull form was one of the subjects of study during the CFD Workshops Gothenburg 2000 [3] and 2010 [4] and the SIMMAN 2008 Workshop [1]. For straight ahead conditions, the flow features and resistance values were measured, see Lee et al. [5] and Kim et al. [6].

Captive model tests for the bare hull KVLCC2 were conducted by INSEAN in preparation for the SIMMAN 2008 Workshop [1], see also Fabbri et al. [7, 8] and Campana and Fabbri [9]. A set of PMM tests comprising amongst others the measurement of the forces and moments for steady drift motion and oscillatory yaw motion was performed. During the tests, the model was free to heave and pitch. For the present work, only the tests with the bare hull form are considered.

4 Viscous Flow Solver and Computational Setup

4.1 ReFRESCO

REFRESCO is a MARIN spin-off of FRESCO [10], which was developed within the VIRTUE EU Project together with Technische Universität Hamburg-Harburg

(TUHH) and Hamburgische Schiffbau-Versuchanstalt (HSVA). ReFRESCO is an acronym for *Reliable and Fast Rans Equations solver for Ships, Cavitation and Offshore*. It solves the multi-phase unsteady incompressible RANS equations, complemented with turbulence models and volume-fraction transport equations for each phase. The equations are discretised using a finite-volume approach with cell-centred collocated variables. The implementation is face-based, which permits grids with elements with an arbitrary number of faces (hexahedrals, tetrahedrals, prisms, pyramids, etc.). The code is targeted, optimised and highly validated for hydrodynamic applications, in particular for obtaining current, wind and manoeuvring coefficients of ships, submersibles and semi-submersibles [11–14]. For all cases presented in this study the y^+ values in the first cell from the wall are below 1, such that the equations are integrated down to the wall.

4.2 Turbulence Closure Models

Several different turbulence closure models are available in ReFRESCO. In this study, the SST version [15] of the two-equation $k - \omega$ turbulence model is used. In the turbulence model, the Spalart correction (proposed by Dacles-Mariani et al. [16]) of the stream-wise vorticity can be activated.

4.3 Implementation of Rotational Motion

For ship manoeuvres, not only oblique flow is of interest, but also the flow around the ship when it performs a rotational (yaw) motion. In RANS, the rotational motion can be modelled in several ways, such as moving the grid in a rotational motion through a stationary flow (inertial reference system), or by letting the flow rotate around the stationary ship (non-inertial reference system). For this work a non-inertial reference system is chosen. Centrifugal and Coriolis forces to account for the rotation of the coordinate system are added to the momentum equation as source terms. More information about the implementation can be found in Toxopeus [14].

4.4 Grid Generation

For best performance of ReFRESCO, multi-block structured O–O grids are used for this study. Grid points have been clustered towards the hull surface and bottom to ensure proper capturing of the boundary layers. The far field boundary is generated as a cylindrical surface, to facilitate the use of a single grid for all computations. An example grid is given in Fig. 1.

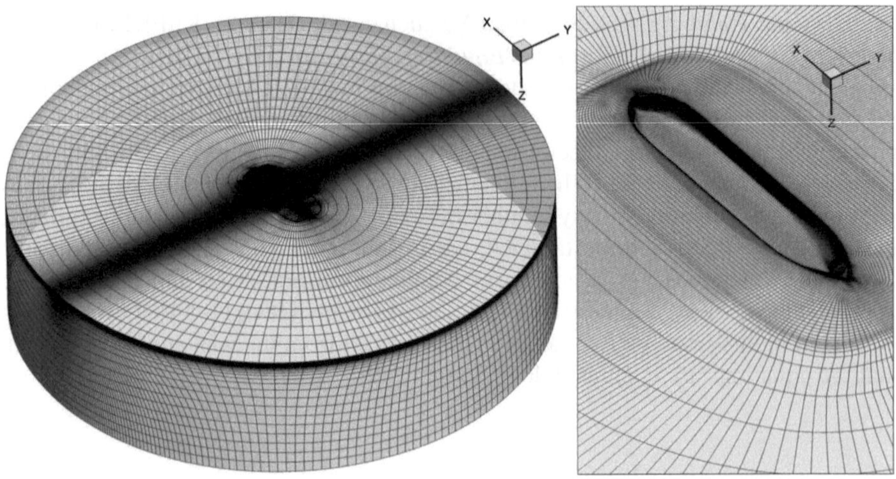

Fig. 1 Example grid, KVLCC2, deep water (coarsened for presentation)

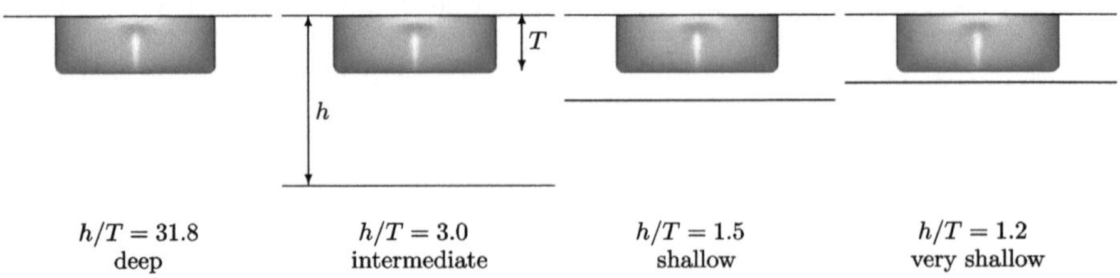

Fig. 2 Water depth to draught ratios considered in this study

Grids were generated with GridPro for four different water depth h to draught T ratios, i.e. $h/T = 31.8$ representing deep water, $h/T = 3.0$ representing an intermediate water depth, $h/T = 1.5$ representing shallow water and $h/T = 1.2$ representing very shallow water, see Fig. 2. Basically, the grid topology around the hull for the four water depths was the same, the only difference being the addition of grid blocks between the bottom of the hull and the sea floor for each water depth.

Based on these grids, geometrically similar grids were generated using GridPro in order to be able to assess the discretisation errors and to accelerate the iterative procedures by using coarse grid solutions as initial flow fields for fine grid computations. Additional grids are obtained by coarsening the finest grid in all directions. Table 1 lists the grid densities used for this study.

4.5 Boundary Conditions

The calculations presented in this study were all conducted without incorporating free-surface deformation. Based on the speeds used during the tests and the range of

Table 1 Overview of grid densities

h/T	Grid cells ($\times 10^{-3}$)
31.8 (Deep)	12721, 8455, 5388, 3340, 2270, 1590, 121
3.0 (Intermediate)	13005, 8597, 5573, 3446, 2374, 1604, 137
1.5 (Shallow)	11659, 7688, 4936, 3106, 2112, 1437, 119
1.2 (Very shallow)	11031, 7270, 4664, 2899, 1999, 1351

drift angles or rotational rates studied, the effects of Froude number and free-surface deformation on the forces on the manoeuvring ship are likely to be reasonably small and assumed to be smaller than the uncertainties due to e.g. discretisation errors or errors in the experimental results. To simplify the calculations, symmetry boundary conditions were therefore applied on the undisturbed water surface and dynamic sinkage and trim was neglected. On the hull surface, no-slip and impermeability boundary conditions are used. For all calculations, even for deep water, the boundary condition on the bottom surface is set to moving-wall/fixed slip ($\overline{u} = \overline{V}_\infty$).

Calculations for ships at drift angles or rotation rates are conducted by setting the boundary conditions at the exterior to the proper inflow velocities. This is done using the `BCAutoDetect` boundary condition, which automatically applies inflow conditions or outflow (Neumann) conditions on the cell faces, depending on the normal velocity at each cell face on the boundary. Therefore, the computational domain does not need to be changed for each new calculation and a single grid for different manoeuvring conditions can be used. Details about `BCAutoDetect` can be found in Toxopeus [14].

4.6 Acceleration of the Calculations

In order to efficiently generate results for many drift angles, a routine was used to automatically increment the drift angle during a single simulation. Simulations begin with a pre-set drift angle, until a specified number of iterations is reached, or when the maximum change in the residuals is less than a specified convergence criterion. Next, the drift angle is incremented by $\Delta\beta$, by changing the inflow conditions, and the solution is continued from the solution from the previous drift angle. Starting the calculations from a converged solution at a slightly different drift angle saves time compared to performing each calculation separately from undisturbed flow. This procedure is repeated until the desired maximum inflow angle is reached. In Toxopeus [14], it is demonstrated that this approach provides the same results as those obtained with multiple single-drift angle calculations.

This procedure was designated *drift sweep* and the application has already been presented in e.g. Toxopeus [14], Vaz et al. [12] and Bettle et al. [17].

Table 2 Overview of computations

Condition	Deep $h/T = 31.8$	Intermediate $h/T = 3.0$	Shallow $h/T = 1.5$	Very shallow $h/T = 1.2$
Drift sweep	5388×10^3 Cells, 0°–32°, 2° incr.	5573×10^3 Cells, 0°–32°, 2° incr.	4936×10^3 Cells, 0°–32°, 2° incr.	4464×10^3 Cells, 0°–38°, 2° incr.
$\beta = 0°$	All 7 grids	All 7 grids	All 7 grids	All 6 grids
$\beta = 4°$	All 7 grids	All 7 grids	All 7 grids	All 6 grids
$\gamma \neq 0$	12721×10^3 Cells, 0.1–0.6, 0.1 incr., 0.65	All 7 grids, 0.1–0.6, 0.1 incr.	All 7 grids, 0.1–0.6, 0.1 incr.	All 6 grids, 0.1–0.6, 0.1 incr.
$\gamma = 0.4$	All 7 grids	All 7 grids	All 7 grids	All 6 grids

5 Programme of Calculations

Most calculations were conducted for a Reynolds number of $Re = 3.7 \times 10^6$ which corresponds to the Reynolds number during the INSEAN model tests (see [7–9]). The conditions are specified in Table 2. Sinkage and trim and free surface deformation were not taken into account. Furthermore, it was assumed that the flow domain was not restricted in horizontal direction, i.e. the basin walls were neglected. During the measurements, the model was free to sink and trim and basin walls were present. Especially for the shallow water conditions, this may lead to differences between the model test results and the computations.

Additional calculations were conducted for the deep water condition, at straight ahead sailing and a Reynolds number of $Re = 4.6 \times 10^6$. This condition was chosen in order to be able to compare the REFRESCO results with wind-tunnel measurements of the flow field around the KVLCC2, see Lee et al. [5] and with towing tank experiments performed by Kim et al. [6]. The calculations comprised the change of Reynolds number and variations in the turbulence modelling.

6 Presentation and Discussion of the Results

6.1 *Iterative Error*

All calculations were run until the maximum normalised residual res_{max} (the so-called L_∞ norm) between successive iterations had dropped well below 1×10^{-5} or when further iterative convergence was not obtained. The changes in the non-dimensional integral quantities (forces and moments) were well below 1×10^{-7}. A representative convergence history of the residuals and the changes in the forces for a calculation on the finest grid is presented in Fig. 3.

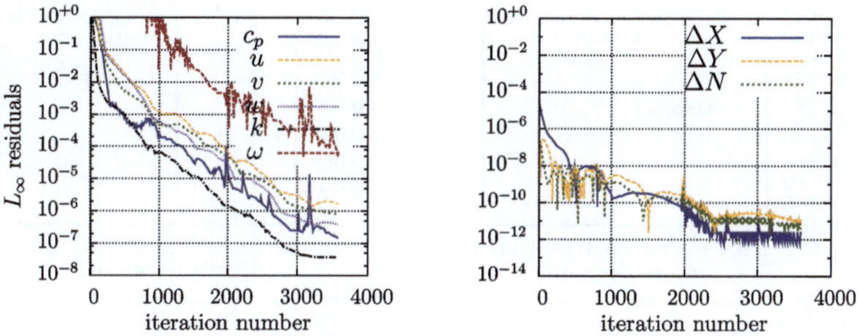

Fig. 3 Iterative convergence, deep, $\beta = 4°$

6.2 Discretisation Error

Using the procedure proposed by Eça et al. [18], the uncertainties in the forces and moments are estimated. Based on an analysis of the results for each grid, it was decided to use the 5 finest grids for the uncertainty analysis. The number of grids n_g used depended on the scatter in the results for the coarsest grids. It was found that for grids with a relative step size h_i of 2 and above, the results are not consistent with the finer grid results. This means that with the present grid layout, grids of more than about 1.6×10^6 cells are required to obtain a reliable solution of the forces and moments. As an example of the verification results, plots are presented for $\beta = 4°$ with deep and very shallow water in Fig. 4.

Verification studies have been performed for all other calculations with $\beta = 0°$, $\beta = 4°$ and $\gamma = 0.4$, but the results have not been included in this paper. These studies have indicated that the uncertainties for the rotational motion cases are higher than for the pure drift cases ($\beta = 0°$ and $\beta = 4°$). This can be attributed to the increased complexity of the flow. Especially for the large rotation rates ($\gamma \geq 0.4$), the uncertainties increase. For rotational motion, the uncertainties in X and Y are large, while the uncertainty in N is reasonable. This is probably caused by the fact that during pure yaw motion, the yaw moment (sum of contributions) is better defined than the longitudinal force or side force (difference between contributions). The theoretical order of convergence should be 2 for REFRESCO. However, due to flux limiters, discretisation of the boundary conditions and other factors, the apparent order of convergence is expected to be between 1 and 2 for geometrically similar grids in the asymptotic range. Considering uncertainty estimates for the various water depths and conditions, the apparent orders do not always follow this expectation. This indicates that either even finer grids are required, or that scatter in the results spoils the uncertainty estimate. A relation between the uncertainties and the water depths was not found.

Additionally, the influence of the grid density on the flow field at the propeller plane has been investigated. It was seen that with increasing grid density, the agreement between the CFD results and the experiments becomes qualitatively slightly better, but the hook shape in the propeller plane as visible in the experiments [5] is not well resolved (see Fig. 6). The sensitivity of the flow field to grid refinement

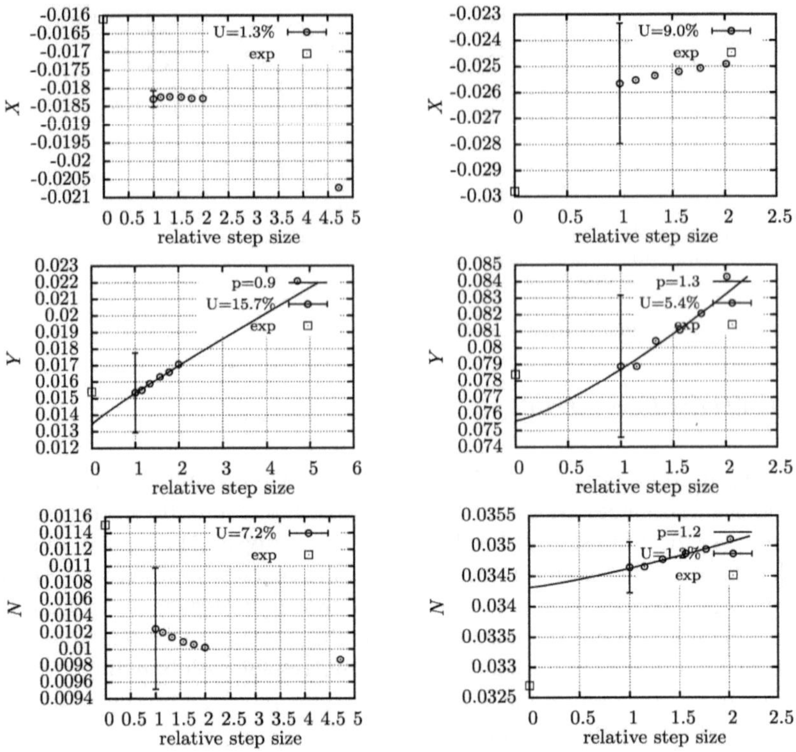

Fig. 4 Uncertainty analysis, deep (*left*) and very shallow (*right*) water, $\beta = 4°$

is judged to be small. Between densities of 3340×10^3 and 12721×10^3, the flow field does not change significantly. Looking at the differences between the solutions obtained on the six grids, it is not thought that the solution will improve upon grid refinement and therefore modelling errors are expected to exist in the CFD results. Alternative turbulence models might improve the results, as was shown for e.g. PARNASSOS during the Gothenburg 2010 CFD Workshop [19].

6.3 Comparison with the Experiments

6.3.1 Manoeuvring Conditions

Comparisons between the CFD results and the experiments are shown in Fig. 5. In general, the agreement is qualitatively reasonable, but quantitatively, considerable differences are seen. In most cases, validation of the solution is not achieved, which indicates that modelling errors are present in the simulations or that the uncertainties claimed for the experimental results are optimistic. When the solution is validated, the level of validation is generally large, e.g. larger than 10 $\%D$. Especially for the

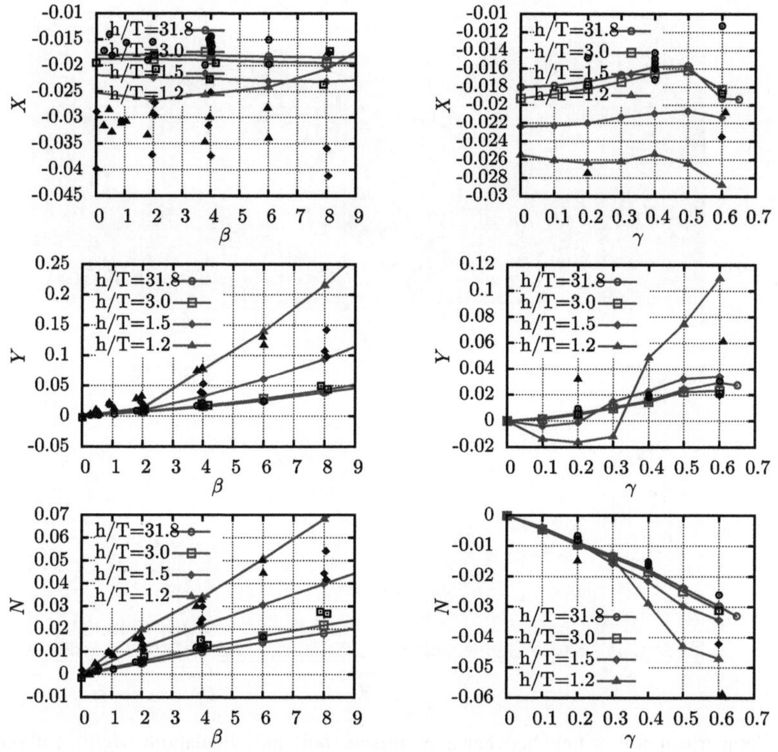

Fig. 5 Comparison between calculations (*blue lines with symbols*) and experiments (*black symbols*), relation with β (*left*) and with γ (*right*)

X force, discrepancies are found and the trends do not appear to be represented well. Similar results and deviations from the trends were obtained by Zou [20]. Reasons for the discrepancies might be the neglect of free surface and dynamic trim and sinkage. However, large scatter exists in the experimental data and therefore the uncertainty in the experiments is expected to be relatively large. Furthermore, it is questioned whether the false bottom used during the tests was sufficiently sealed at all sides to correctly model shallow water conditions and whether blockage of the basin walls influences the results, see also Simonsen et al. [21]. This should be investigated further.

6.3.2 Straight Ahead Sailing

For the calculation for deep water, $\beta = 0°$ and $Re = 4.6 \times 10^6$, comparisons were made with the experiments performed by Lee et al. [5] and Kim et al. [6]. The agreement between the CFD results and the experiments is good for the two most upstream planes ($x = -0.35L_{\text{pp}}$ and $x = -0.4L_{\text{pp}}$, not presented here). Further downstream, differences appear which increase when going aft. The comparison

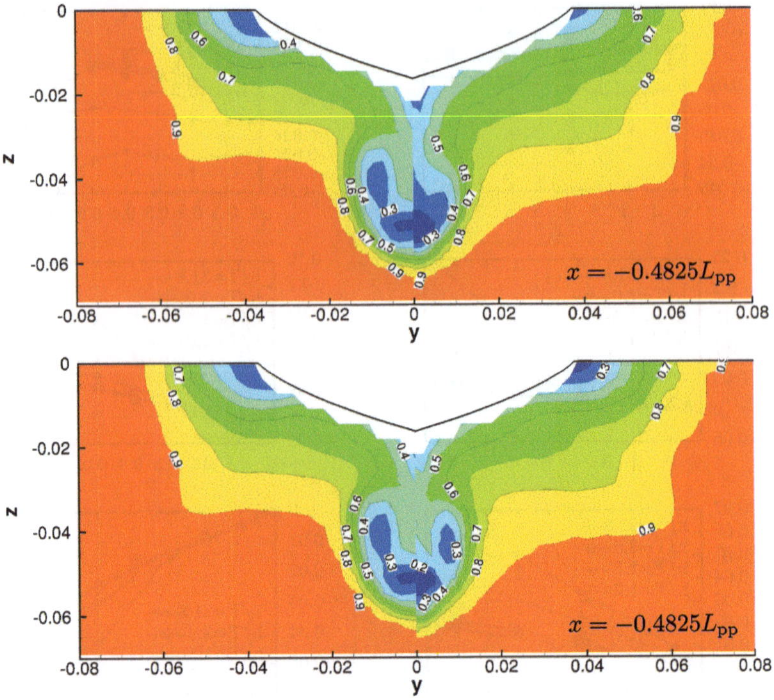

Fig. 6 Comparison of flow field between experiments (*left*) and calculations (*right*), without (*top*) and with (*bottom*) correction for vorticity, deep water, $\beta = 0°$

for $x = -0.4825L_{\mathrm{pp}}$ is shown in the top half of Fig. 6. As discussed in Sect. 6.2, it was expected that improvements can be made by selecting alternative turbulence modelling. This assumption was tested using additional calculations in which the Spalart correction of the streamwise vorticity according to Dacles-Mariani et al. [16] was activated. With this correction, the strain rate S is replaced by $S_c = \Omega + C \cdot \min(0, S - \Omega)$ in which the constant C should be chosen, see Eça [22]. In the present calculation with active correction, C was set to 10 (which is far outside the normal range of application), and the results are given in the lower half of Fig. 6. It is seen that with this correction, the agreement between the computations and the experiments improves considerably, which demonstrates the sensitivity of the results to the turbulence modelling. These findings correspond to those of Eça [22] and Eça et al. [19], based on results obtained with the viscous-flow solver PARNASSOS.

In Table 3, the predicted resistance in deep water is compared to previous calculations and to the experiments[1] (indicated by D, with $U_D = 1.0\,\%D$) performed by MOERI [6]. Additionally, results from tests by Kume et al. [23] with the KVLCC2M

[1]During the Gothenburg 2010 CFD workshop, it was concluded that the MOERI experiments were performed with the rudder attached to the model. Therefore, the G2010 KVLCC2 case contribu-

Table 3 Resistance prediction, deep water, $\beta = 0°$, $\gamma = 0$

	$n_{\text{cells}} \times 10^{-3}$ (Half ship)	$C_T \times 10^3$	$C_F \times 10^3$	$C_{PV} \times 10^3$	E (%D)
Experiments [6]	–	4.110	–	–	–
Experiments (KVLCC2M) [23]	–	4.152	–	–	1.0
Mean G2010 [4]	–	4.180	–	–	1.7
PARNASSOS [19]	6000	4.077	3.325	0.752	−0.8
REFRESCO	6361	4.185	3.457	0.728	1.8
REFRESCO (DM $C = 10$)	6361	4.109	3.280	0.829	0.0

hull form (identical to the KVLCC2 except for some fairing of the propeller shaft) are added ($U_{D'} = 3.3$ %D'). The KVLCC2M resistance value has been corrected for the difference in Reynolds number, using a form factor $(1+k) = 1.2$. In order to compare with other resistance predictions for the KVLCC2 found in literature, the values in the table have been made non-dimensional using the wetted surface. The total resistance is indicated with C_T, the friction component by C_F and the pressure component by C_{PV}.

The resistance predicted by REFRESCO is about 1.8 % higher than the value found in the experiments. Assuming that the numerical uncertainty U_{SN} is the same for $Re = 4.6 \times 10^6$ as for $Re = 3.7 \times 10^6$, i.e. $U_{SN} = U_I + U_G \approx U_G = 1.3$ %S, the validation uncertainty is about $U_V = \sqrt{U_D^2 + U_{SN}^2} \approx 1.6$ %D. The validation uncertainty and the comparison error for the uncorrected REFRESCO results are found to be of similar orders of magnitudes, although strictly validation of results is not obtained.

The Spalart streamwise vorticity correction used to improve the wake field also leads to a reduction of the comparison error. However, although the streamwise vorticity correction appears to improve the results, care should be taken with this modification since it is not guaranteed that the correction will lead to better results for other conditions as well. It just demonstrates that the turbulence modelling can have a large impact on the computational results.

6.4 Influence of Water Depth

6.4.1 Forces and Moments

In Fig. 7 and Fig. 8 the forces and moments as a function of the drift angle and rotation rate are shown for each water depth. The influence of the water depth is already

tions comprised calculations including the rudder, while the present calculations are for the bare hull. The KVLCC2M test results were obtained for the bare hull.

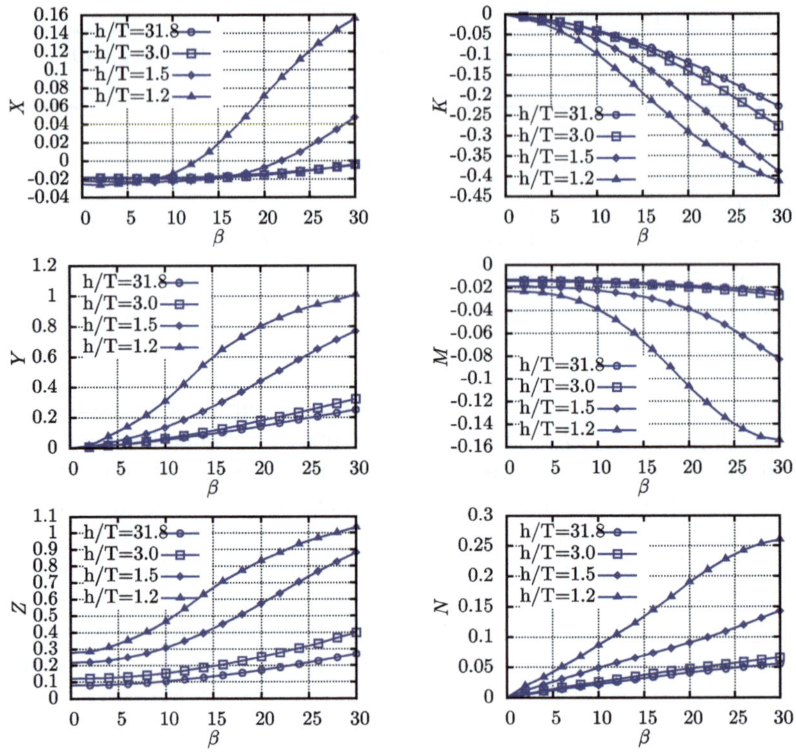

Fig. 7 Influence of drift angle and water depth

evident in these graphs at $h/T = 3.0$. For the forces and moments as a function of the drift angle β, consistent trends are found. The relation between the forces and moments and the rotation rate is much more complex, however. Noteworthy is the trend observed in the yaw moment N: for yaw rates below $\gamma = 0.4$, the influence of the water depth is hardly visible, while a considerable increase in yaw moment is found for the larger yaw rates when the water depth is reduced. Unfortunately, the influence of yaw rate on forces and moments in shallow water has hardly been published by other authors and therefore it cannot be determined whether this is a modelling error or a physical feature. Therefore, more attention to rotational motion in future studies is strongly recommended.

During free sailing tests with the KVLCC2, see Quadvlieg and Brouwer [24], the maximum drift angle and non-dimensional yaw rate during the manoeuvres were respectively $\beta \approx 20°$ and $\gamma \approx 0.8$. Figure 7 and Fig. 8 indicate that for these conditions the influence of the water depth on the forces and moments is much larger for a fixed drift angle than for a given yaw rate: e.g. the yaw moment for $\beta = 20°$ increases by a factor of 4.7 between deep and very shallow water, while for $\gamma = 0.6$ the increase is only by a factor of 1.6.

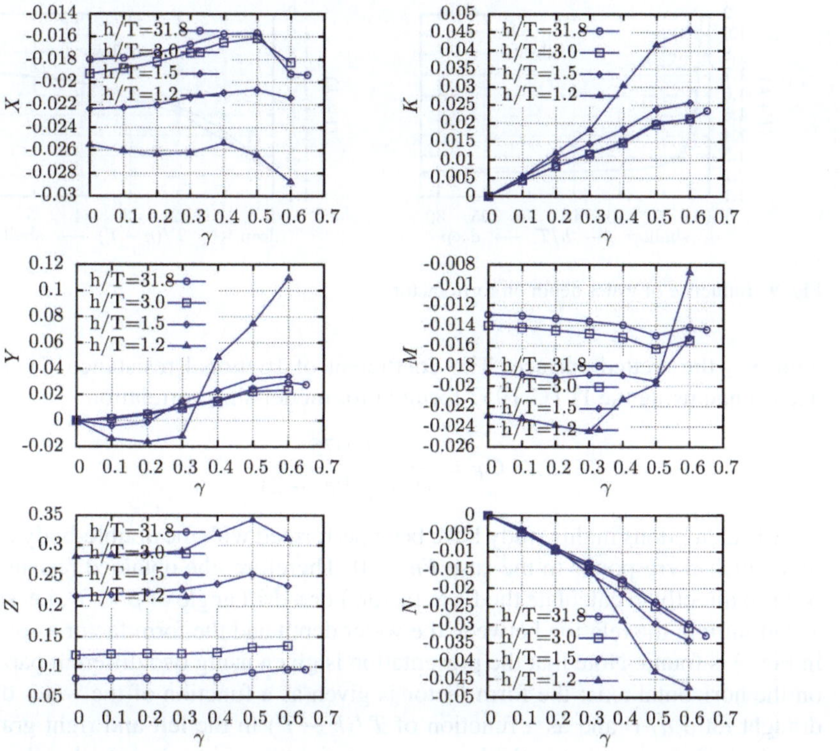

Fig. 8 Influence of yaw rate and water depth

6.4.2 Form Factor

To extrapolate the resistance obtained during model tests to full scale values, use is commonly made of a parameter called the form factor, see e.g. Larsson and Raven [25]. The form factor $(1+k)$ is the ratio between the viscous resistance of the hull and the frictional resistance of a flat plate with the same length and wetted surface area. The form factor is assumed to be the same for model scale as for full scale. In experiments, this factor is determined for each individual hull form from low speed resistance measurements where the wave resistance components are supposed to vanish according to a certain rule:

$$1+k = \lim_{Fn \to 0} \frac{R}{R_F}$$

The flat plate resistance R_F is determined by the formula

$$R_F = \frac{1}{2}\rho V^2 S_{wa} C_F,$$

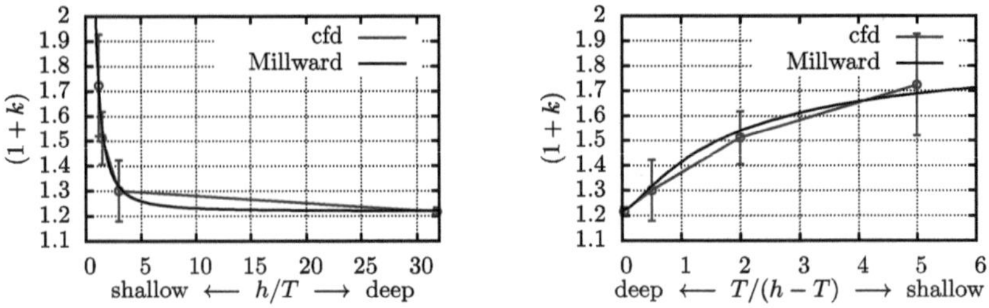

Fig. 9 Influence of water depth on form factor

with S_{wa} the wetted surface. The coefficient of frictional resistance C_F is often determined using the ITTC-1957 formula for model-ship correlation:

$$C_F = \frac{0.075}{(^{10}\log(Re) - 2)^2}$$

All calculations in this study have been performed with the double-body assumption which corresponds to the case $Fn \rightarrow 0$. Therefore, the obtained resistance can be used directly to calculate the form factor. For a drift angle of $\beta = 0°$, i.e. straight-ahead sailing, the relation between the water depth and the form factor as presented in Fig. 9 is found. Note that the presentation is given using two different parameters on the horizontal axis: the form factor is given as a function of the water depth to draught ratio h/T and as a function of $T/(h-T)$ in the left and right graphs respectively. It is seen that in the latter presentation the points do not all collapse near the $h/T = 0$ axis and a clearer relation between the influence of the water depth on the form factor is found.

The form factor as calculated with the viscous flow calculations changes from about 1.22 in deep water to 1.73 in very shallow water. Assuming the same form factor on model scale and full scale, this clearly shows that resistance extrapolation to full scale for shallow water conditions cannot be done using a form factor that was obtained in deep water. Millward [26] published a method to correct form factors found for deep water to shallow water conditions. According to Millward, the increase of the form factor due to shallow water is:

$$\Delta k = 0.644 \left(\frac{T}{h} \right)^{1.72}$$

For $h/T = 1.2$, this amounts to $\Delta k = 0.471$, such that the form factor becomes $(1+k) = 1.22 + 0.471 = 1.69$ which is reasonably close to the form factor found in the CFD calculation for this water depth. The relation between the form factor based on Millward's formula and the water depth has been plotted in the graph. The agreement between Millward's formula and the form factors predicted by CFD is encouraging. It is therefore recommended to investigate whether the form factor correction proposed by Millward can be applied to other ships as well. Furthermore, studies are currently conducted at MARIN to investigate whether the form factor on model scale is the same as the form factor on full scale.

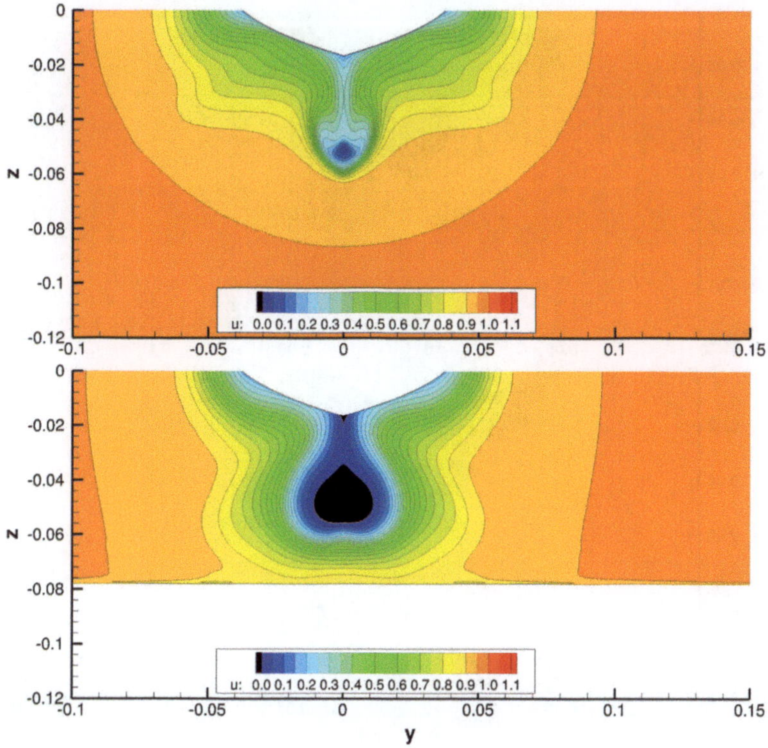

Fig. 10 Axial velocity in deep (*top*) and very shallow (*bottom*) water, $\beta = 0°$

6.5 Local Quantities

In Figs. 10, 11, and 12, flow fields around the hull are presented for deep and very shallow water, with $\beta = 0°$, $\beta = 4°$ and $\gamma = 0.4$. After examination of the flow fields for these conditions and for the other drift angles and rotation rates, it is observed that for reducing water depth, the pressures below the hull decrease, while the velocities increase (which is according to expectations). Flow separation is detected in restricted water, just below the propeller hub, see Fig. 10. The flow separation at the aft ship increases and grows upward for decreasing water depths. For straight ahead condition, the flow separation is much more pronounced than for the drift or rotation condition. The wake near the water surface does not change significantly when changing the water depth. However, the wake around the propeller hub becomes thicker as the water depth reduces. Furthermore, the wake starts to extend to the sea floor, resulting in bulging contour lines. The vortices developing around the aft body, responsible for the famous hook shape in the propeller wake plane, are less concentrated at reduced water depth, which is demonstrated in more detail in Fig. 13.

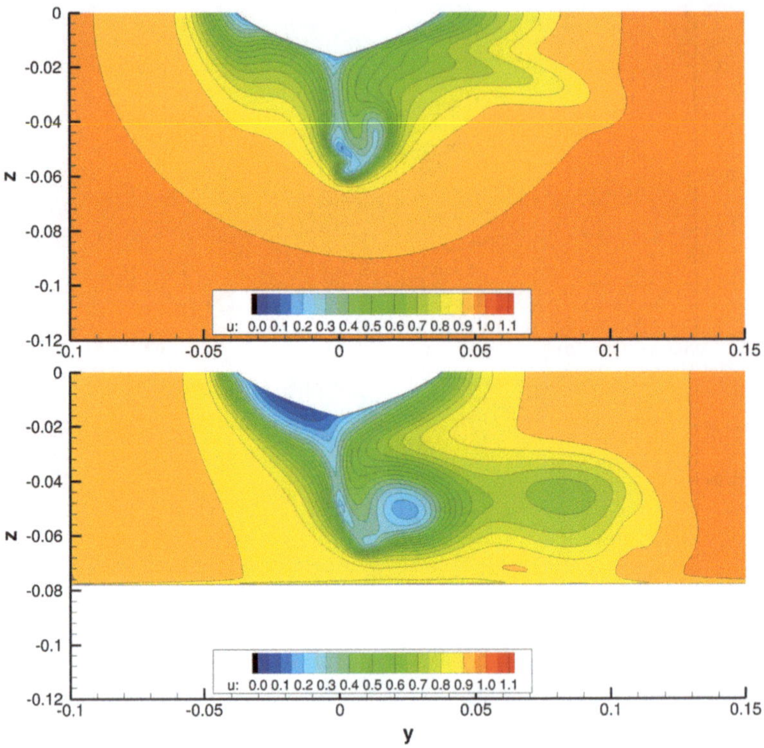

Fig. 11 Axial velocity in deep (*top*) and very shallow (*bottom*) water, $\beta = 4°$

7 Conclusions

An elaborate numerical study has been conducted for the KVLCC2 hull form sailing in various water depths. The calculations give clear insight into the forces and moments acting on the ship as a function of drift angle, yaw rate and water depth. Uncertainty estimates were made for the various calculations. In general, the uncertainties increase with increased flow complexity, i.e. for larger drift angles or yaw rates. A dependency of the uncertainty on the water depth was not found.

The correspondence between the experiments and the calculations is qualitatively reasonable. Quantitatively, it is difficult to draw conclusions because of the scatter in the INSEAN experimental data. Additional measurements, with a larger range of drift angles and rotation rates and combinations thereof, are very much desired. Comparisons between resistance measurements by MOERI and the computations show a good correlation, with a comparison error of the same order of magnitude as the validation uncertainty. The resistance predicted by REFRESCO is therefore judged to be good. By varying the settings of the turbulence modelling, the comparison error could be cancelled completely. This clearly demonstrated the sensitivity of the results to the turbulence modelling.

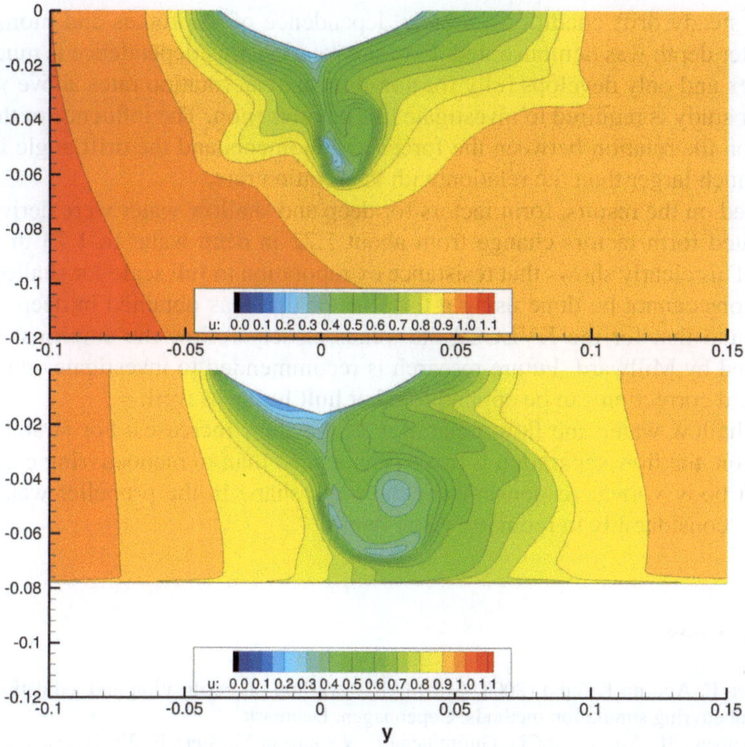

Fig. 12 Axial velocity in deep (*top*) and very shallow (*bottom*) water, $\gamma = 0.4$

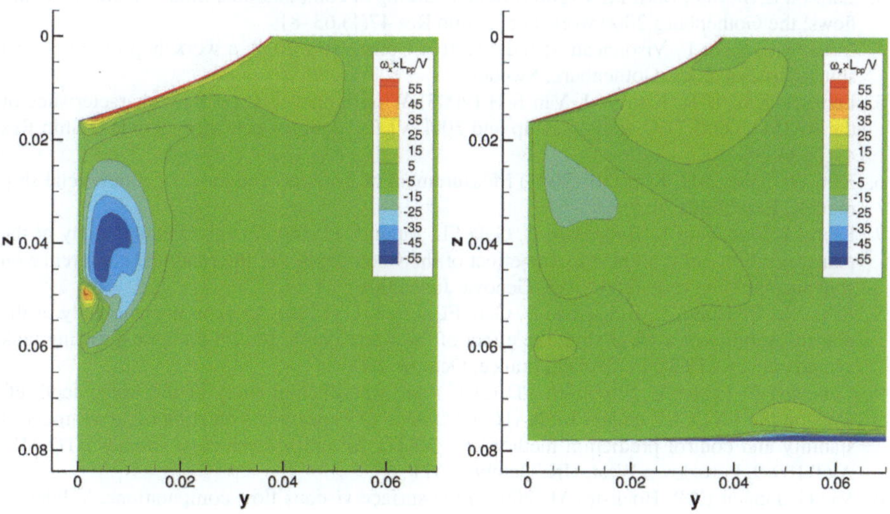

Fig. 13 Axial vorticity in deep (*left*) and very shallow (*right*) water, $\beta = 0°$

For steady drift conditions, a clear dependence of the forces and moments on the water depth was demonstrated. For pure rotation, this dependence is much more complex and only develops fully for non-dimensional rotation rates above $\gamma = 0.3$. Further study is required to investigate this phenomenon. The influence of the water depth on the relation between the forces and moments and the drift angle is found to be much larger than the relation with the rotation rate.

Based on the results, form factors for deep and shallow water were derived. The calculated form factors change from about 1.22 in deep water to 1.73 in shallow water. This clearly shows that resistance extrapolation to full scale for shallow water conditions cannot be done using a form factor that was obtained in deep water. It was found that for the KVLCC2 the trends closely follow the empirical formula proposed by Millward. Future research is recommended to investigate whether the Millward correction can be applied to other hull forms as well.

In shallow water, the flow separation at the stern increases. For straight ahead condition, the flow separation is more pronounced than in manoeuvring conditions. The aft body vortex, responsible for the hook shape in the propeller wake plane, reduces considerably in reducing water depth.

References

1. Stern F, Agdrup K (eds) (2008) SIMMAN workshop on verification and validation of ship manoeuvring simulation methods, Copenhagen, Denmark
2. Toxopeus SL, Simonsen CD, Guilmineau E, Visonneau M, Stern F (2011) Viscous-flow calculations for KVLCC2 in manoeuvring motion in deep and shallow water. In: NATO RTO AVT-189 specialists meeting on assessment of stability and control prediction methods for NATO air and sea vehicles, number RTO-MP-AVT-189-10, Portsdown West, UK, October 2011
3. Larsson L, Stern F, Bertram V (2003) Benchmarking of computational fluid dynamics for ship flows: the Gothenburg 2000 workshop. J Ship Res 47(1):63–81
4. Larsson L, Stern F, Visonneau M (eds) (2010) Gothenburg 2010: a workshop on numerical ship hydrodynamics. Gothenburg, Sweden
5. Lee S-J, Kim H-R, Kim W-J, Van S-H (2003) Wind tunnel tests on flow characteristics of the KRISO 3,600 TEU containership and 300k VLCC double-deck ship models. J Ship Res 47(1):24–38
6. Kim W-J, Van S-H, Kim DH (2001) Measurement of flows around modern commercial ship models. Exp Fluids 31(5):567–578
7. Fabbri L, Benedetti L, Bouscasse B, Gala FL, Lugni C (2006) An experimental study of the maneuverability of a blunt ship: the effect of the water depth. In: International conference on ship and shipping research (NAV), Genova, June 2006
8. Fabbri L, Benedetti L, Bouscasse B, Gala FL, Lugni C (2006) An experimental study of the maneuverability of a blunt ship: the effect of the water depth. In: 9th numerical towing tank symposium (NuTTS), Le Croisic, France, October 2006
9. Campana E, Fabbri L, Simonsen CD (2011) An experimental study of the water depth effects on the KVLCC2 tanker. In: NATO RTO AVT-189 specialists meeting on assessment of stability and control prediction methods for NATO air and sea vehicles, number RTO-MP-AVT-189-9, Portsdown West, UK, October 2011
10. Vaz G, Jaouen FAP, Hoekstra M (2009) Free-surface viscous flow computations. Validation of URANS code FreSCo. In: 28th international conference on ocean, offshore and arctic engineering (OMAE), number OMAE2009-79398, Honolulu, Hawaii, May 31–June 5, 2009

11. Vaz G, Waals O, Fathi F, Ottens H, Le Souef T, Kwong K (2009) Current affairs—model tests, semi-empirical predictions and CFD computations for current coefficients of semi-submersibles. In: 28th international conference on ocean, offshore and arctic engineering (OMAE), number OMAE2009-80216, Honolulu, Hawaii, May 31–June 5, 2009
12. Vaz G, Toxopeus SL, Holmes S (2010) Calculation of manoeuvring forces on submarines using two viscous-flow solvers. In: 29th international conference on ocean, offshore and arctic engineering (OMAE), number OMAE2010-20373, Shanghai, China, June 2010
13. Koop AH, Klaij CM, Vaz G (2010) Predicting wind shielding for FPSO tandem offloading using CFD. In: 29th international conference on ocean, offshore and arctic engineering (OMAE), number OMAE2010-20284, Shanghai, China, June 2010
14. Toxopeus SL (2011) Practical application of viscous-flow calculations for the simulation of manoeuvring ships. PhD thesis, Delft University of Technology, Faculty Mechanical, Maritime and Materials Engineering
15. Menter FR (1994) Two-equation eddy-viscosity turbulence models for engineering applications. AIAA J 32(8):1598–1605
16. Dacles-Mariani J, Zilliac GG, Chow JS, Bradshaw P (1995) Numerical/experimental study of a wingtip vortex in the near field. AIAA J 33:1561–1568
17. Bettle MC, Toxopeus SL, Gerber AG (2010) Calculation of bottom clearance effects on Walrus submarine hydrodynamics. Int Shipbuild Prog 57(3–4):101–125
18. Eça L, Vaz G, Hoekstra M (2010) A verification and validation exercise for the flow over a backward facing step. In: Fifth European conference on computational fluid dynamics, ECCOMAS, Lisbon, Portugal, June 2010
19. Eça L, Hoekstra M, Raven HC (2010) Calculation of the flow around the KVLCC2 tanker: case 1.1.a. In: Gothenburg 2010: a workshop on CFD in ship hydrodynamics, Gothenburg, Sweden, December 2010, pp 547–552 (Preprints)
20. Zou L (2011) CFD predictions including verification and validation of hydrodynamic forces and moments on a ship in restricted waters. Licentiate thesis, Chalmers University of Technology, Gothenburg, Sweden
21. Simonsen CD, Stern F, Agdrup K (2006) CFD with PMM test validation for manoeuvring VLCC2 tanker in deep and shallow water. In: International conference on marine simulation and ship manoeuvring (MARSIM), number M04, Terschelling, The Netherlands, June 2006
22. Eça L (2007) Calculation of the flow around the KVLCC2 tanker at model and full scale Reynolds numbers with eddy-viscosity models. IST report D72-43, Lisbon, Portugal
23. Kume K, Hasegawa J, Tsukada Y, Fujisawa J, Fukasawa R, Hinatsu M (2006) Measurements of hydrodynamic forces, surface pressure, and wake for obliquely towed tanker model and uncertainty analysis for CFD validation. J Mar Sci Technol 11(2):65–75
24. Quadvlieg FH, Brouwer J (2011) KVLCC2 benchmark data including uncertainty analysis to support manoeuvring predictions. In: IV international conference on computational methods in marine engineering (MARINE), Lisbon, Portugal, September 2011
25. Larsson L, Raven HC (2010) Ship resistance and flow. In: The principles of naval architecture, SNAME
26. Millward A (1989) The effect of water depth on hull form factor. Int Shipbuild Prog 36(407):283–302

Computational Investigation of Non-body-of-Revolution Hull Form Maneuvering Characteristics

Keegan P. Delaney

Abstract Maneuvering characteristics for traditional body-of-revolution (BOR) hull forms are well understood through decades of testing, while less traditional non-body-of-revolution (Non-BOR) hull forms are not nearly as well understood. A computational investigation on the maneuvering capability for two Non-BOR hull forms (Single and Twin Tail stern configurations) and an equivalent BOR hull form is completed, and compared to prior experimental tests. Initially, global forces and moments on the hull forms over various static angles of attack are computed and studied. Next, steady turn computations are used to determine global forces and moments on the hull forms over various turning radii. Force and moment information from the static and steady turn computations are combined to compute indices of stability for each hull form. Stability indices are used as a means of quantifying maneuvering stability for all of the hull forms in the vertical and horizontal planes. Computationally predicted stability indices displayed the same maneuvering characteristics as previous experiments. The Twin Tail hull form was shown to be highly stable in both the vertical and horizontal planes. The Single Tail hull form was unstable in the vertical plane and highly stable in the horizontal plane. The BOR proved to be marginally stable to stable in both planes. The results of this study show that the Twin Tail hull form would be more suitable from a maneuvering standpoint than the Single Tail hull form in future Non-BOR design studies.

Keywords RANS · Maneuvering · Underwater body · Non-BOR

1 Introduction

In this study, Reynolds Averaged Navier-Stokes (RANS) computations are used to shed insight into maneuvering characteristics for the DTMB Series 66 non-body-of-revolution (Non-BOR) hull forms. Non-BOR hull forms are of interest due to

K.P. Delaney (✉)
Carderock Division, Naval Surface Warfare Center, 9500 MacArthur Blvd., West Bethesda, MD 20817, USA
e-mail: keegan.delaney@navy.mil

L. Eça et al. (eds.), *MARINE 2011, IV International Conference on Computational Methods in Marine Engineering*, Computational Methods in Applied Sciences 29,
DOI 10.1007/978-94-007-6143-8_10, © Springer Science+Business Media Dordrecht 2013

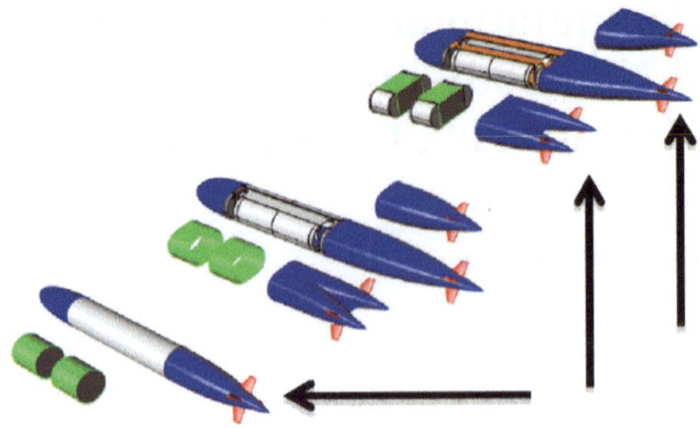

Fig. 1 Series 66 hull forms with configurations used in this study marked with *arrows*

Table 1 Series 66 geometric properties

	Model length (ft)	Beam to-height	Surface area (ft^2)
BOR	10.4125	1.0	30.161
Twin tail	10.4125	2.5	57.587
Single tail	12.2751	2.5	61.118

their potential for increased storage capacity and midbody modularity vice traditional body-of-revolution (BOR) hull forms. Maneuvering characteristics of BOR hull forms are well understood from decades of testing, while less traditional Non-BOR hull forms are not nearly as well understood.

Series 66 is a systematic sequence of bodies: Single Tail, Twin Tail, and an equivalent BOR hull forms. Figure 1 shows the complete Series 66 hull form family. Only the BOR, wide-body long Single Tail, and wide-body Twin Tail configurations (marked with arrows in Fig. 1) are investigated in this study. Other Series 66 hull forms will be investigated in future studies. The wide-body Single and Twin Tail hull forms share the same bow and midsection, which are essentially the BOR body with added beam at constant height, while only the sterns are different. Table 1 shows the geometric hull form characteristics for the three Series 66 hull forms used in this study.

This computational investigation of the Series 66 Non-BOR hull forms is complementary to experimental tests performed by Roddy et al. [4] at the Naval Surface Warfare Center, Carderock Division (NSWCCD). All experimental data presented in this study is from the work of Roddy et al.

Initially, global forces (vertical and horizontal) and moments on the hull forms over various angles of attack are studied. Both bare hull and X-configuration stern appended hull forms are compared to analyze the effect of stern appendages for each of the hull forms. Figure 2 shows larger views of the Single and Twin Tail stern appended hull forms.

Next, steady turn computations are performed for all of the stern appended configurations over varied turning radii. Vertical and horizontal plane steady turn ma-

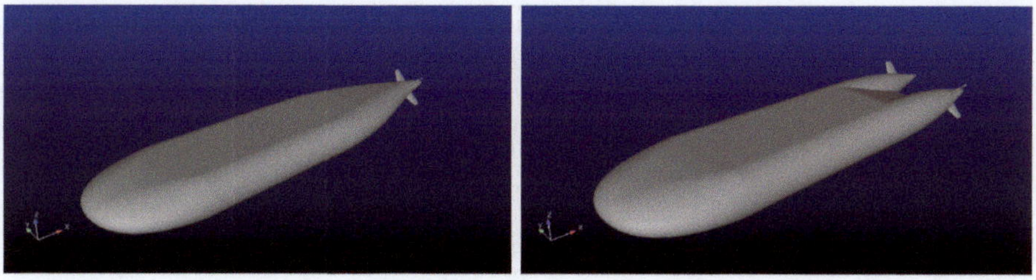

Fig. 2 Series 66 Single (*left*) and Twin (*right*) Tail stern appended hull forms

Table 2 Index of stability (G) definitions

G	Degree of stability
Negative	Unstable
Positive	Stable
0 to 0.2	Marginally stable (satisfactory in the horizontal plane)
0.5 to 0.7	Stable (satisfactory in the vertical plane)
>0.8	Highly stable

neuver computations are used to compute rotary derivatives from the resulting force and moment curves.

Finally, information from the force and moment computations is combined with rotary derivatives to compute indices of stability. Indices of stability in the vertical (G_v) and horizontal (G_h) planes quantify the maneuvering stability of the underwater body. Table 2 shows the relationship between stability indices and maneuvering stability. Typical applications that the Series 66 hull forms would be used for target G_v and G_h values of 0.6 and 0.2, respectively. It is typically desired that the underwater body be slightly more stable in the vertical plane. It is important for an underwater body to not have too high of an index of stability or else the body will not turn and maneuver readily enough. Conversely, it is important for the hull form's index of stability to not be too low, or else the body will have a hard time maintaining steady course in turning maneuvers.

2 Numerical Methods

NavyFOAM V1.0's single-phase RANS solver is used for this work. NavyFOAM V1.0 [5] was developed from the extended version of OpenFOAM (www.wikki.co.uk), which is extended from the standard release (www.openfoam.com). The solver employs a cell-centered finite-volume method that permits use of arbitrary polyhedral elements including quadrilateral, hexahedral, triangular, tetrahedral, pyramidal, prismatic, and hybrid meshes. The single-phase solver is run in parallel using domain decomposition and message passing.

For incompressible flows the continuity (mass conservation) and momentum equations can be written as Eqs. (1) and (2), respectively.

$$\frac{\partial U_i}{\partial x_i} = 0 \tag{1}$$

$$\frac{DU_i}{Dt} = -\frac{1}{\rho}\frac{\partial p}{\partial x_i} + \frac{\partial}{\partial x_j}\left(\nu\frac{\partial U_i}{\partial x_j} - \overline{u'_i u'_j}\right) \tag{2}$$

Where ρ is the density of the fluid, ν is the kinematic viscosity, p is the pressure, U_i is the mean velocity component and u'_i is the fluctuating velocity component. The term $\overline{u'_i u'_j}$ is the Reynolds stress. The Boussinesq approximation is used to relate the Reynolds stress to velocity gradients by Eq. (3).

$$\overline{u'_i u'_j} = \nu_t\left(\frac{\partial U_i}{\partial x_j} + \frac{\partial U_j}{\partial x_i}\right) - \frac{2}{3}k\delta_{ij} \tag{3}$$

Where k is the turbulent kinetic energy and ν_t is the turbulent viscosity. Closure for the RANS equations is achieved by modeling ν_t using Menter's shear stress transport (SST) turbulence model, which is a two-equation $k - \varepsilon/k - \omega$ hybrid model [3].

3 Grid Generation

Due to the necessity for quick turnaround time and relatively complex geometry around the various appendages and stern sections unstructured gridding methods are preferred over structured gridding methods for this study. Multi-element unstructured grids are developed using SolidMesh [1], a suite of tools developed at Mississippi State University. SolidMesh provides tools for geometry preparation and surface grid generation. The volume grid is generated by using an advancing normal methodology for the boundary layer prism elements and an Advancing Front/Local Reconnection technique [2] to develop isotropic elements. Using these tools, multi-element unstructured grids can be created around complex geometries, like those presented in this work, significantly faster than traditional structured grids. Typical unstructured grids used in this study contained 5 to 25 million cells. All grids were created to solve down to the wall, with a non-dimensional normal wall spacing ($y+$) of 1. Figure 3 shows typical unstructured surface meshes used for these computations.

4 Force and Moment Computations

Results from pitch (vertical) and drift (horizontal) angle computations are presented for the various hull forms under bare hull and stern appended configurations. Computational results for all hull forms are compared to experimental data up to 20

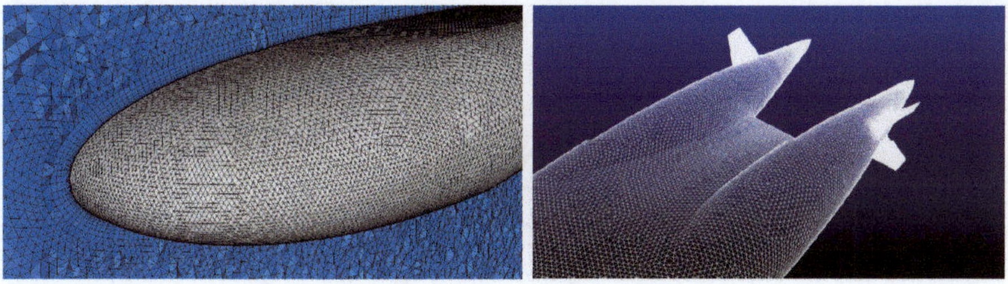

Fig. 3 Views of the unstructured mesh on the Twin Tail bow and a stream-wise centerline plane (*left*) and on the stern (*right*)

degrees angle of attack as a means of comparison between the solver and experiment. However, for maneuvering purposes of this study only the forces and moments from $+/-4$ degrees are important as stability indices include the slope of force and moment curves over this range. All forces and moments are presented in non-dimensional form according to Eqs. (4) and (5).

$$F' = \frac{F}{\frac{1}{2}\rho U^2 L^2} \tag{4}$$

$$M' = \frac{M}{\frac{1}{2}\rho U^2 L^3} \tag{5}$$

Where L is the characteristic length of the hull form, F and M are the dimensional forces and moments, and F' and M' are the non-dimensional forces and moments. All moments for computation and experiment are taken about the center of mass.

Figure 4 shows the Single Tail hull form forces and moments plotted over a range of pitch and drift angles. As an important note, all of the experimental results at large negative drift angles used in this study should be considered unreliable. In the experimental tests two struts were connected to the body, and at negative drift angles the body was turned into the wake of the struts, thus altering experimental results. Nevertheless, Fig. 4 shows that for the appropriate conditions computational predictions match experimental measurements very well over a wide range of pitch and drift angles for both bare hull and stern appended hull form configurations.

One interesting note from the pitch angle results is that the Single Tail bare hull and stern appended cases yield similar lifting force and pitching moment results for both computation and experiment. This indicates that the stern appendages add little control in the vertical plane. In the horizontal plane there is a noticeable difference in forces and moments for the bare hull and appended cases. Also, in general the forces and moments in the vertical plane are an order of magnitude larger than in the horizontal plane. This is to be expected, as the body appears as a well-rounded BOR in the horizontal plane, whereas there is a large flat midsection that is exposed to the flow in the vertical plane.

As previously mentioned for stability index purposes the slope of the force and moment curves around 0 angle of attack is important. Figure 5 shows pitch and drift

Fig. 4 Single Tail forces and moments for the bare hull (BH) and stern appended (StAp) hull forms

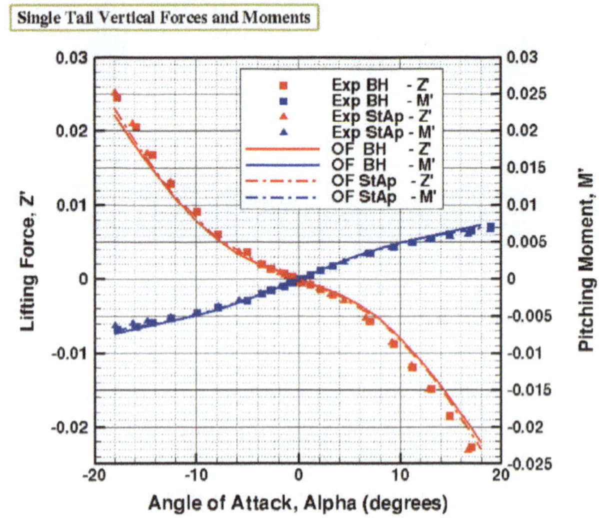

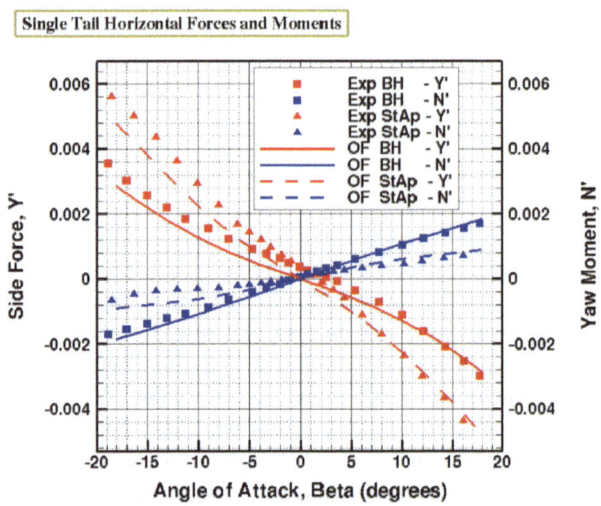

forces around 0 degrees angle of attack for the Single Tail hull form. At times there is a slight offset in the experimental data as the drift angle force plot shows, where the force and/or moment curve does not cross through 0 exactly at 0 degrees angle of attack. Despite the slight offset in force for the drift angle case the experimental and computational slopes line up very well.

Figure 6 shows the Twin Tail hull form force and moment results. There are no Twin Tail bare hull form experimental drift angle results to compare to, so only computational predictions are presented. As in the Single Tail case the experimental measurements and computational predictions agree very well over a wide range of attack angles. The stern appendages seem to alter the force and moment results slightly more than in the Single Tail case, but again the vertical plane difference

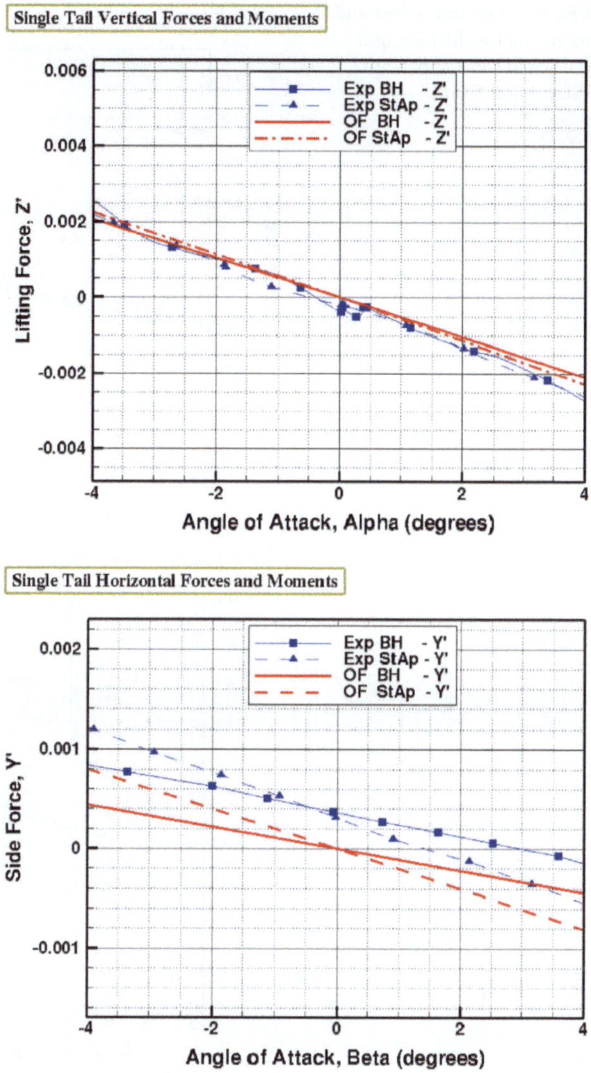

Fig. 5 Single Tail forces and moments around 0 degrees angle of attack

is minimal compared to the horizontal plane difference. Again, the forces and moments in the vertical plane are an order of magnitude larger than in the horizontal plane, because the flat midsection is directly exposed to the incoming flow in the vertical plane.

Lastly, Fig. 7 shows the forces and moments for the equivalent BOR hull form. The plot on the left is presented with the experimental data in original form, and the plot on the right is presented with experimental results shifted by an offset to make forces and moments cross through 0 at 0 degrees angle of attack. The vertical and horizontal plane results are presented in one plot, as the body is symmetric in both panes. The forces and moments of the BOR are of the same order of magnitude as

Fig. 6 Twin Tail forces and moments for the bare hull (BH) and stern appended (StAp) hull forms

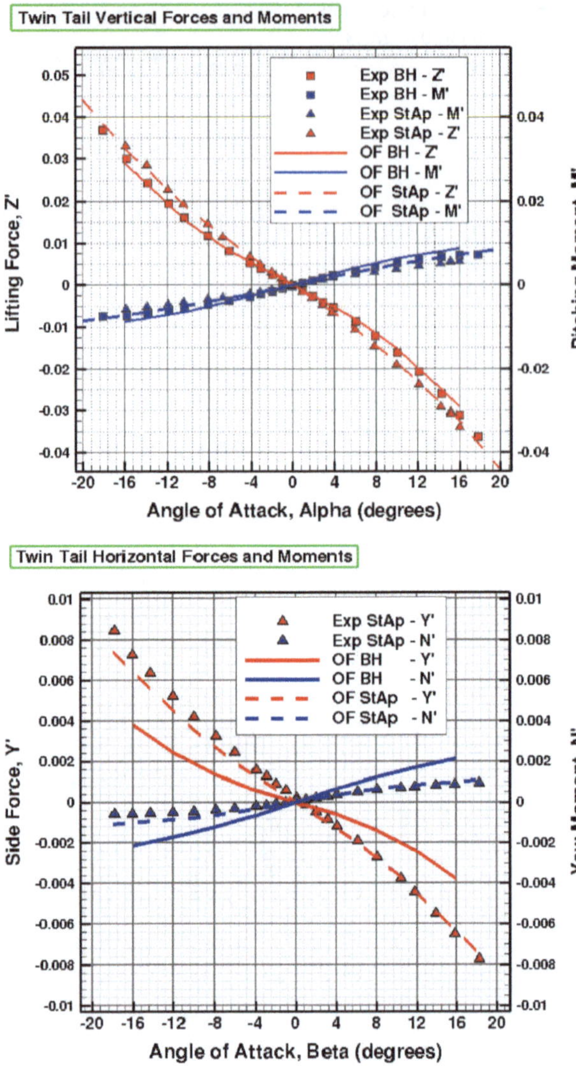

the Single and Twin Tail drift angle results. This makes sense, as the Non-BOR hull forms appear as a BOR in the horizontal plane profile. There is a slight disagreement in computed and experimental results that is exaggerated by the offset, but generally the results and slopes agree well.

5 Steady Turn Results

Steady turn computations are completed for all of the stern appended hull forms with the intention of computing rotary derivatives that are needed for stability indices.

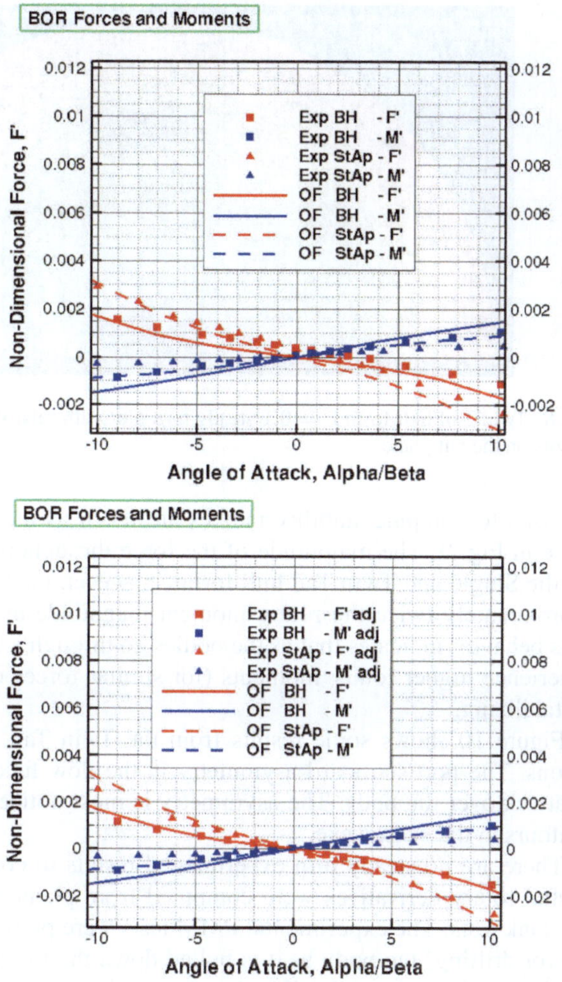

Fig. 7 Body-of-Revolution forces and moments without (*top*) and with (*bottom*) adjusted experimental data

The various hull forms undergo a steady turn maneuver (similar to a rotating arm tow tank experiment) at various radii. Computations are done for vertical steady maneuvers where the simulated body undergoes a constant pitch, and for horizontal steady maneuvers (the body is rotated 90° from the vertical turn) where simulated body undergoes a constant drift. Equation (6) defines the non-dimensional radii used for this work.

$$r' = \frac{L}{R} \tag{6}$$

Where R is the dimensional radius of the steady turn. Figure 8 shows the computational domain for the Twin Tail hull form at r' of 0.10. The relative velocity varies with the radius as the contours and vectors show.

Figure 9 shows computed rotary forces and moments for various turning radii in the vertical plane for the Single and Twin Tail hull forms. Rotary derivatives that

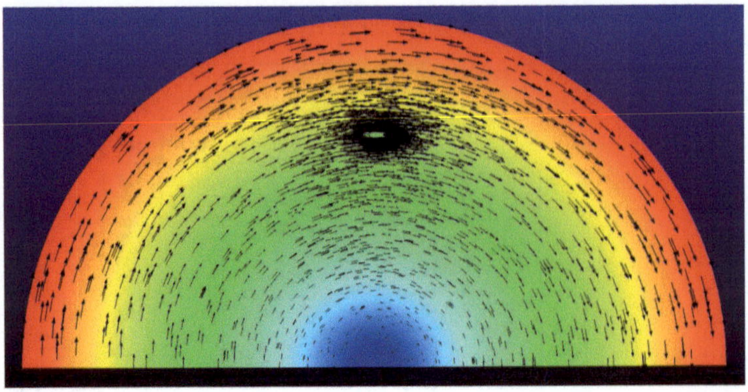

Fig. 8 Twin Tail hull form $r' = 0.1$ steady turn plot with relative velocity contours and velocity vectors on the cut plane

are used to compute stability indices are taken as the slope of rotary curves like those in Fig. 9. The magnitude of the force throughout the range tested is similar for the Single and Twin Tail hull forms; however, the Single Tail hull form creates approximately twice the rotary moment magnitude than the Twin Tail hull form. This behavior indicates that as the bodies go through a turn the Single Tail hull will experience higher rotary moments (for similar forces on the body), which can be destabilizing.

Figure 10 shows some results from the Twin Tail hull form $r' = 0.1$ computations. The port/starboard asymmetry in the flow field is due to the varied angle of attack over the body. The asymmetry is most noticeable in the relative velocity contours in the stern wake.

There are no steady turn experimental results to compare with; instead experimental rotary derivatives were computed from planar motion mechanism (PMM) tow tank tests. The experimental PMM tests were performed by dynamically pitching (or drifting) the body as it is pulled down the tow tank. These differences will be discussed in greater detail in the next section.

6 Stability Indices

The stability indices from Table 2 are computed from slopes of the translational (pitch and drift) and rotary (vertical and horizontal) curves as discussed in the previous sections. Equations (7) and (8) show the vertical (G_v) and horizontal (G_h) indices.

$$G_v = 1 - \frac{M'_w \cdot (Z'_q + m')}{Z'_w \cdot M'_q} \tag{7}$$

$$G_h = 1 - \frac{N'_v \cdot (Y'_r - m')}{Y'_v \cdot N'_r} \tag{8}$$

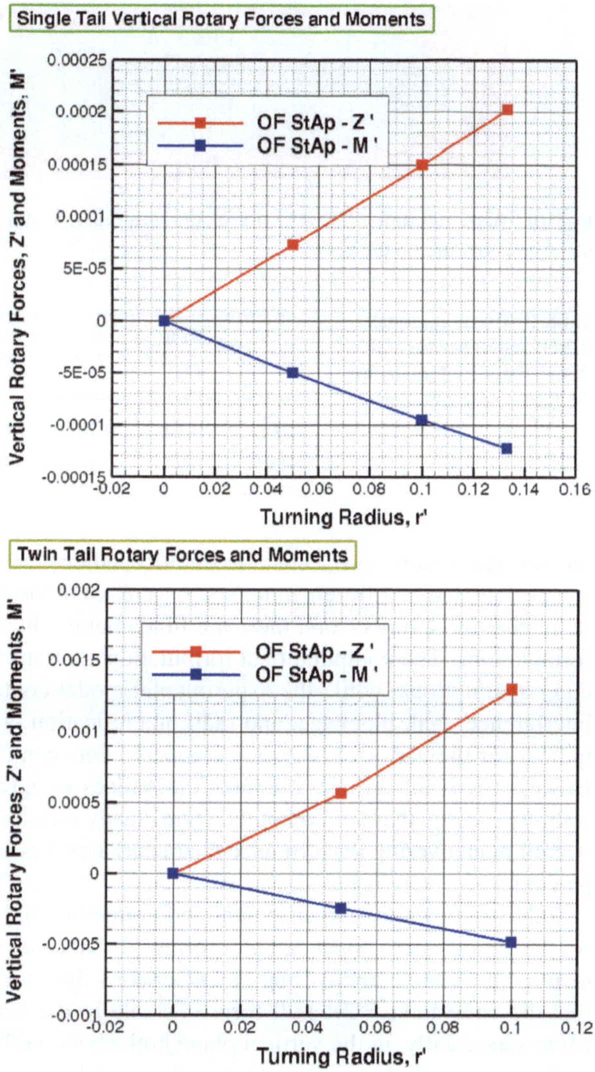

Fig. 9 Computed rotary vertical force and moment curves for the Single (*top*) and Twin (*bottom*) Tail hull forms

Where m' is the dimensionless mas, $m/(0.5\rho L^3)$, Z'_w and M'_w are the slopes of the straight ahead vertical force and moment curves, respectively. Z'_q and M'_q are the slopes of the vertical rotary force and moment curves, respectively. Y'_v and N'_v are the slopes of the straight ahead horizontal force and moment curves, respectively. Y'_r and N'_r are the slopes of the horizontal rotary force and moment curves, respectively.

Computational rotary derivatives were calculated from the steady turn runs above; however, experimental rotary derivatives were determined from dynamic pitch and drift PMM tests in a straight ahead tow tank. Given the scope of this project it was not feasible to perform computational PMM predictions or experi-

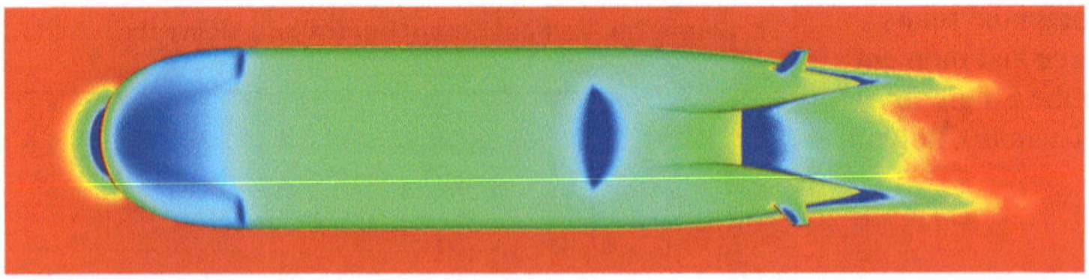

Fig. 10 Twin Tail hull form $r' = 0.1$ steady turn contour plot with relative velocity on the cut plane and pressure on the body

Table 3 Computational (CFD) and experimental (EXP) Stability Indices for the Series 66 hull forms

	BOR		Twin Tail		Single Tail	
	G_v	G_h	G_v	G_h	G_v	G_h
CFD	0.14	0.14	1.66	1.48	−0.50	1.35
EXP	0.37	0.37	0.94	2.63	−2.08	2.70

mental steady turn runs. Thus, the stability indices for computation and experiment were calculated in different manners. As a result of the different rotary tests and their individual modeling errors, there are fundamental differences in the stability index formulations. The computational formulation contains RANS and steady turn modeling assumptions; while the experimental model contains free surface, strut wake interference, tank blockage, and radii quantification effects. Nevertheless, stability indices for the various bodies are compared between experiment and computation. However, it is unrealistic to expect the indices to agree completely. Instead if the stability indices that are computed from steady turn and PMM tests show the same overall maneuvering performance characteristics it is considered a successful comparison.

Table 3 shows the experimental and computational stability index comparisons for the various stern appended hull forms. For all cases experiment and computation display the same maneuvering characteristics, but in differing degrees. Both show that the equivalent BOR hull form is stable in the vertical and horizontal planes. More specifically, in the vertical plane both show the BOR to be marginally stable. In the horizontal plane computations show the body to be marginally stable (which is satisfactory), while experiments show the body to be more stable. Both experiment and computation indicate that additional control surfaces or modifications to existing surfaces need to be added to the body to achieve the acceptable amount of vertical stability.

Both experiment and computation show the Twin Tail hull form to be highly stable (to varying degrees) in both the vertical and horizontal planes. The Twin Tail hull form is different from the BOR in that a ship designer would need to remove stability from the body for typical applications. If the hull form were too stable then the body would be difficult to readily turn and maneuver.

Meanwhile experiment and computations show that the Single Tail hull form is highly stable in the horizontal plane and unstable in the vertical plane. Thus, new control surfaces or modification to existing surfaces would be needed to improve vertical plane stability performance, while reducing the horizontal plane stability.

Between the two Non-BOR hull forms it would appear that the Twin Tail hull would be a better candidate to select for future design based on the greater inherent stability. The Twin Tail hull form's superior stability as compared to the Single Tail hull form is most likely due to the wedge-like section in between the two tails which acts as a vertical stabilizer.

7 Conclusions

In conclusion, a computational investigation on the maneuvering capability for two non-body-of-revolution (Non-BOR) hull forms and an equivalent body-of-revolution (BOR) hull form was completed. This study was complimentary to previous experimental tests done by Roddy et al. at NSWCCD. Initially, global force and moment predictions over various pitch and drift angles were analyzed. Computational predictions on the Single Tail hull form showed that the stern appendages did not seem to affect control in the vertical plane, but had a significant effect in the horizontal plane. The stern appendages seem to make a relatively larger difference for the Twin Tail configuration in the vertical plane, and show the same significant effect in the horizontal plane. For both Non-BOR hull forms the forces and moments in the vertical plane are an order of magnitude larger than horizontal forces and moments. This is to be expected as the large flat midsection is exposed to the flow in the vertical plane. Meanwhile, in the horizontal plane the bodies appear more as a fared traditional BOR in profile, thus resulting in lower forces and moments. Equivalent BOR computations confirmed that the forces and moments are of similar magnitude to Non-BOR horizontal plane results.

Next, steady turn computations were completed with the intention of using the results to compute rotary derivatives that are used in stability indices. Experimental results used planar motion mechanism (PMM) tests to determine rotary derivatives. Each method for computing rotary derivatives includes their own uncertainties and modeling errors.

Finally, stability indices were computed for all of the stern appended bodies in the vertical and horizontal planes. Computationally predicted stability indices displayed the same maneuvering characteristics as experimental measurements. The Twin Tail hull form was shown to be highly stable in both the vertical and horizontal planes. The Single Tail hull form was unstable in the vertical plane and highly stable in the horizontal plane, indicating that additional control surfaces would be necessary to add stability in the vertical plane for typical applications. The wedge-like surface that connects the twin tails likely added vertical stability that made the Twin Tail hull form stable, while the Single Tail hull form lacked any such surface and suffered a lack of vertical stability. The BOR proved to be marginally stable to stable in both planes. The results of this study show that the Twin Tail hull form would be more suitable from a maneuvering standpoint than the Single Tail hull form in future Non-BOR design studies.

Future work will center around altering the two Non-BOR configurations and analyzing the effect on maneuvering. The length-to-beam ratios of the bodies will be altered, and the resulting vertical and horizontal stability indices will be compared.

Acknowledgements The U.S. Office of Naval Research sponsored this work, and the program monitor is Dr. Ronald D. Joslin, Code 331. The author is also grateful to Mr. Robert Roddy who provided the experimental data for the present study.

References

1. Gaither AJ (1997) A solid modeling topology data structure for general grid generation. Master's thesis, Mississippi State University
2. Marcum DL, Gaither AJ (1999) Mixed element type unstructured grid generation for viscous flow applications. AIAA paper No 99-3252
3. Menter FR (1994) Two-equation eddy viscosity turbulence models for engineering applications. AIAA J 32:1598–1605
4. Roddy RF, El-Taher DJ, Hess DE, Junghans KA, Jiang L (2010) An investigation into the performance of non-body-of-revolution submarines—DTMB series 66. In: 28th symposium on naval hydrodynamics, September 2010
5. Shan H, Delaney K, Kim S-E, Rhee B, Gorski J, Ebert M (2011) Guide to NavyFOAM V1.0. NSWCCD-50-TR-2011/025, April 2011

Part VI
Propulsion and Cavitation

Numerical Prediction of Erosive Collapse Events in Unsteady Compressible Cavitating Flows

Michael S. Mihatsch, Steffen J. Schmidt, Matthias Thalhamer, and Nikolaus A. Adams

Abstract The objective of the present investigation is the numerical prediction of the potential of a flow to inflict surface damage by cavitation. For this purpose, physical criteria are derived that detect and quantify relevant flow phenomena. In particular, we present a numerical approach for tracing isolated collapses of vapor clouds during the numerical simulation of the flow. The suggested "collapse detector" provides the frequency of collapses, their positions, and resulting maximum pressures, as well as the maximum condensation rate of each event. This data, together with the maximum wall pressure, allow for an automatic indication of erosion-sensitive areas.

The employed flow solver CATUM (CAvitation Technische Universität München) is a density-based 3-D finite volume method equipped with a Low-Mach-number consistent flux function. All fluid components (liquid, vapor, saturated mixture) are modeled by closed form equations of state.

To assess this novel approach we simulate an experimentally investigated nozzle-target flow. A comparison of numerically predicted collapse events with the experimentally observed areas of cavitation erosion substantiates the proposed methodology. The obtained data represents a time-history of collapse events together with their position and strength, and may be used to estimate erosion rates.

Keywords Cavitation · Erosion · Numerical simulation · Multiphase flow

1 Introduction

Hydrodynamic cavitation and cavitation erosion are well-known phenomena in naval engineering [1, 2]. Distinctive examples include cavitation on suction sides of propeller blades and cavitating tip vortices. The basic mechanisms of hydrodynamic cavitation are flow-induced evaporation due to the pressure drop of an ac-

M.S. Mihatsch (✉) · S.J. Schmidt · M. Thalhamer · N.A. Adams
Institute of Aerodynamics and Fluid Mechanics, Technische Universität München, Boltzmannstr. 15, 85748 Garching, Germany
e-mail: michael.mihatsch@aer.mw.tum.de
url: http://www.aer.mw.tum.de/en/home/

L. Eça et al. (eds.), *MARINE 2011, IV International Conference on Computational Methods in Marine Engineering*, Computational Methods in Applied Sciences 29,
DOI 10.1007/978-94-007-6143-8_11,

celerated liquid, and inertia controlled recondensation of liquid-embedded vapor bubbles due to pressure recovery. The collapse-like recondensation results in the generation of shock waves that are supposed to constitute the main mechanism of cavitation erosion. These shocks propagate approximately with the speed of sound $c_l = O(10^3)$ m/s of the liquid and their intrinsic time scale is about 2 orders of magnitude smaller than the time scale of the underlying shear flow.

Since the simulation of erosive collapses in cavitating flows requires the resolution of both time scales, time step sizes of the order of nanoseconds are inevitable and millions of time steps have to be computed to cover the characteristic time interval of the shear flow. It is neither reasonable to store the enormous amount of data generated by such a simulation, nor is it feasible to evaluate manually the data with respect to collapse characteristics and erosion aggressiveness. Instead, new approaches have to be developed that allow for automatic detection and evaluation of aggressive collapse events predicted by 3-D unsteady simulations of cavitating flows.

2 Underlying Numerical Model

We apply our flow simulation code CATUM (CAvitation Technische Universität München [3]), which is a density based finite volume method employing a Low-Mach-number consistent flux function and an explicit time marching procedure. The spatial reconstruction of the velocity field is a WENO-3 procedure; density and internal energy are reconstructed by monotonic TVD limiters ("minmod"). Time marching is performed through an explicit low storage 4-step Runge-Kutta method with an optimized stability region. The combination of both methods results in a 2nd order numerical approach in space and time for smooth flow, and it ensures a sharp representation of discontinuous flow features such as shocks and contact waves. In this investigation we focus on the simulation of inertia-driven effects and wave dynamics and we neglect viscous effects. Hence, the governing equations are the compressible Euler equations.

In the present investigation the thermodynamic properties of the working fluid, in this case water, are characterized by closed-form equations of state: the liquid phase is specified by a modified Tait law, saturated two-phase regions are modeled by polynomial fits of the IAPWS [4] data, and pure vapor is modeled by an modified ideal gas law. In order to allow for the simulation of shock formation and wave propagation, the compressibility of the fluids (liquid and vapor) is taken into account. Therefore, the numerical time step is necessarily proportional to the ratio of the smallest length scale (minimum grid size) and the fastest signal speed ($\sim$speed of sound of the liquid).

The phase transition model is based on local equilibrium assumptions for pressure, temperature, and specific Gibbs functions. We have observed that this assumption is justified since most of technically relevant hydrodynamic cavitation processes occur close to the equilibrium vapor pressure. Although this observation is questionable for incipient cavitation, it applies to developed cavitating flows, such as the one investigated within the presented research work. Previous investigations show that CATUM is able to predict even delicate flow features, such as irregular break-up patterns of partial cavities of developed cavitating flows [5].

We adopt body-fitted hexahedral grids to discretize the flow field. The data structure is based on local blocks, and a pre-processor load balance strategy is utilized to obtain approximately linear scaling on multi-processor systems.

3 Modeling and Implementation of Erosion Indicators and Collapse Detectors

Shock waves initiated by violent cloud collapses can generate intense pressure peaks on the order of several thousand bars. If such a shock wave is emitted sufficiently close to a solid surface, its impact on the solid material may result in dynamic loads that lead to material fatigue. Thus, instantaneous maximum loads may provide an indication of locations where erosion is likely to occur. Therefore monitoring the maximum pressure arising in each computational cell during the whole simulation is one efficient approach for detecting erosion sensitive areas within the flow field, and especially along solid walls. Although this procedure was applied successfully in previous investigations [5], it has several obvious drawbacks: The approach neither provides information about the frequency of collapse events, nor does it allow to distinguish between stagnation pressures and instantaneous maximum loads with durations of only a few microseconds. However, even this limited amount of achievable information allows for the indication of erosion-sensitive areas.

Our recent development focuses on a more detailed approach that we denote as "collapse detector". The concept involves the derivation of a set of physical criteria to detect collapses of isolated clouds and to characterize the strength of the generated shock waves as predicted by our numerical simulations. Figure 1 shows a schematic of consecutive stages of a cloud-collapse. In the lower part of this figure the average divergence of the velocity field around the collapsing cloud is depicted.

The onset of condensation results in the formation of a velocity field that is approximately directed towards the center of the cloud—similar to an elementary sink flow. Thus the divergence of the velocity field is negative and decreasing. Since we assume that the specific Gibbs functions remain in equilibrium, and due to disregarding solved or dissolved gas, the pressure within the cloud remains approximately constant. Hence the surrounding liquid is further accelerated towards the center of the cloud until the vapor completely condenses at the last stage of the collapse. At that point the inertia of this sink flow results in a massive increase of the pressure. The maximum pressure is reached when the mass flow towards the center vanishes. At that point, the divergence changes its sign and the negative radial pressure gradient leads to an outward directed velocity field. Consequently, the pressure at the center decreases and may even lead to a re-evaporation of the liquid (rebound of the cavity). These considerations motivate the following definition of a numerical "collapse detector":

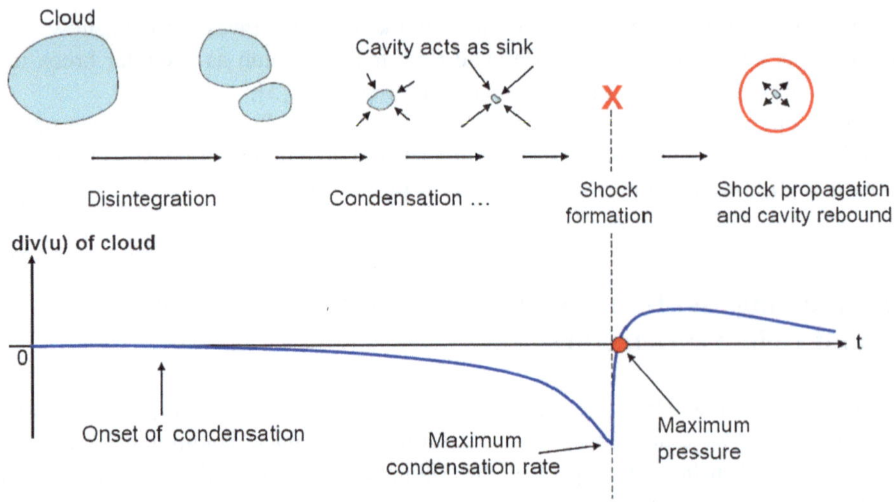

Fig. 1 Sketch of consecutive stages of the collapse of a vapor cloud together with the temporal evolution of the divergence of the velocity field

- We denote as "candidates" such computational cells where the vapor volume content condenses completely during the last time step. If the surrounding cells of a "candidate" contain liquid only, an isolated collapse is detected.
- Once a collapse is detected, the maximum pressure is generated at that instant in time when the divergence of the velocity field changes its sign.
- The strength of the collapse is characterized by its maximum (negative) divergence and its maximum pressure.

The main advantages of the "collapse detector" as compared to the maximum pressure approach are as follows:

- The collapse detector automatically distinguishes between collapse-induced maximum pressures and high pressures at stagnation points or due to wave interaction.
- The number of collapse events, as well as their position, and their strength provide important information about a possible stress profile the material is exposed to. This information can be used to estimate erosion rates.

An apparent deficiency of the "collapse detector" could be the missing information of the collapse intensity at the material surfaces. To overcome this drawback, an efficient projection method is proposed. We adopt the linear decay law of spherical waves to estimate the resulting pressure at the wall p_{wall} from the known maximum pressure in the collapse center $p_{collapse}$ and from the distance of the collapse to the wall r_{wall}. The linear decay law states that the amplitude of a linear spherical pressure wave is inversely proportional to the radial distance of the wave front measured from its origin, hence $p(r) \sim 1/r$. We assume that the initial radius of the wave front

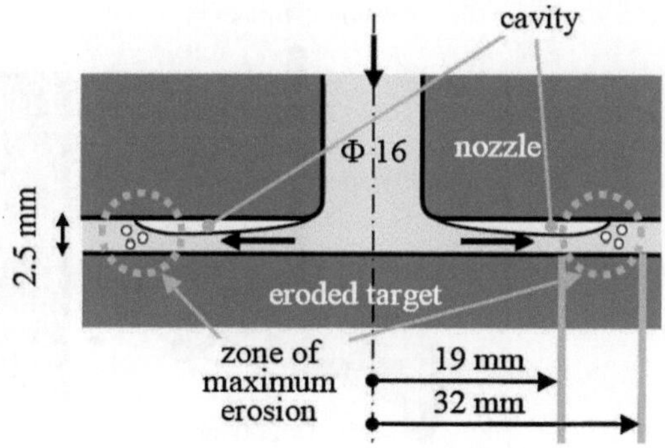

Fig. 2 Sketch of the experimental setup [8]. Reprint with permission of ASME

is proportional to the cube root of the volume V_{cell} of the cell where the collapse was detected. Thus, we obtain

$$p_{wall} \sim \frac{\sqrt[3]{V_{cell}}}{r_{wall}} p_{collapse}. \tag{1}$$

This concept allows for the reduction of grid dependencies of the predicted maximum pressures as well. Instead of the maximum pressure, the value $p_{collapse} \cdot \sqrt[3]{V_{cell}}$ may be used to characterize the strength of a collapse. Preliminary investigations demonstrate the applicability of this approach [6, 7].

4 Results

We simulate an experimentally investigated nozzle-target flow [8] to evaluate the numerical predictability of erosion-sensitive areas. The experimental setup results in an axisymmetric stagnation flow as shown in Fig. 2. At the exit of the nozzle the fluid accelerates along a small radius and forms a radial cavitation pocket. Collapse-induced erosion is observed in the experiment within an annulus at the surface of the target disc. The remaining parts are manufactured out of highly resistant material.

Figure 3 shows a cut through the computational domain. In accordance with the experiment, water at approximately room temperature enters the nozzle with the inlet-velocity $u_{in} = 31$ m/s. An asymptotic pressure boundary condition $p_e = 10.1$ bar is imposed at the exit of a circular reservoir attached at a radial distance of 100 mm. All solid boundaries are treated as inviscid adiabatic walls. The computational grid consists of $5 \cdot 10^5$ body-fitted hexahedrons structured in 56 blocks.

Figure 4 shows the time history of the vapor volume fraction within the computational domain. The time intervals 1 and 2 correspond to a grid-sequencing-technique where the numerical solution is initially obtained on a coarse grid (interval 1) and interpolated to the fine grid. Only a short time interval (interval 2) is required until the solution on the fine grid establishes. The analyzed part of the simulation (interval 3) consists of $2.5 \cdot 10^6$ time steps with a time step-size of $\Delta t \approx 2.8 \cdot 10^{-8}$ s. This corresponds to a simulated physical time interval of $7 \cdot 10^{-2}$ s. The computation

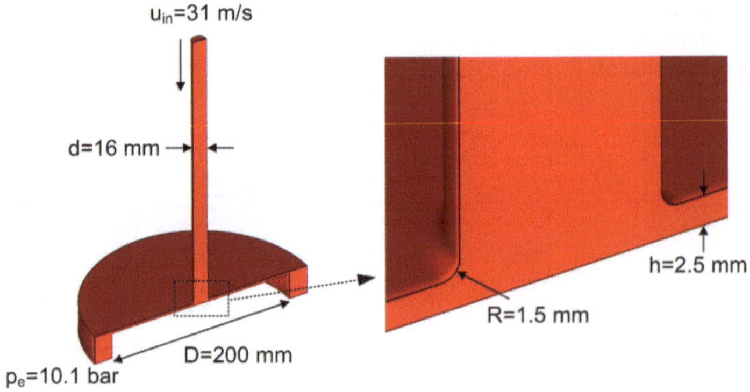

Fig. 3 Cut through the numerical domain and imposed boundary conditions. The grid consists of $5 \cdot 10^5$ cells in 56 blocks and the numerical time step size is $\Delta t \approx 2.8 \cdot 10^{-8}$ s. The simulation is performed for the whole 360° domain

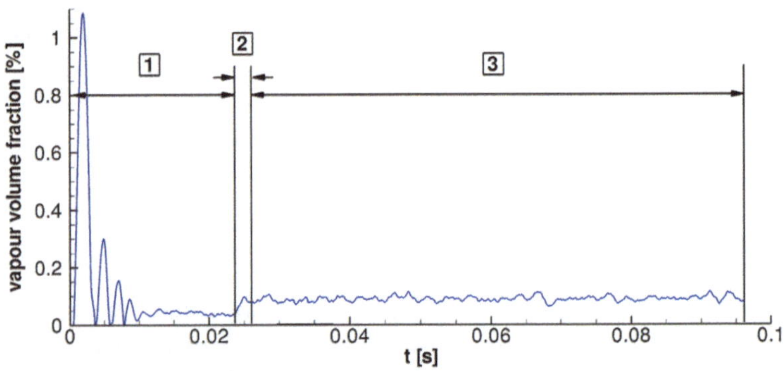

Fig. 4 Time history of the vapor volume fraction referred to the volume of the whole computational domain. *Interval 1*: Preliminary calculation on a coarse grid. *Interval 2*: Interpolation to the fine grid. *Interval 3*: Analyzed part of the simulation on the fine grid

required 3920 CPU-hours (70 hours on 56 cores, Intel Nehalem-EP based 8-way nodes).

Figure 5 shows instantaneous vapor structures at 4 equidistant instants in time ($\Delta t = 8.4 \cdot 10^{-5}$ s). Blue iso-surfaces correspond to a vapor volume fraction of $\alpha = 0.1$. We observe the onset of sheet-cavitation at the rounding of the nozzle outlet. The fragmentation of the sheet and its transient shedding are strongly non-uniform in circumferential direction. The collapse of the vapor structures is enforced by the positive pressure gradient in radial direction.

Figure 6 depicts the spectrum of the integrated vapor volume. The dominant frequency is 408 Hz (arrow), which corresponds to the radial shedding. Lower frequencies might be circumferential modes. The highest well-defined frequencies at 1139 Hz and 1182 Hz are supposed to be related to the collapse of small clouds.

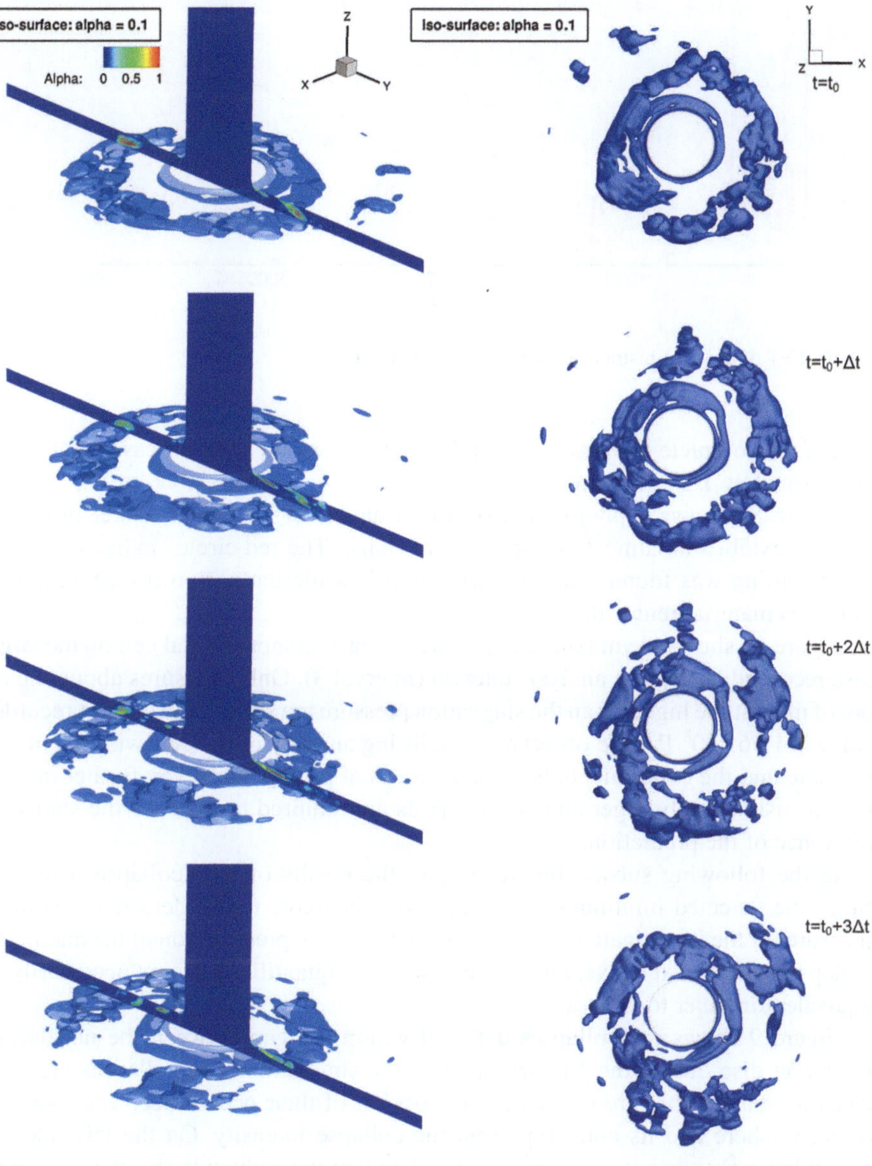

Fig. 5 Iso-surfaces of the vapor volume fraction ($\alpha = 0.1$) at 4 equidistant instants in time ($\Delta t = 8.4 \cdot 10^{-5}$ s). *Left side*: Perspective view with an additional axial cut plane showing the contour of the vapor volume fraction. *Right side*: Top view

Figure 7 shows a perspective view of two consecutive instants in time ($\Delta t = 2.8 \cdot 10^{-5}$ s). On the left side, a couple of vapor structures marked in red are observed at a radial position of $\approx 2.6 \cdot 10^{-2}$ m. On the right side, most of these vapor structures have already collapsed and a shockwave has formed. Hence, the dura-

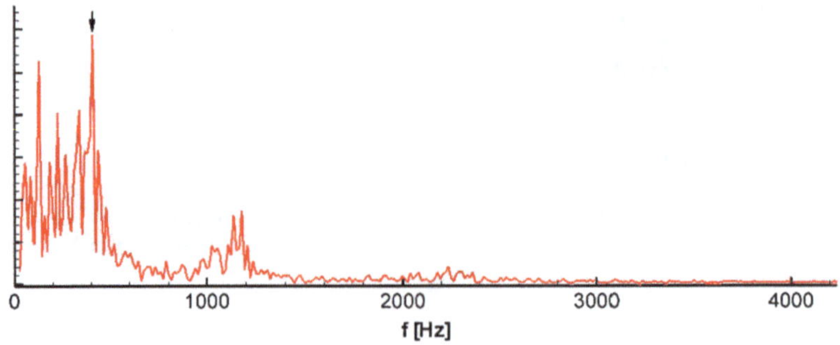

Fig. 6 FFT of the time history of the vapor volume fraction

tion of the complete collapse is about $2.8 \cdot 10^{-5}$ s, which is in our case resolved by approximately 1000 time steps.

Figure 8a shows a photograph (from an angle) of an eroded target disc. The damage exhibits an almost perfect circular shape. The red circles indicate the area where pitting was found in the experiment [8], while the narrower dark gray ring indicates material removal.

Figure 8b shows the maximum pressure for each computational cell on the target disc recorded during the analysis interval (interval 3). Only pressures about one order of magnitude higher than the stagnation pressure are shown, the highest recorded value is $1.56 \cdot 10^8$ Pa. We observe a convincing analogy of the cell-wise maximum pressure and the experimentally detected areas of pitting. However, further investigations using much larger analysis intervals are required to improve the statistical relevance of the prediction.

In the following subsection we present the results of the "collapse detector". Since the detected minimum divergence is comparable to the detected maximum pressure for the investigated test cases, we focus on the presentation of the maximum collapse pressure exclusively. However, these two quantities may not necessarily be equivalent in other test cases.

Figure 9 shows the collapses detected within the gap between the nozzle and the target disc during the analyzed part of the simulation. The collapses are indicated as spheres that are plotted at the position of their occurrence. The diameter of each sphere and its color represent the collapse intensity. On the left side (a), the collapse pressure $p_{collapse}$ is visualized and on the right side (b), the maximum pressure referred to the target wall p_{wall} (Eq. (1)) is depicted. In both cases most of the collapses are found within the marked area.

Compared to Fig. 9a, in Fig. 9b the agreement with the experiment even improves when the collapse pressure is projected onto the target.

Figure 10 shows the radial positions of all collapses plotted against the time of their occurrence. According to Fig. 9b, size and color of each sphere represent the collapse pressure referred to the target wall (p_{wall}). One observes that all violent collapses (red and yellow spheres) are located within the marked area. The lower

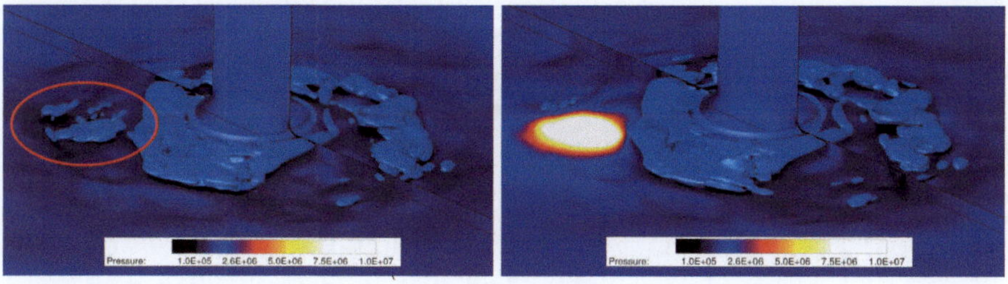

Fig. 7 Perspective view of two time instants ($\Delta t = 28$ μs) showing the collapse of a vapor structure and the resulting shock. Vapor structures are indicated by iso-surfaces ($\alpha = 0.1$). The color on the target wall and on the axial cut plane shows the static pressure

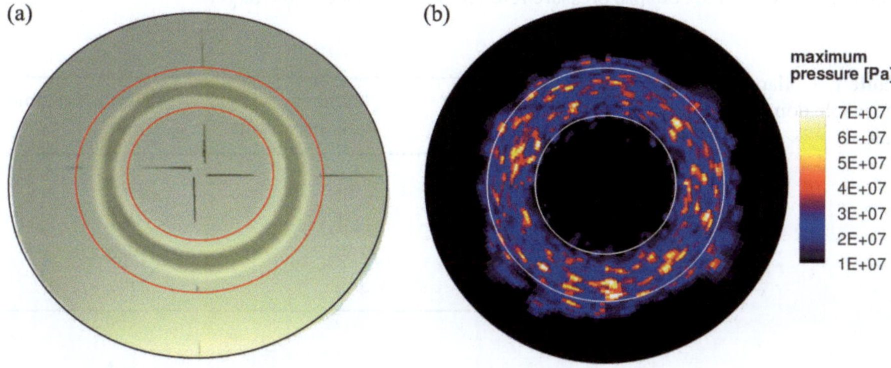

Fig. 8 (**a**) Photograph of an eroded target from the experiment—slightly perspective view [8]. According to experimental observation, erosion damage occurs within $r = 19$ mm and $r = 32$ mm (*red circles*) [8]. Reprint with permission of ASME. (**b**) Maximum pressure observed in the simulation on the target during the analysis interval [9]

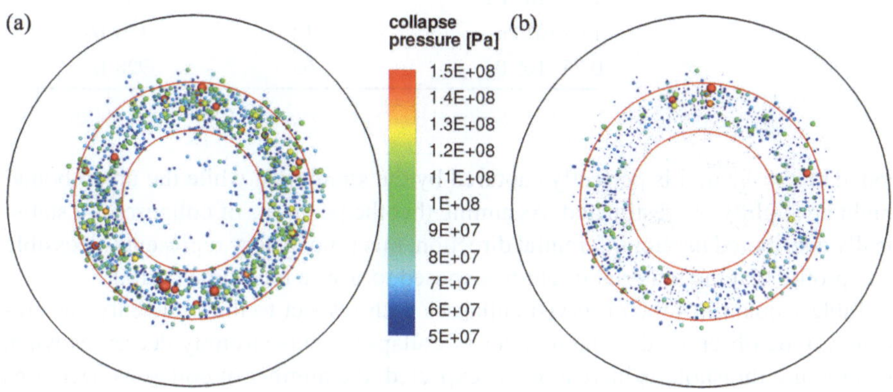

Fig. 9 Visualization of all collapses detected during the analysis interval ($\Delta t = 7 \cdot 10^{-2}$ s). Each collapse is represented by a sphere at the position of its occurrence. *Size* and *color* represent the collapse intensity. On the *left side*, the maximum pressure at the collapse center is shown—on the *right side*, the collapse pressure referred to the surface of the target is depicted [9]

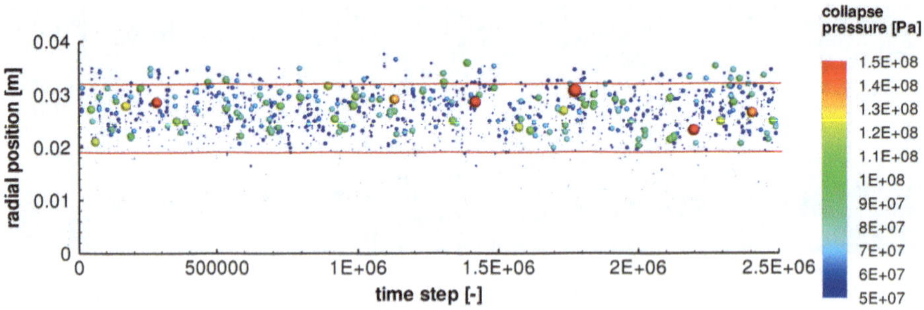

Fig. 10 Time history of collapses detected during the analysis interval. The diagram shows the radial position of each detected collapse against the time of its occurrence. *Size* and *color* of the collapses represent the collapse pressure referred to the surface of the target

Table 1 Collapse events in the whole domain per time unit

Collapse pressure	Number of collapse events	Events per time unit
$1.5 \cdot 10^8$ Pa	11	157 1/s
$1.25 \cdot 10^8$ Pa	31	442 1/s
$1.0 \cdot 10^8$ Pa	139	1986 1/s
$0.75 \cdot 10^8$ Pa	498	7114 1/s

Table 2 Collapse events in the whole domain per time unit. The collapse pressure is referred to the surface of the target

Collapse pressure referred to the surface of the target	Number of collapse events	Events per time unit
$1.5 \cdot 10^8$ Pa	2	29 1/s
$1.25 \cdot 10^8$ Pa	4	58 1/s
$1.0 \cdot 10^8$ Pa	12	170 1/s
$0.75 \cdot 10^8$ Pa	65	928 1/s

bound at $r = 19$ mm is perfectly captured by the simulation while the upper bound might be slightly overestimated. Assuming that the positions of collapses are statistically distributed in circumferential direction, this time history represents a possible load profile that the material might be exposed to over a long time period.

Table 1 quantifies the observed collapses with respect to their intensity and frequency. One observes that the number of collapse events strongly decreases when the pressure threshold is increased. As expected, the number of collapses decreases significantly as well when the collapse pressure is referred to the surface of the target (Table 2). This data may offer a possibility for calibrating thresholds for the collapse pressure by making a comparison of the numerically predicted loads to experimental pitting tests and pit counts.

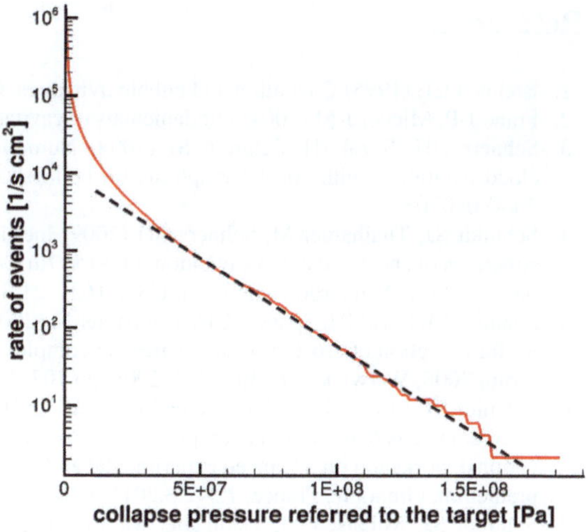

Fig. 11 Cumulative spectrum of collapse events. The diagram shows the collapse pressure referred to the surface of the target against the rate of collapse-events per time unit and area. *Red solid line*: simulation. *Black dashed line*: exponential decay

Figure 11 shows the cumulative spectrum of collapse events as a diagram of the collapse pressure referred to the surface of the target against the rate of collapse events per time unit and area.

One observes an exponential decrease of the rate of events with increasing collapse pressure. This behavior has been reported from experiments of the same setup [10, 11] when plotting the rate of events against a measure for the intensity of the causal collapse event (e.g. diameter of a pit [10], or maximum force of a peak on a pressure sensor [11]).

5 Conclusions

We present a novel approach referred to as "collapse detector" for the numerical prediction of cavitation erosion. Based on a set of physical criteria we detect and characterize isolated collapses of vapor structures within unsteady cavitating flows. The collapse detector is implemented into our flow solver CATUM, which enables the simulation of cloud collapses and the subsequent formation and propagation of shocks due to the compressible treatment of the working fluid. By comparing our results with experimental data we demonstrate the ability of the proposed methodology to predict the position of cavitation erosion. Additionally we provide information about the relative strength and the time history of collapse events. Currently the significance of solved gas and its effect on cavitation erosion are being investigated. Further research will focus on the comparison of our numerical predictions to experimental pitting tests, as well as the application of our collapse data to a material law as proposed by Karimi [12].

Acknowledgements We would like to thank Jean-Pierre Franc for providing detailed information on his experimental setup as well as for fruitful discussions and suggestions. All computational resources have been provided by the Leibniz Supercomputing Center.

References

1. Brennen CE (1995) Cavitation and bubble dynamics. Oxford University Press, New York
2. Franc J-P, Michel J-M (2004) Fundamentals of cavitation. Kluwer Academic, Dordrecht
3. Schnerr GH, Sezal IH, Schmidt SJ (2008) Numerical investigation of three-dimensional cloud cavitation with special emphasis on collapse induced shock dynamics. Phys Fluids 20(4):040703
4. Schmidt SJ, Thalhamer M, Schnerr GH (2009) Inertia controlled instability and small scale structures of sheet and cloud cavitation. In: Proc 7th CAV 2009—7th international symposium on cavitation, Ann Arbor, Michigan, USA, 16.8.–21.8.2009. Paper 17, CD-ROM publication
5. Schmidt SJ, Sezal IH, Schnerr GH, Thalhamer M (2008) Numerical analysis of shock dynamics for detection of erosion sensitive areas in complex 3-D flows. In: Proc WIMRC cavitation forum 2008, Warwick, UK, July 7–9, 2008, pp 107–120
6. Schmidt SJ, Mihatsch M, Thalhamer M, Egerer C, Hickel S, Adams NA (2011) Assessment of erosion sensitive areas via compressible simulation of unsteady cavitating flows. In: International workshop on advanced experimental and numerical techniques for cavitation erosion prediction, Grenoble, France, 1.3–2.3.2011. Post-workshop CD
7. Schmidt SJ, Mihatsch M, Thalhamer M, Adams NA (2011) Assessment of the prediction capability of a thermodynamic cavitation model for the collapse characteristics of a vapor-bubble cloud. In: Proc WIMRC cavitation forum 2011, Warwick, UK
8. Franc JP (2009) Incubation time and cavitation erosion rate of work-hardening materials. J Fluids Eng 131
9. Schmidt SJ, Mihatsch M, Thalhamer M, Adams NA (2013) Assessment of erosion sensitive areas via compressible simulation of unsteady cavitating flows. In: Chahine GL, Franc JP, Karimi A, Kim KH (eds) Advanced experimental and numerical techniques for cavitation erosion prediction. Springer, Berlin (in press)
10. Franc JP, Riondet M, Karimi A, Chahine GL (2012) Material and velocity effects on cavitation erosion pitting. In: Wear, pp 248–259
11. Franc JP, Riondet M, Karimi A, Chahine GL (2011) Impact load measurements in an erosive cavitating flow. J Fluids Eng 133
12. Karimi A, Leo WR (1987) Phenomenological model for cavitation erosion rate computation. Mater Sci Eng 95:1–14

Numerical and Experimental Investigation into Propulsion and Cavitation Performance of Marine Propeller

Nobuhiro Hasuike, Shosaburo Yamasaki, and Jun Ando

Abstract This paper discusses the application of the CFD to transitional and cavitating flow around marine propellers. Especially the emphasis is put on the adaption of the 3 equations turbulence model for the non-cavitating flow and tip vortex resolution on cavitating flow. This research finds 3 equations turbulence model is effective for prediction of the propeller open characteristics in the wide range of Reynolds number. Next, numerical simulation of cavitating flow with tip vortex resolution is validated. Furthermore, simple cavitation erosion index is applied and shows good agreement with experimental results. Finally, effect of tip vortex resolution on the pressure fluctuation at the top of the propeller is discussed.

Keywords CFD · Transitional flow · Cavitation · Pressure fluctuation

1 Introduction

In the design of marine propeller, it is necessary for designers to achieve higher propeller open efficiency and good cavitation performance. Therefore, the higher accuracy prediction tool is expected for estimation of propeller open characteristics (POC in short), erosion risk and pressure fluctuation caused by cavitation at initial design stage.

For realizing high performance in full scale, it is necessary for propeller designers to estimate the scale effects of POC in higher accuracy. Further, in wake flow, pressure distribution on the blade surface and cavity void fraction effect on erosion risk and cavity volume time variation gratefully effects on pressure fluctuation ampli-

N. Hasuike (✉) · S. Yamasaki
Nakashima Propeller Co., Ltd., 688-1, Joto-Kitagata, Higashi-ku, Okayama 709-0625, Japan
e-mail: nobuhiro@nakashima.co.jp
url: http://www.nakashima.co.jp/

J. Ando
Faculty of Engineering, Kyushu University, 744 Motooka Nishi-ku, Fukuoka 819-0395, Japan
e-mail: ando@nams.kyushu-u.ac.jp
url: http://www.eng.kyushu-u.ac.jp/e/index.html

L. Eça et al. (eds.), *MARINE 2011, IV International Conference on Computational Methods in Marine Engineering*, Computational Methods in Applied Sciences 29,
DOI 10.1007/978-94-007-6143-8_12, © Springer Science+Business Media Dordrecht 2013

tude. Therefore, numerical tool which gives detail flow characteristics of cavitating flow is highly expected.

In this paper, newly developed 3 equations $k - k_L - \omega$ model is applied to the transitional flow around model propellers and compared with other typical turbulence models and experimental results. Secondly, POC in the wide range of Reynolds number between model and full-scale is predicted and the boundary layer characteristics and variation of propeller efficiency is discussed.

Further, detailed cavitation simulation including tip vortex cavitation is conducted. Simple cavitation erosion index is introduced and validated with high speed video observation and paint erosion test result.

Finally, simulated pressure fluctuation and variation of cavity volume is compared with the experimental results and the effectiveness of numerical simulation is investigated.

2 Numerical Model

In this research, transitional and cavitating flow around propeller were simulated using SOFTWARE CRADLE SCRYU/Tetra V9 software, which was based on a finite volume method with an unstructured grid. The $k - \varepsilon$ model, the Shear-Stress Transport $k - \omega$ model [1, 2] were applied to the transitional flow. In addition to these widely used turbulence models, newly developed 3-equations $k - k_L - \omega$ model was also applied. For the non-uniform cavitating flow simulations, the $k - \varepsilon$ model and the full-cavitation model was applied and the simple erosion index was introduced.

2.1 Transitional Flow Simulation

It is important to predict the transition point of a flow around a propeller in operating in low-Reynolds-number. LKE (Laminar Kinetic Energy) model [3] was developed to simulate the transitional flow.

In the LKE model, the disturbance energy in a pre-transitional region of a boundary layer is represented as Laminar Kinetic Energy (k_L), while the turbulence energy is as k. The transport equation of k_L is solved by using two equations of fully turbulent model. SC/Tetra introduces the following $k - k_L - \omega$ model [4] which was developed based on the $k - \omega$ model:

$$\frac{\partial \rho k_T}{\partial t} + \frac{\partial \rho u_i k_T}{\partial x_i} = \rho(P_{k_T} + R_{BP} + R_{NAT} - \omega k_T - D_T) + \frac{\partial}{\partial x_i}\left(\left(\mu + \frac{\rho \alpha_T}{\sigma_k}\right)\frac{\partial k_T}{\partial x_i}\right) \quad (1)$$

$$\frac{\partial \rho k_L}{\partial t} + \frac{\partial \rho u_i k_L}{\partial x_i} = \rho(P_{k_L} - R_{BP} - R_{NAT} - D_L) + \frac{\partial}{\partial x_i}\left(\mu \frac{\partial k_L}{\partial x_i}\right) \quad (2)$$

$$\begin{aligned}\frac{\partial\rho\omega}{\partial t}+\frac{\partial\rho u_i\omega}{\partial x_i}&=\rho\left[C_{\omega1}\frac{\omega}{k_T}P_{k_T}+\left(\frac{C_{\omega R}}{f_w}-1\right)\frac{\omega}{k_T}(R_{BP}+R_{NAT})-C_{\omega2}f_W^{\frac{4}{3}}\omega^2\right.\\&\left.+f_\omega\alpha_T f_W^2\frac{\sqrt{k_T}}{Y^3}\right]+\frac{\partial}{\partial x_i}\left(\left(\mu+\frac{\rho\alpha_T}{\sigma_\omega}\right)\frac{\partial\omega}{\partial x_i}\right)\end{aligned}\tag{3}$$

The parameter P_{kT} and P_{kL} are both production terms of k_T and k_L .

The parameter $\nu_{T,s}$ and $\nu_{T,l}$ are the eddy viscosities of small scale and large scale, respectively. The sum of these values ($\nu_t = \nu_{T,s} + \nu_{T,l}$) is used for the eddy viscosity of the momentum equation. The parameter R_{BP} and R_{NAT} are the contributions of the bypass transition and the natural transition, respectively. The contribution of the bypass transition increases as the turbulent intensity in the external flow increases.

2.2 Cavitating Flow Simulation

The full-cavitation model [5] accounts for all first-order effects, i.e., phase change, bubble dynamics, turbulent pressure fluctuations and non-condensable gases contained in the liquid.

The fluid density is a function of vapor mass fraction f, which is computed by solving a transport equation coupled with the mass and momentum conservation equations. Suffixes v, g, l denote the vapor phase, non-condensable gas (NCG in short), liquid respectively. $\vec{v}$ is velocity vector. γ is diffusion coefficient of the vapor mass fraction. Net phase change rate is expressed as $R_e - R_c$. The source terms R_e and R_c denote vapor evaporation and condensation rates. Where, k and σ are the turbulent kinetic energy and the surface tension. The phase change rate R_e and R_c are derived from the Rayleigh-Plesset equation and limiting bubble size (interface surface area per unit volume of vapor) is considered. C_e and C_c are the empirical components. It is assumed that saturated vapor pressure P_v depends on the turbulence kinetic energy. ((8), (9)) Subscripts *turb* denotes the turbulence.

$$\frac{1}{\rho}=\frac{f_v}{\rho_v}+\frac{f_g}{\rho_g}+\frac{1-f_v-f_g}{\rho_l}\tag{4}$$

$$\frac{\partial}{\partial t}(\rho f)+\nabla\cdot(\rho\vec{v}f)=\nabla\cdot(\gamma\nabla f)+R_e-R_c\tag{5}$$

$$R_e=C_e\frac{\sqrt{k}}{\sigma}\rho_l\rho_v\sqrt{\frac{2(p_v-p)}{3\rho_l}}(1-f_v-f_g)\tag{6}$$

$$R_c=C_c\frac{\sqrt{k}}{\sigma}\rho_l\rho_l\sqrt{\frac{2(p-p_v)}{3\rho_l}}f_v\tag{7}$$

$$P_v=\left(P_{sat}+\frac{P'_{turb}}{2}\right)\tag{8}$$

$$P'_{turb}=0.39\rho k\tag{9}$$

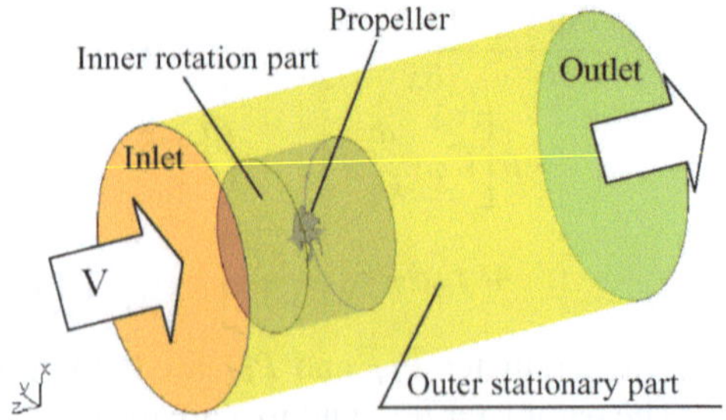

Fig. 1 Computational domain

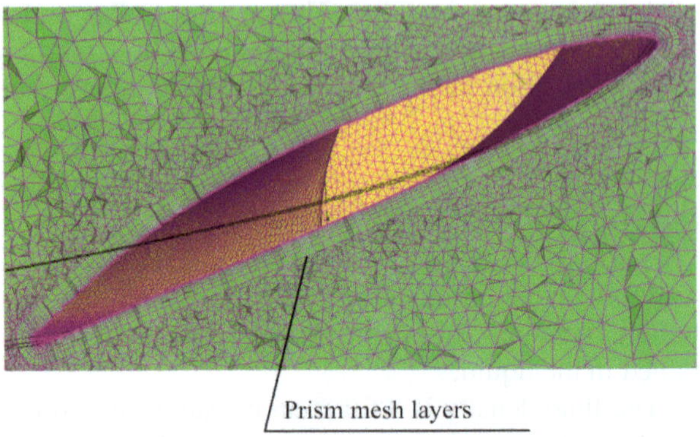

Fig. 2 Prism mesh arrangement near blade surface

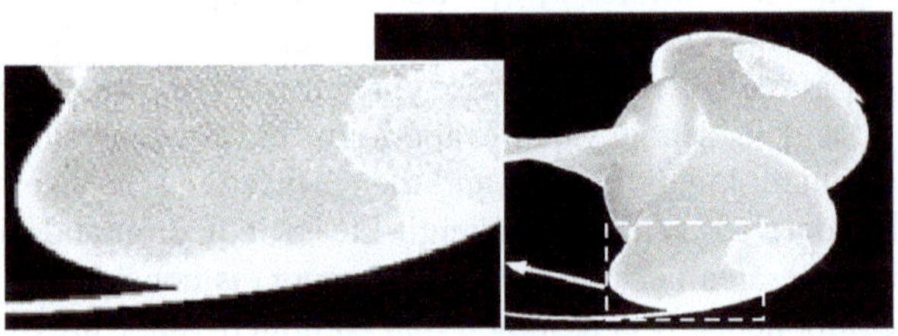

Fig. 3 Mesh near the blade surface

2.3 *Numerical Grids*

The computational domain is composed of the inner rotational part including the propeller and the outer stationary part. The stationary part and the rotational part are connected discontinuously. Constant velocity and zero pressure are prescribed at the inlet and the outlet boundary. Figure 1 shows the computational domain. The numerical mesh is an unstructured grid, and basic cells are tetrahedral and prismatic cells are applied to near the blade surface for resolving the boundary layer (Fig. 2). The first layer thickness of the prism layer was set to a non-dimensional wall distance for a wall-bounded flow (y^+ in short) $= 50$ in the case of the $k - \varepsilon$ model. $y^+ = 50$ was set in the case of the SST $k - \omega$ model and the $k - k_L - \omega$ model. In this research, fine meshes are applied near the tip region for the resolution of tip vortex (Fig. 3).

Table 1 Principal particulars of DTMB P4119 propeller [6, 7]

	DTMB P4119
Number of blades	3
Diameter	0.3048 m
Pitch ratio (0.7R)	1.084
Skew angle	0°

Fig. 4 DTMB P4119 propeller [6, 7]

3 Numerical Simulation of Propeller Open Characteristics

3.1 Propeller Open Characteristics

POC at model scale gratefully depends on mixture ratio of laminar and turbulent flow. DTMB P4119 propeller [6, 7] was selected for the benchmark. This propeller's experimental dataset was used as 22nd ITTC benchmark of RANS calculations. Table 1 and Fig. 4 show the principal particulars and propeller shape. The $k-\varepsilon$ model, the SST $k-\omega$ model with low Reynolds correction [2] (SST $k-\omega$ (Low) in short), the SST $k-\omega$ model without low Reynolds correction (SST $k-\omega$ (w/o corr.) in short) and the $k-k_L-\omega$ model were applied. Turbulent intensity T_{uin} at inlet was set to 1 % for the $k-\varepsilon$ model and the SST $k-\omega$ models and set to 5 % for the $k-k_L-\omega$ model.

First of all, POC in a different operation point is compared in Fig. 5. In the case of the $k-\varepsilon$ model, thrust coefficient K_T was overall small compared with the experiment, and torque constant K_Q was large excluding advancement ratio $J=1.1$. The SST $k-\omega$ (Low) predicted higher K_T and corresponding K_Q value compared with experiment. Predicted open efficiency η_o was overestimated at design point $J=0.833$. On the other hand, the $k-k_L-\omega$ model gave corresponding K_T and little smaller K_Q value. Propeller open efficiency η_o was more corresponding well than the other turbulence models at design point $J=0.833$.

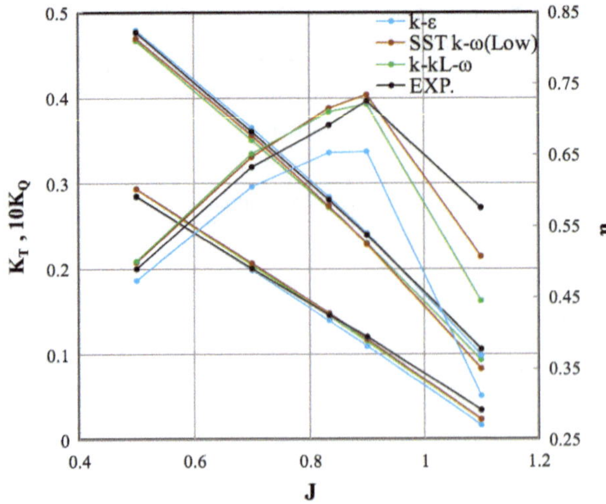

Fig. 5 Propeller open characteristics

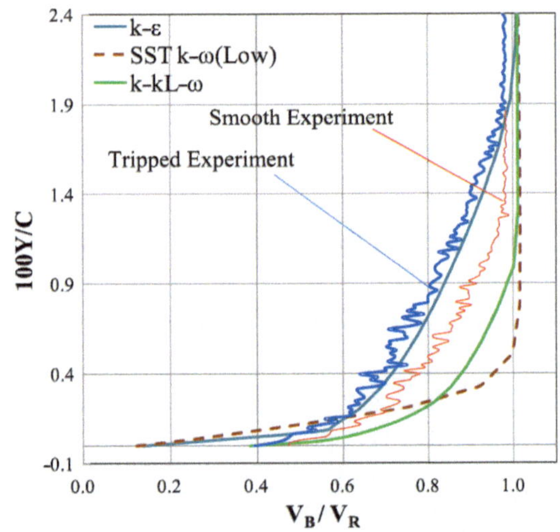

Fig. 6 Boundary layer profiles at $r/R = 0.7$, back side, $x/C = 0.9$

Next, the velocity distributions in boundary layer are compared in Fig. 6. Y/C and V_B/V_R denote non-dimensional distance from wall and velocity in the boundary layer divided by outer flow respectively. The $k-\varepsilon$ model shows good agreement with the experimental measurement results shown in figure as "Tripped" which means the state of with roughness on the leading edge of the propeller. The estimated boundary layer thickness by the SST $k-\omega$ (Low) model is thinner than the experimental measurement results shown in figure as "Smooth" which means the state of without roughness on the leading edge of the propeller. The $k-k_L-\omega$ model gave better result than the SST $k-\omega$ (Low) in case of "Smooth".

Turbulence kinetic energy distributions around the blade section at 70 % radius are compared in Figs. 7(a)–(d). The $k-\varepsilon$ and the SST $k-\omega$ (w/o corr.) predicted

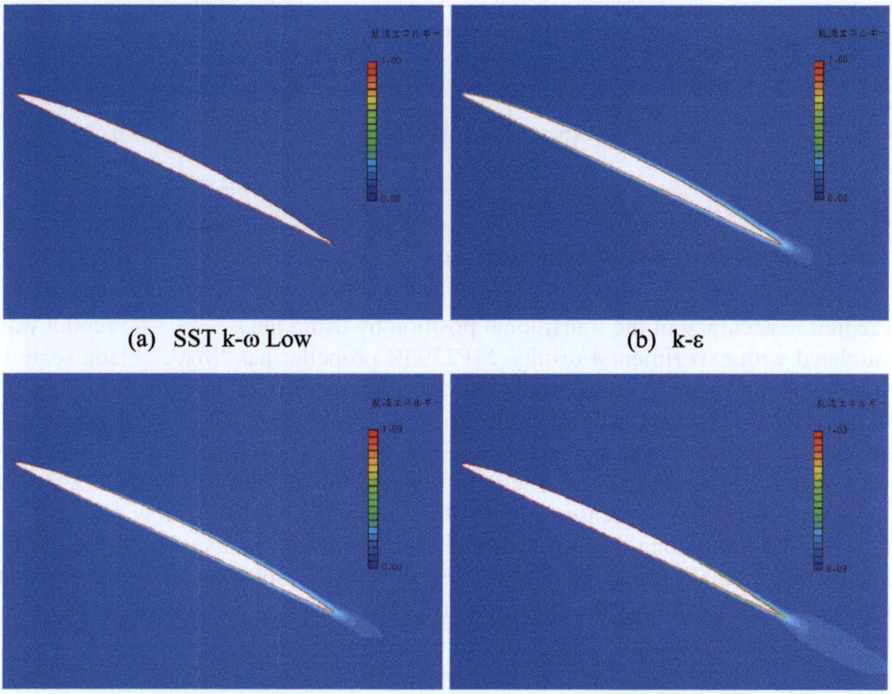

(a) SST k-ω Low (b) k-ε (c) SST k-ω w/o.corr (d) k-k_L-ω

Fig. 7 Turbulent kinetic energy at $r/R = 0.7$

the fully turbulent flow. Underestimation of η_o was mainly due to overestimation of viscous component of K_Q. On the other hand, the flow field simulated by the SST $k-\omega$ (Low) was fully laminar and which causes underestimation of viscous component of K_Q. In the case of the $k-k_L-\omega$ model, flow field was mainly composed by laminar flow and partially composed by turbulent flow. In propeller open water test, leading edge roughness was not applied. Therefore, the $k-k_L-\omega$ model gave more corresponding η_o for the POC prediction at model scale.

Table 2 Principal particulars of model propellers

	MP2293R	MP0193R
Number of blades	5	5
Diameter	250 mm	250 mm
Pitch ratio (0.7R)	0.703	0.7506
Expanded area ratio	0.4	0.4
Skew angle	20°	20°

Table 3 Principal particulars of model propellers [1]

Number of blades	5
Diameter	400(950) mm
Pitch ratio	0.95
Expanded area ratio	0.65
Skew angle	10.5°

3.2 Transitional Flow Simulation

Prediction accuracy of the transitional position by using the $k - k_L - \omega$ model was validated with experimental results. MP2293R propeller has 'MAU' blade section which was traditionally used in past decades. MP0193R has laminar flow type of blade section. Principal particulars of propellers are shown in Table 2. Experimental and simulated results are shown in Figs. 8 and 9. Back side transitional line was located at about 60 % chord length from leading edge in case of MP2293 and located at about 70 % in case of MP0193. Further, face side oil flow pattern was largely different among these two propellers. Simulated results of both MP2293R and MP0193 show good agreement with experimental results. It was confirmed that the $k - k_L - \omega$ model predicted the effects of blade section on transitional region.

3.3 Scale Effects of Propeller Open Characteristics

Seiun-Maru conventional model propellers (CP, Table 3), which experimental data includes open water test results in the wide range of Reynolds number were selected for the calculation. Experimental results were referred from Funeno's literature [8]. Calculation results are compared with experimental results in Figs. 10, 11, 12. Subscripts p and v denote pressure component and viscous component respectively. Figure 13 shows typical representative of flow pattern. Turbulent flow is located at outer side of critical radius (C.R.). Figure 14 shows calculation results of flow pattern.

(1) SST $k - \omega$ (w/o corr.), SST $k - \omega$ (Low), $T_{uin} = 5$ %
Simulated flow field was fully turbulent. Pressure component of K_T increased and frictional component of K_T decreased when Reynolds number increased. Pressure component of K_Q increased and frictional component of K_Q decreased when Reynolds number increased. K_Q much depended on frictional component and decreased totally. As a result, η_o simply increased, however η_o was lower than experimental results at lower range of Reynolds number because of the overestimation of the frictional component.

(2) SST $k - \omega$ (Low), $T_{uin} = 1$ %
Simulated flow field was mixture of laminar and turbulent flow and its critical radius was fixed in the range of $R_n(K) = 2.5 \times 10^5 \sim 3.7 \times 10^5$. In this range of Reynolds number, η_o increased by decrease of boundary layer thickness. In the range of $R_n(K) = 3.7 \times 10^5 \sim 1 \times 10^6$, critical radius moved toward

Fig. 8 Surface streamlines of MP2293

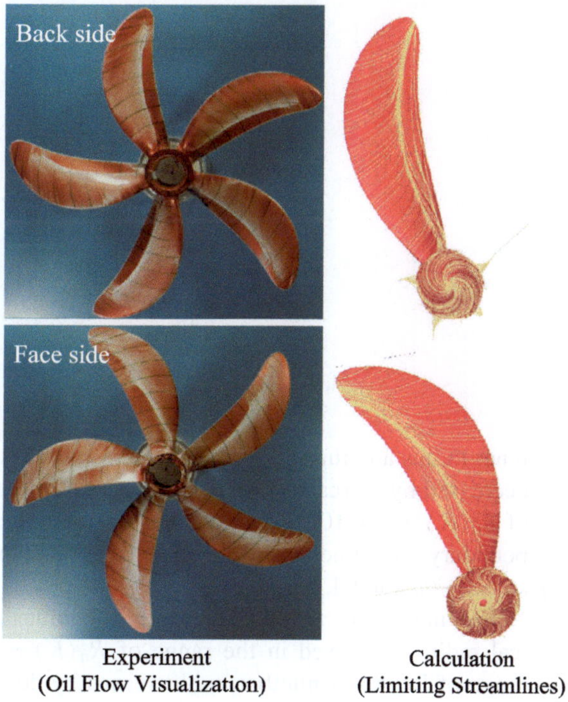

Fig. 9 Surface streamlines of MP0193

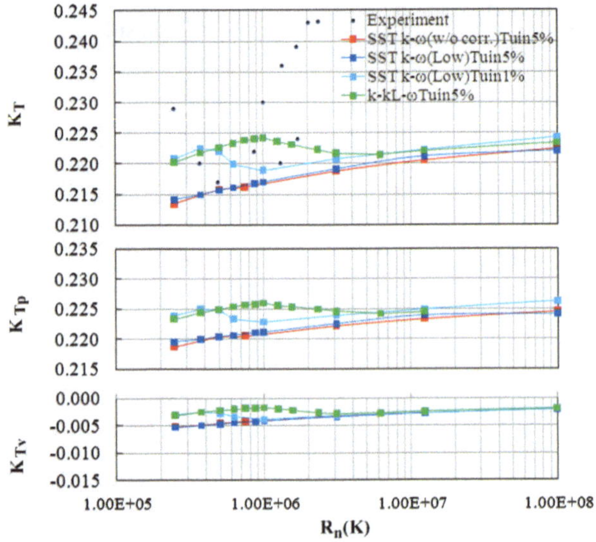

Fig. 10 Effect of Reynolds number on K_T

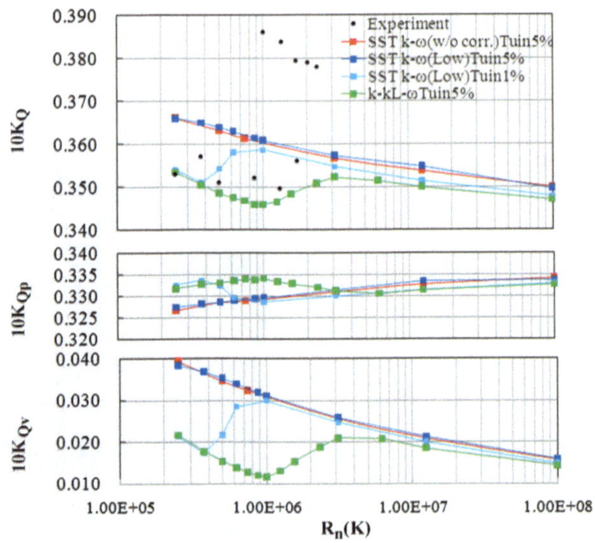

Fig. 11 Effect of Reynolds number on K_Q

inner radius and turbulent area increased when Reynolds number increased. η_o decreased by increasing wall shear stress of turbulent flow part. Over the range of $R_n(K) = 1 \times 10^6$, flow field was fully turbulent. By decreasing of turbulent boundary layer thickness, η_o increased at this range of Reynolds number.

(3) $k - k_L - \omega$ model, $T_{uin} = 5\ \%$

Simulated flow field was mixture of laminar and turbulent flow and its critical radius was fixed in the range of $R_n(K) = 2.5 \times 10^5 \sim 1 \times 10^6$. In this range of Reynolds number, η_o increased by decrease of boundary layer thick-

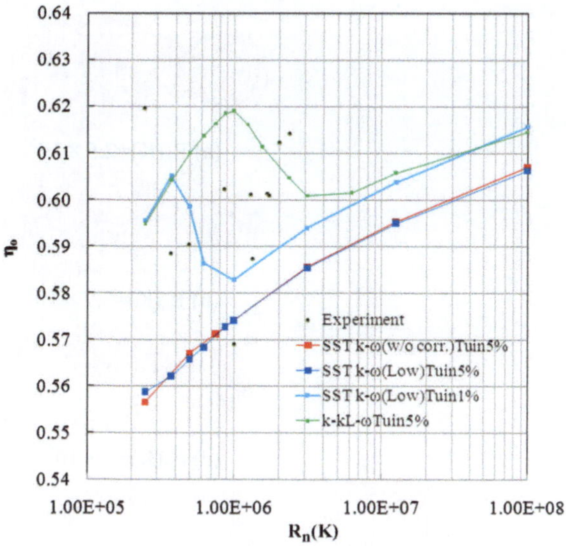

Fig. 12 Effect of Reynolds number on η_o

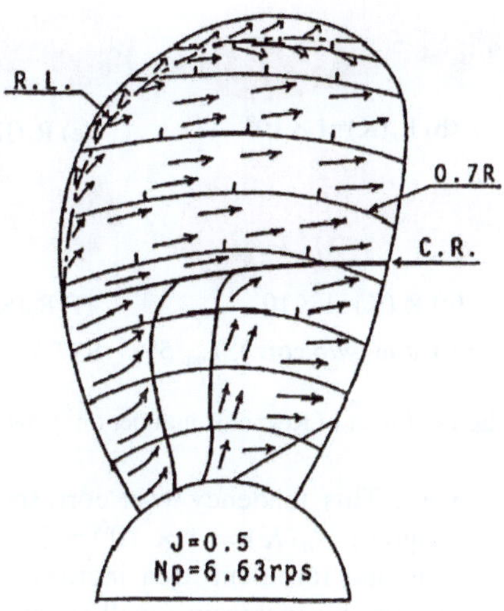

Fig. 13 Oil flow visualization in the Experiment ($R_n(K) = 6.75 \times 10^5$) [9]

Table 4 Principal particulars of model propellers [10]

Number of blades	4
Diameter	241.8 mm
Pitch ratio	0.69
Expanded area ratio	0.49
Skew angle	15°

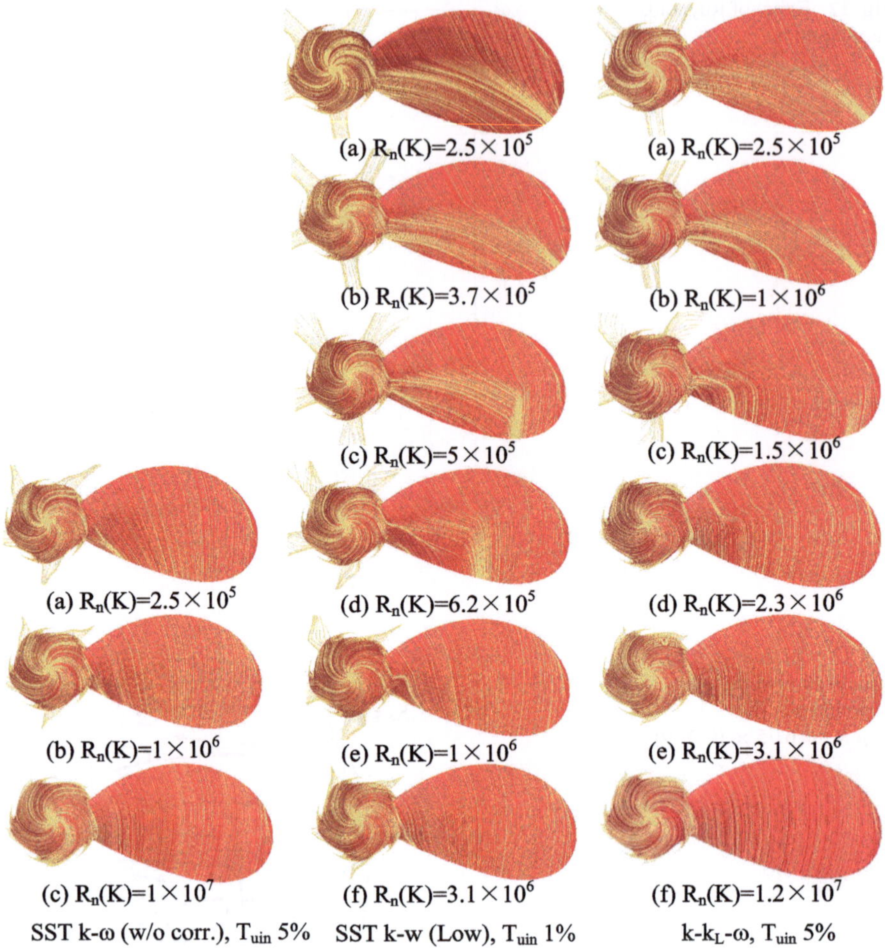

Fig. 14 Effect of Reynolds number on surface streamline

ness. This tendency was corresponding to propeller open water test. In the range of $R_n(K) = 1 \times 10^6 \sim 3 \times 10^6$, critical radius moved toward inner radius and turbulent area increased when Reynolds number increased. η_o decreased by increasing wall shear stress of turbulent flow part. Over the range of $R_n(K) = 3 \times 10^6$, flow field was fully turbulent.

4 Numerical Simulation of Cavitating Flow

4.1 Experimental Data of Partially Eroded Propeller

Cavitation performance test of a propeller for large bulk carrier was conducted. Principal particulars of model propeller and wake pattern are shown in Table 4 and

Fig. 15 Wake patten [10]

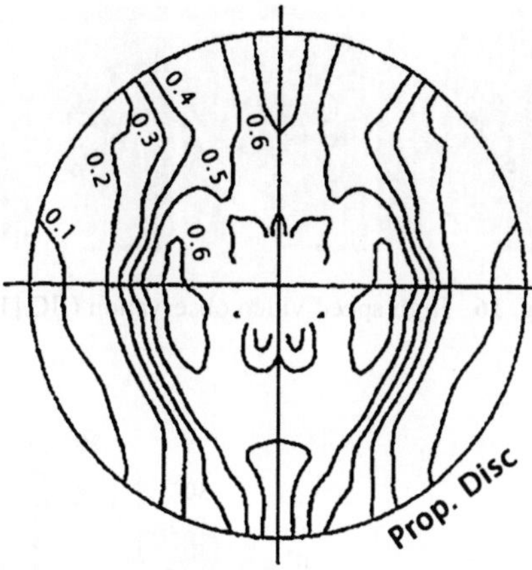

Fig. 15. The wake distribution was simulated by wire mesh. High speed video observation was conducted at IHI Corporation R&D Center (IIC in short). K_T and σ_n were set to 0.16 and 1.68 respectively. Observation result is shown in Fig. 16. Cavitation remained around trailing edge like a patch and collapse drastically at the 67° in the rotating position. Paint erosion test was conducted at the University of Tokyo (UT in short). K_T and σ_n were set to 0.16 and 1.53 respectively. Lower σ_n was applied for promoting cavitation erosion. Test result is shown in inset of Fig. 17. Paint removal was confirmed at trailing edge near the tip.

4.2 Numerical Evaluation of Cavitation Erosion Risk

Cavitation erosion is a fatigue phenomenon caused by the impact load generated when the cavitation bubble collapses acts repeatedly on the solid surface in neighborhood.

Nowadays, the generation of impact load is predictable in the state of the single bubble. However, flow analysis for actual turbo machinery needs to treat many bubbles collapse and needs to treat comparatively long time phenomenon. However, such detail simulation is not practical for actual design procedure in current computer ability.

Therefore, it is thought to be practicable that the physical index obtained from the global flow analysis is related to the strength of cavitation erosion. Nohmi et al. [11] conducted the analysis of cavitating hydrofoil by the barotropy model in combination with the bubble dynamics, and suggested the simple four indexes [11] related the strength of cavitation erosion from their finding.

In this research, one index of them was applied to the cavitating flow around propeller and the evaluation of the erosion risk was conducted. The index is shown in expression (10). Expression is the product of the numerical value of the cavity void fraction α at the surface position and the pressure recovery.

$$\text{Index} = \alpha \cdot \max\big[(p - p_v), 0\big] \tag{10}$$

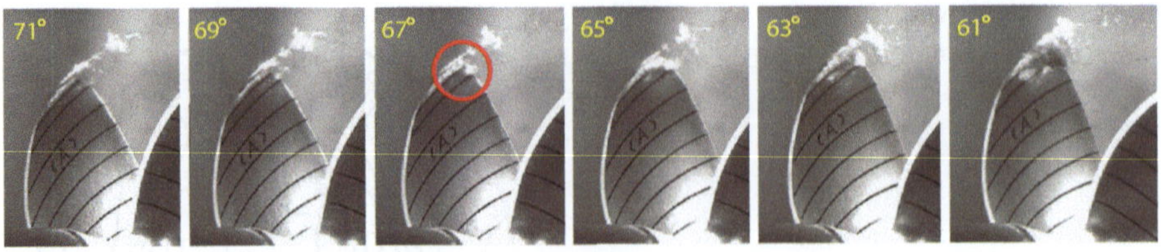

Fig. 16 High speed video observation (IIC [10])

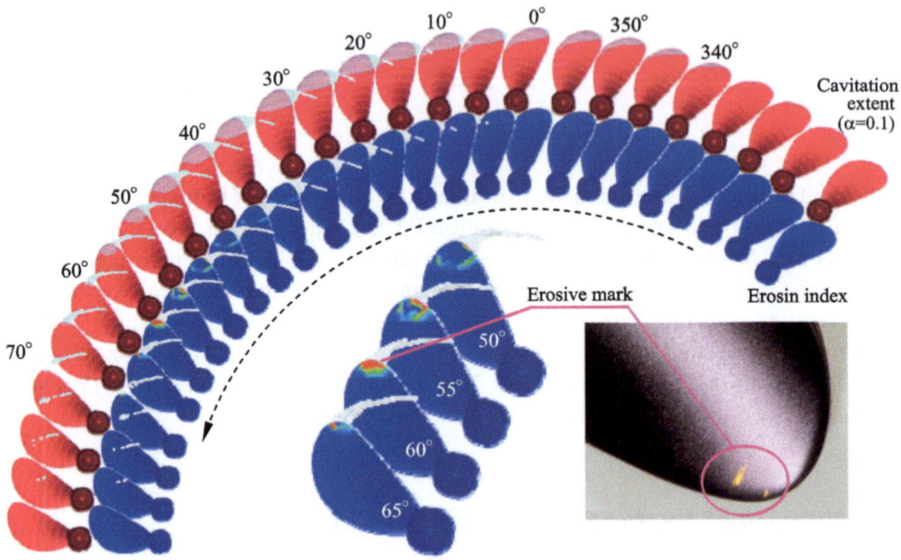

Fig. 17 Cavitation extent and contour of erosion index. *Inset*: Paint erosion test result (UT [10])

The adaptive mesh refinement methodology was used for the resolution of tip vortex. To calculate efficiently, the flow field around only one blade out of four blades was divided by fine meshes. This blade is called "Key blade" here and focused.

The propeller is rotated 1800° by 1° per cycle to develop flow field, after that 0.3° per cycle was adopted during the final time step. The second order accuracy of the convective term was adapted. The full-cavitation model was used in the calculation. NCG was set to 1 ppm and the empirical constants C_e and C_c were set to 0.02 and 0.01 respectively. K_T and σ_n were set to 0.16 and 1.68 respectively.

Simulated cavity extent (cavity void fraction $\alpha = 0.1$) and contour of erosion index are shown in Fig. 17. Red color in contour map of erosion index means higher erosion risk.

In this case, cavitation remained like a patch at the 50–70°. This type of cavitation pattern is thought to be erosive. Simulated results show higher value of erosion index at the 55–65° and correspond well with experimental results.

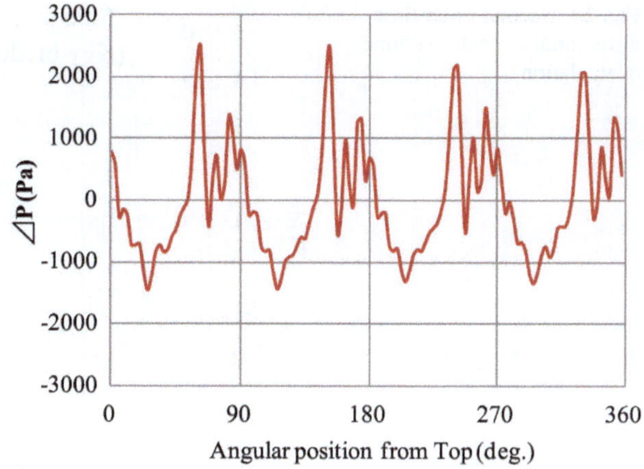

Fig. 18 Pressure pulse (Experiment) (IIC [10])

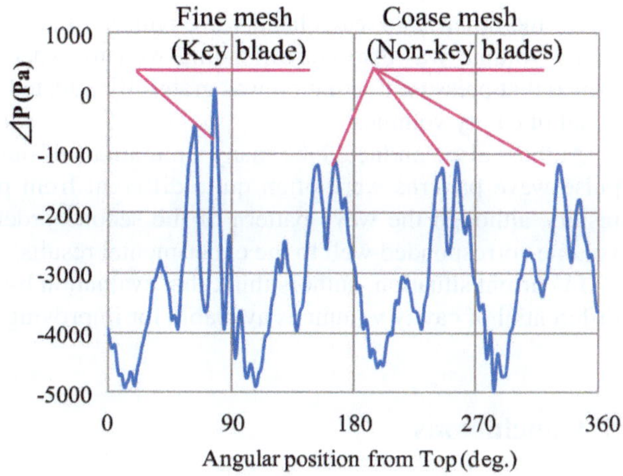

Fig. 19 Pressure pulse (Calculation)

4.3 Pressure Pulse Evaluation

Pressure pulse at the propeller top position was measured at IIC as shown in Fig. 18. Pressure pulse amplitude is thought to be gratefully related to the second order time differential of cavity volume. In this research, pressure pulse and cavity volume variation were compared with experimental results in Figs. 19 and 20. Experimental pressure pulse was measured with transducer in flat plate. However, simulation were conducted without plate.

For considering this difference, illustrated pressure pulse in Fig. 19 is product of simulated pressure and solid boundary factor 2.

By using fine mesh for tip vortex resolution, higher order blade frequency component increased. Simulated pressure pulse wave agrees comparatively well with experimental results in this case.

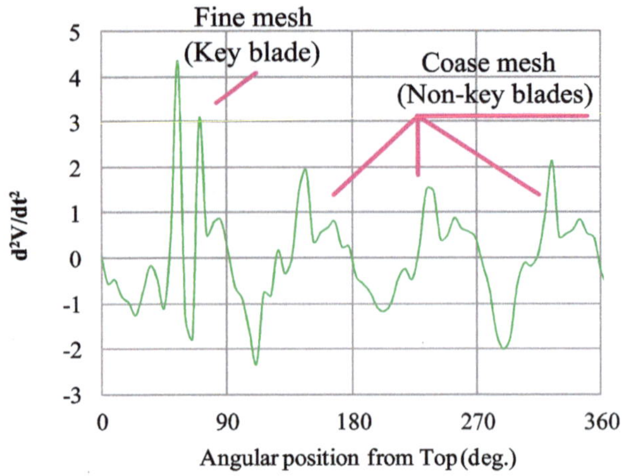

Fig. 20 Second order time differential of cavity volume (Calculation)

On the other hand, calculated wave pattern of second order time differential of cavity volume was also corresponding well to experimental pattern. It was confirmed that pressure fluctuation was gratefully affected by second order time differential of cavity volume.

Authors also conducted the many simulations of other cases, simulated pressure pulse wave patterns were often quite different from pressure pulse measurement results, although the wave pattern of the second order time differential of cavity volume corresponded well to the experimental results.

In current situation, authors think that evaluation by using the second order time differential of cavity volume is available for improving cavitation performance.

5 Conclusions

- The $k-\varepsilon$ model predicted the boundary layer thickness in good agreement with experimental results in the fully turbulent flow case.
- The SST $k-w$ model underestimated the boundary layer thickness in model scale and predicted Reynolds number of transition to fully turbulent flow was lower than open water test tendency.
- The $k-k_L-\omega$ model predicted boundary layer thickness and POC in comparatively good accuracy in the range of transitional Reynolds number.
- The cavitation simulation of the eroded propeller was conducted and erosion index was introduced. RANS predicted the erosive cavitation pattern. Predicted erosive position showed good agreement with experimental results.
- Cavitation simulation with tip vortex resolution is necessary for analyzing the higher order component of pressure pulse.
- Prediction accuracy of pressure pulse was not enough. Much improvement of discretizing accuracy is necessary. However, the evaluation by second order time differential of cavity volume gives valuable information for improving the pressure level of designing propeller.

References

1. Menter F (1993) Zonal two equation $k - \omega$ turbulence models for aerodynamic flows. AIAA 93-2906
2. Wilcox D (1998) Turbulence modeling for CFD, 2nd edn. CDW Industries, Essex
3. Walters DK, Leylek JH (2004) A new model for boundary layer transition using a single-point RANS approach. J Turbomach 126:193–202
4. Walters DK, Cokljat D (2008) A three-equation eddy-viscosity model for Reynolds-averaged Navier-Stokes simulations of transitional flow. J Fluids Eng 130:121401
5. Singhal AK, Athavale MM, Li H, Jiang Y (2002) Mathematical basis and validation of the full cavitation model. J Fluids Eng 124:617–624
6. Gindroz B, Hoshino T, Pylkkanen V (1998) Propeller RANS/panel method workshop. In: 22nd ITTC propulsion committee propeller RANS/panel method workshop, Grenoble, April
7. Jessup SD (1998) Experimental data for RANS calculations and comparisons (DTMB P4119). In: 22nd ITTC propulsion committee propeller RANS/panel method workshop, Grenoble, April
8. Funeno I (2002) On viscous flow around marine propellers—Hub vortex and scale effect. J Kansai SNA 238:17–27
9. Takei Y, Kakugawa A, Ukon Y (1987) Flow visualization on the blades of marine propellers. Flow Vis 7(26):7–10
10. Yamasaki S, Hasuike N, Himei K, Okazaki A, Mishima T, Fujii K, Yamaguchi H, Kawanami Y, Ukon Y (2011) Cavitation erosion of a propeller for large bulk carrier. In: Proc of conference. Proceedings the Japan society of naval architects and ocean engineers, vol 12, pp 415–418
11. Nohmi M, Iga Y, Ikohagi T (2008) Numerical prediction method of cavitation erosion. In: Conference proceeding of turbomachinery society of Japan, vol 59, pp 49–54

References

[illegible]

[illegible] A [illegible] model [illegible] propulsion [illegible]
[illegible] Tech [illegible]

[illegible] A [illegible] simulation [illegible]

[illegible]

Advanced Lagrangian Approaches to Cavitation Modelling in Marine Applications

Sergey Yakubov, Bahaddin Cankurt, Thierry Maquil, Patrick Schiller, Moustafa Abdel-Maksoud, and Thomas Rung

Abstract The paper scrutinizes different approaches to cavitation modelling in the framework of volume of fluid based marine engineering Navier-Stokes simulations. Traditional Eulerian cavitation models compute the vapor content based on computationally efficient, semi-empirical mass-transfer models for cavitation. In conjunction with Lagrangian cavitation models, separate equations for the bubble size and momentum are solved for each individual bubble/nuclei of a dispersed vapor phase, and a subsequent mapping procedure provides the vapor-volume fraction of the Eulerian mixture. The paper aims to advocate the benefits of a combined approach, which reduces the computational effort of the Lagrangian approach whilst maintaining its enhanced predictive realm in critical flow regimes. Validation examples refer to 2D hydrofoils and outline the strong parameter dependency for the Eulerian cavitation models as well as its insensitivity to water-quality aspects. On the contrary, Lagrangian cavitation models return an improved accuracy and capture the influence of water quality. Results of an open-water propeller flow investigation confirm these findings and display a fair predictive agreement in conjunction with a combined modelling approach which allows to perform accurate cavitation predictions at reasonable cost.

Keywords Computational methods · Marine engineering · Cavitation · Euler-Euler · Euler-Lagrange

1 Introduction

In many hydrodynamic engineering applications liquid-embedded vapor cavities occur due to the evaporation of liquid in low pressure regions of an accelerated flow. These cavities—which are usually formed by a number of vapor bubbles—may violently collapse owing to re-condensation once they reach a region where the ambient

S. Yakubov (✉) · B. Cankurt · T. Maquil · P. Schiller · M. Abdel-Maksoud · T. Rung
Institute of Fluid Dynamics and Ship Theory (M-8), University of Technology Hamburg-Harburg, Schwarzenbergstraße 95C, 21073 Hamburg, Germany
e-mail: sergey.yakubov@tu-harburg.de
url: http://www.tu-harburg.de/fds

L. Eça et al. (eds.), *MARINE 2011, IV International Conference on Computational Methods in Marine Engineering*, Computational Methods in Applied Sciences 29,
DOI 10.1007/978-94-007-6143-8_13,

pressure recovers. Rational models for this complex phenomenon have been attracting a vast amount of researchers for decades. The modelling fundamentals of bubble dynamics, i.e. bubble growth and collapse, trace back to the work of Rayleigh [17] and later on the work of Plesset [16]. Hence, the equation describing the bubble dynamics is commonly referred to as the Rayleigh-Plesset equation.

When attention is directed to computational modelling of cavitation three main categories, i.e. barotropic cavitation models (based on barotropic equation of state), Euler-Euler models and Euler-Lagrange techniques, are employed in Navier-Stokes based engineering simulations. All approaches employ two-phase flow models for a vapor and a liquid phase. Euler-Euler and barotropic approaches are mostly employed in conjunction with volume of fluid methods, which assume that the two phases share the kinematic field. In conjunction with the barotropic model, all transport effects are neglected and the vapor volume fraction is directly computed from local pressure Using an Euler-Euler model the vapor volume fraction is advanced by a transport equation along with a source term, describing the mass transfer between liquid and vapor. Accordingly, cavitation physics is reduced to modelling the source term which refers to a simplified Rayleigh-Plesset equation. For some flows these approaches are shown to be quite efficient. However, they both ignore local (inhomogeneous) water properties and are restricted to simplified bubble dynamics which often requires a questionable, case-dependent calibration of model parameters.

Within the Euler-Lagrange approach the non-dispersed field properties follow from Eulerian conservation equations whereas the vapor part is governed by Newtonian motion of individual, spherical bubbles driven by the surrounding flow field. Several authors considered different forces acting on a bubble such as drag, lift, buoyancy, volume variation, etc. An overview of the forces acting on spherical bubbles and the influence of them on a bubble's trajectory can be found in [2, 15]. Euler-Lagrange approaches allow to consider bubble deformation, bubble splitting, and bubble-bubble or bubble/wall interactions. Examples refer to Apte et al. [20], who have implemented the collision of bubbles via a standard collision model, Chahine [3], who has modelled non-spherical deforming and splitting bubbles, and Lauterborn et al. [8] focusing upon acoustic cavitation based on a Lagrangian approach. Using different bubble number spectra for the approaching water provides access to water-quality influences which is considered to be a major advantage of Lagrangian cavitation models. The information exchange between liquid and vapor components can be performed in various ways. According to Elghobashi [4] and Sommerfeld [22], dispersed flows can be classified by the interaction between the bubbles and the carrier flow. Using a one-way-coupling, the solution of the Eulerian field is imposed on the Lagrangian part in order to move the bubbles under the influence of pressure and velocity fields. This strategy is suitable to investigate cavitation inception, as demonstrated by Oweis [15] and Chahine [3], who employed a one-way-coupling to capture cavitation inception in vortical and propeller flows. However, for developed cavitating flows the interaction between the liquid and vapor parts should be taken into account in both ways. In this case the bubbles have a significant influence on the flow field but bubble-bubble interactions still remain small. Implementations of two-way couplings have been only recently been reported

[2, 5, 20]. The main drawback of the Euler-Lagrange approach is the large computational effort due to large number of bubbles required for the simulation of complex 3D applications. This motivates the development of non-trivial parallelization techniques and hybrid cavitation modelling strategies.

The present paper aims to convey the merits of Lagrangian cavitation models. Supplementary, means to enhance the computational efficiency of Euler-Lagrange models based upon hybrid MPI/OpenMP parallelization techniques are outlined and a combined Euler-Euler/Euler-Lagrange model is presented. Investigated examples are confined to a 2D hydrofoil and a propeller in open-water for different operating conditions. The paper is organised as follows: Sect. 2 briefly describes the numerical method. Section 3 is devoted to the rationale of the Euler-Euler model and the related results. The 4th section describes the Euler-Lagrange cavitation model, outlines the associated hybrid parallelisation method and displays the predictive performance for the investigated examples. Section 5 is concerned with the performance of the combined Euler-Euler/Euler-Lagrange model. Final conclusions follow in Sect. 6.

2 Numerical Method

Results of the present study were obtained from the Navier-Stokes procedure FreSCo$^+$, a joint development of Hamburg University of Technology (TUHH) and the Hamburgische Schiffbau-Versuchsanstalt (HSVA) [18]. The procedure uses a segregated algorithm based on the strong conservation form of the momentum equations. It employs a cell-centered, co-located storage arrangement for all transport properties. Structured and unstructured grids, based on arbitrary polyhedral cells or hanging nodes, can be used. The implicit numerical approximation is second-order accurate in space and time. Integrals are approximated using the conventional mid-point rule. The solution is iterated to convergence using a pressure-correction scheme. Various turbulence-closure models are available with respect to statistical (RANS) or scale-resolving (LES, DES) approaches. Since the data structure is generally unstructured, suitable pre-conditioned iterative sparse-matrix solvers for symmetric and non-symmetric systems (e.g. GMRES, BiCG, QMR, CGS or BiCGStab) can be employed. FreSCo$^+$ is efficiently parallelized for several hundred (or more) processes. Multi-phase flows are modelled using a VOF-type mixture-fraction approach, which can be coupled to a number of Euler-Euler or Euler-Langrange models for cavitation.

2.1 Governing Equations

The fluid mixture of liquid and vapor is described by the standard isothermal Navier-Stokes equations

$$\frac{\partial \rho}{\partial t} + \nabla \cdot (\rho \mathbf{u}) = 0 \tag{1}$$

$$\frac{\partial \rho \mathbf{u}}{\partial t} + (\mathbf{u} \cdot \nabla)(\rho \mathbf{u}) = -\nabla p + \nabla \cdot \boldsymbol{\tau} + \mathbf{f}, \tag{2}$$

where $\boldsymbol{\tau}$ is the viscous stress tensor which is composed from modelled turbulent and laminar contributions. The two phases of the mixture are considered to be immiscible and changes of the mixture fraction are attributed to cavitation. The mixture density ρ and mixture viscosity μ are computed as a sum of partial densities and viscosities of the liquid (l) and vapor (v), viz.

$$\begin{aligned} \rho &= \alpha \rho_v + (1-\alpha)\rho_l \\ \mu &= \alpha \mu_v + (1-\alpha)\mu_l \end{aligned} \tag{3}$$

Here α is the vapor-volume fraction defined as the ratio between the vapor volume and the total volume of a control volume

$$\alpha = \frac{V_v}{V_v + V_l} \tag{4}$$

A control volume (CV) filled with liquid yields $\alpha = 0.0$, with vapor $\alpha = 1.0$, respectively. Values out of $\alpha \in [0.0, 1.0]$ describe non-realizable situations. The prediction of the vapor-volume fraction follows different routes for the Euler-Euler and Euler-Lagrange model.

Examples included were all based upon fully-turbulent Reynolds-averaged flow simulations using a $k - \omega$ turbulence model and monotonicity preserving QUICK scheme for the approximation of convective terms.

3 Euler-Euler Cavitation Model

Within the Euler-Euler approach, the vapor-volume fraction is computed by an additional transport equation:

$$\frac{\partial \alpha}{\partial t} + \nabla \cdot (\alpha \mathbf{u}) = S_{\text{cav}} \tag{5}$$

Several empirical Eulerian cavitation models, which express the source term S_{cav} differently, are available from the literature. The present work employs the source term reported by Zwart et al. [25] and Sauer [19] . The Zwart-model uses different formulations for vaporisation and condensation, i.e.

$$S_{\text{cav}} = \begin{cases} F_{\text{vap}} \cdot \frac{3\,\alpha_{\text{nuc}}}{R_0} \cdot \sqrt{\frac{2}{3}\frac{|p_v^* - p|}{\rho_l}}(1-\alpha), & p_v^* < p \\ F_{\text{cond}} \cdot \frac{3}{R_0} \cdot \sqrt{\frac{2}{3}\frac{|p_v^* - p|}{\rho_l}}\alpha, & p_v^* > p \end{cases} \tag{6}$$

based on two empirical constants F_{vap} and F_{cond} which allow to distinguish between vaporisation and condensation. Two additional model parameters are the initial nucleation-site volume fraction α_{nuc} and the corresponding initial nucleation-site radius R_0. Mind that all parameters are assigned to constant values and do not

Table 1 Investigated cavitation model coefficients of the Zwart model [22]

	F_{vap}	F_{cond}
Small	5	0.001
Medium	25	0.005
Default [25]	50	0.010

vary in space or time. They are thus not designed to account for an inhomogeneous water quality. The vaporisation and condensation parameters F_{vap} and F_{cond} are deliberately varied in the present study in order to enhance the predictive accuracy. Accordingly, three parameter sets are investigated as displayed in Table 1. Mind that the investigated parameter range is in the range of frequently employed combinations.

The Sauer-model uses the same formulations for vaporisation and condensation

$$S_{cav} = 4\pi R_b^2 n_0 \sqrt{\frac{2}{3}\frac{|p_v^* - p|}{\rho_l}}(1-\alpha)\,\text{sign}(p_v^* - p), \tag{7}$$

where n_0 denotes the bubble density and R_b refers to a representative local bubble radius.

Note that the original vapor pressure used in both models is replaced by a modified pressure p_v^* to take into account turbulence influence on the cavitation inception as suggested by Singhal et al. [21]

$$p_v^* = p_v + 0.195\rho k, \tag{8}$$

with k being the turbulent kinetic energy. The latter proved to be adequate for sheet cavitation prediction in case of using adjusted internal model constants [1].

3.1 Cavitating Hydrofoil Flow

The cavitating flow over a NACA 6-series hydrofoil has been experimentally investigated by researchers of the University of Rostock together with the Hamburgische Schiffbauversuchsanstalt (HSVA) in the K22 cavitation tunnel of HSVA under the aegis of the *KonKav-I* project funded by the German Ministry of Economics and Technology (BMWi). The tunnel features a test section with a 0.57 m wide square cross section and a length of 2.20 m. The investigated hydrofoil is a span two-dimensional cambered NACA66_2-415 foil with a chord length of $c = 0.2025$ m. Experimental data for various air contents, angles of attack, cavitation numbers, inlet velocities are measured as well as bubble distributions upstream the profile. The present simulations are confined to two test cases with the flow conditions outlined in Table 2. The two cases are hydrodynamically not very different and thus suitable for a sensitivity study.

Table 2 2D foil flow conditions

	Case 1	Case 2
V_{inlet}	5.5 [m/s]	5.5 [m/s]
Angle of attack	10.4°	10.0°
Air content	47 %	50 %
Cavitation number	3	2.5

Unsteady RANS simulations were performed with a 2D unstructured grid of 30,000 cells. Fluid properties were set to water properties at 24°. At the inlet a constant velocity of $V_{\text{inlet}} = 5.5$ m/s was applied. The outlet is assigned to a uniform pressure boundary condition, where the outlet pressure matches the experimental cavitation number σ, i.e. $p_{\text{outlet}} = \frac{1}{2}\rho_l V^2 \sigma + p_v$. The tunnel walls are considered as slip walls and a no-slip boundary condition is used along the hydrfoil. The Zwart et al. [25] cavitation model was used for this sensitivity study and model parameters were varied as outlined in Table 1.

The contour plots of the predicted vapor-volume fraction for both cases are depicted by Figs. 1 and 2. Supplementary to the predictions, the experimentally reported cavitation regime is added to the results in Fig. 1 (Case 1; black line) and the measured cavitation thickness in Fig. 2 (Case 2; black line). The figures reveal a strong dependency of the predictive accuracy on the employed model coefficients. For the higher cavitation number of Case 1, the cavitation extent is underpredicted with all three parameter sets, whereas for Case 2, the cavitation length is over-predicted when using the default set and agrees quite well for the medium set.

3.2 Cavitating Flow over the PPTC Propeller

Cavitation tests for a five-blade controllable pitch propeller were conducted in the cavitation tunnel K15A of the SVA Potsdam. The propeller diameter refers to $D = 250$ mm. It features a chord length at $r/R = 0.7$ of $c/D = 0.417$, a pitch ratio of 1.635 at $r/R = 0.7$, a hub-diameter ratio of 0.3, an area ratio of 0.779 and a skew angle of 18.8°. The operating conditions for the two selected cases are given in Table 3.

The cylindrical computational domain extends 2D upstream of the propeller and 4D in downstream direction. The radius of the domain denotes 1.34 D, providing the same cross-sectional area size as in the experimental tunnel. Two nonconformal body-fitted fully hexahedral unstructured grids have been generated using NUMECA HEXPRESS™ software. One has $4 \cdot 10^6$ cells with refinement for one blade in order to assess the blade resolution dependency and another grid has $12 \cdot 10^6$ cells with additionally refined region in tip vortex (Fig. 3). The fine mesh has been generated in line with the experience that Euler-Euler models require a very fine resolution in the tip vortex region [24].

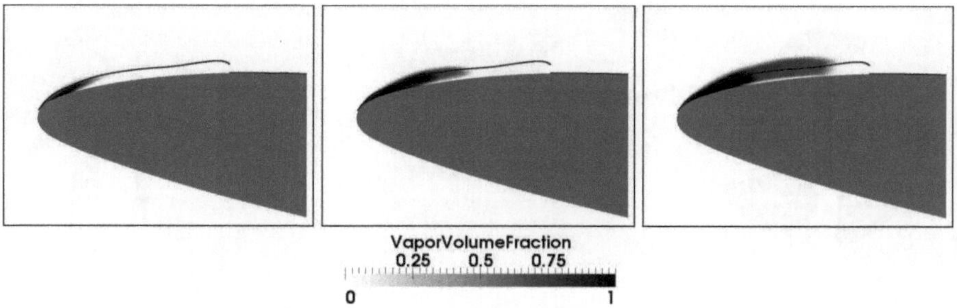

Fig. 1 Cavitating flow over the NACA66$_2$-415 2D hydrofoil at $\sigma = 3$. The figure illustrates the sensitivity to modelling coefficients using $F_{\text{vap}}/F_{\text{cond}} = 5/0.001$ (*left*), 25/0.005 (*middle*), 50/0.01 (*right*)

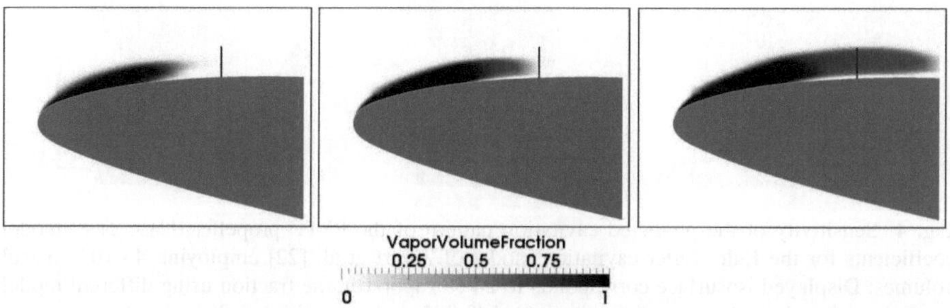

Fig. 2 Cavitating flow over the NACA66$_2$-415 2D hydrofoil at $\sigma = 2.5$. The figure illustrates the sensitivity to modelling coefficients using $F_{\text{vap}}/F_{\text{cond}} = 5/0.001$ (*left*), 25/0.005 (*middle*), 50/0.01 (*right*)

Table 3 PPTC propeller flow conditions

	Case 1	Case 2
V_{inlet}	5.3 [m/s]	8.1 [m/s]
Advance coefficient	1.01	1.269
Number of revolutions	21 [1/s]	25 [1/s]
Cavitation number based on number of revolutions	1.96	1.424
Air content (% of saturation)	80 %	53.5 %

At the outlet boundary a uniform pressure was specified matching the given cavitation number. No-slip walls with wall functions were assigned to the hub and propeller blades. A slip-wall boundary condition was employed along the outer circumference.

At first results of this more practical problem are scrutinised with respect to their sensitivity for the choice of Zwart et al. [25] cavitation model parameters. Therefore, a parameter study has again been performed with $4 \cdot 10^6$ cells grid for the Case 2

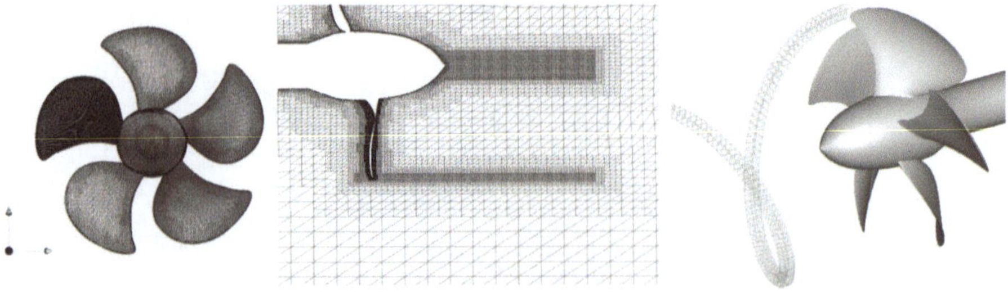

Fig. 3 Illustration of the employed PPTC propeller grids: $4.7 \cdot 10^6$ cells grid (*left*, *centre*) and refinement box for the $12 \cdot 10^6$ cells grid (*right*)

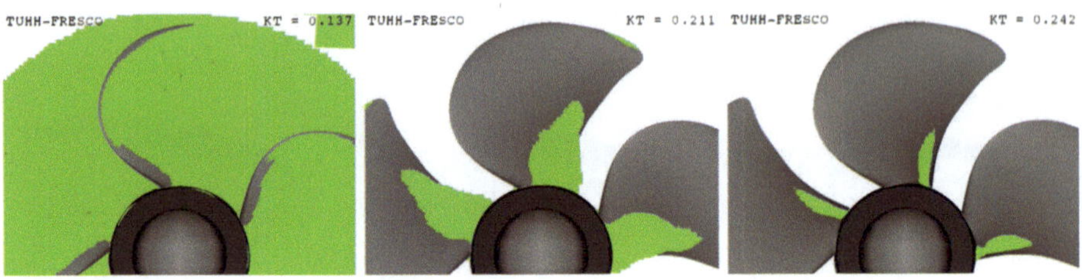

Fig. 4 Sensitivity of the predicted cavitation pattern of the PPTC propeller (Case 2) to model coefficients for the Euler-Euler cavitation model of Zwart et al. [22] employing $4 \cdot 10^6$ control volumes; Displayed isosurface corresponds to 20 % vapor-volume fraction using different model coefficients, i.e. $F_{\text{vap}}/F_{\text{cond}} = 25/0.005$ (*left*), 5/0.005 (*centre*) and 0.1/0.001 (*right*)

Table 4 Predicted thrust for the PPTC propeller (Case 2) using different coefficients of the Zwart model [25] ($4 \cdot 10^6$ control volumes)

$F_{\text{vap}}/F_{\text{cond}}$	K_{T}
25/0.005	0.137
5/0.005	0.211
0.1/0.001	0.242
Non-cavitating measurements	0.245

to investigate the respective influence of F_{vap} and F_{cond} on the predicted thrust coefficient $K_t = T/(n^2 D^4 \rho)$. Three different parameter sets as displayed in Table 4 were used. Mind that they are well within the range of recommended values. The table reveals, that the computed thrust coefficients might differ substantially from the non-cavitating case, depending on the choice of coefficients. Experienced variations reach from 50 % reduction for high parameter values to virtually no changes for small values. The result can be attributed to different predictions of the cavitation volume, which is clearly seen from Fig. 4. For $F_{\text{vap}} = 25$, $F_{\text{cond}} = 0.005$, an excessive vapor volume is produced in the propeller regime. It displaces the primary flow and is convected over a large portion of the domain downstream of the propeller which is of course unphysical (but the solution is converged). In conjunction with

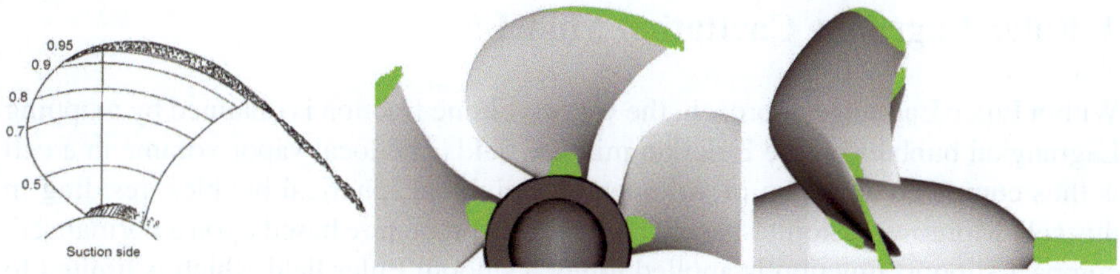

Fig. 5 Euler-Euler simulations of PPTC propeller (Case 1), employing $4 \cdot 10^6$ control volumes. Experimentally observed cavitation pattern (*left*), snapshot of predicted vapor volume fraction ($\alpha = 0.2$, *centre*, *right*)

Fig. 6 Euler-Euler simulations of the PPTC propeller (Case 1) employing $12 \cdot 10^6$ control volumes. Experimentally observed cavitation pattern (*left*), snapshot of predicted vapor volume fraction ($\alpha = 0.2$, *centre*, *right*)

smaller values cavitation exists only in regions near the propeller hub and tip and with the smallest values cavitation region gets very small.

This study shows that Euler-Euler cavitation models may require case-dependent calibration of constants which might be inappropriate for their industrial use. The conclusion applies not only to the employed Zwart model but also any other mass-transfer model of this type.

Finally, Case 1 from Table 3 was computed with Sauer [19] model for both computational grids in attempt to resolve tip vortex cavitation. Figure 5 shows isosurface of vapor volume fraction equal to 0.2 obtained on the $4 \cdot 10^6$ cells grid by the Sauer Euler-Euler model. As expected, sheet cavitation on the blade suction side and hub vortex cavitation are predicted quite well, but no cavitation is observed in the tip vortex cavitation. The resolved vortex is not strong enough to cause the required pressure reduction in it's core which is the only mechanism in the Euler-Euler approach responsible for the cavitation inception. Similar results were obtained with use of other Euler-Euler models [1].

With grid refinement in the tip vortex region, computed pressure drops enough to start cavitation. As one can see on Fig. 6 cavitation pattern is now reproduced quite successfully. It required about eight million additional nodes to resolve tip vortex correctly with a current model. Less dissipative turbulence models like LES/DES or special techniques aimed at reduction of turbulent viscosity in the vortex region i.e. vorticity confinement methods [9] might reduce such severe requirements to the grid quality. That is the subject of further studies.

4 Euler-Lagrange Cavitation Model

Within Euler-Lagrange approach, the vapor-volume fraction is obtained by mapping Lagrangian bubbles on the Eulerian mixture field. The local vapor volume in a cell is thus computed as a sum of volumes of individual spherical bubbles, residing in this cell. Moreover, a kernel-based interpolation procedure based upon a normalised, compact-support function is applied to get a smooth Euler field which is limited to its physical bounds.

The trajectory of a bubble can be described by the equations of motion [2, 14, 15]:

$$\begin{aligned} &\frac{d\mathbf{x}}{dt} = \mathbf{v}; \\ &m_b \frac{d\mathbf{v}}{dt} = (m_b - m_f)\mathbf{g} + m_f \frac{D\mathbf{u}}{Dt} - \frac{1}{2} m_f \left(\frac{d\mathbf{v}}{dt} - \frac{D\mathbf{u}}{Dt} \right) + \mathbf{F}_D + \mathbf{F}_L + \mathbf{F}_V, \end{aligned} \tag{9}$$

where m_b is the bubble mass, m_f is the equivalent mass of the mixture phase. The Euler-Lagrange approach features two velocity fields, i.e. the bubble velocities $\mathbf{v}$ and the mixture velocities $\mathbf{u}$ (interpolated to the present centre of the bubble). The terms on the right-hand side refer to the forces due to buoyancy, fluid acceleration and added mass. The three last terms—$\mathbf{F}_D, \mathbf{F}_L, \mathbf{F}_V$—denote to the drag, lift, and volume variation forces.

Assuming $m_b \ll m_f$, replacing m_f with $\frac{4}{3}\pi R^3 \rho_f$ (where R is the bubble radius) and using empirical correlations for lift, drag and volume variation forces [6, 13, 15], Eq. (9) can be written as:

$$\frac{d\mathbf{v}}{dt} = -2\mathbf{g} + 3\frac{D\mathbf{u}}{Dt} + \frac{3}{4}\frac{C_D}{R}|\mathbf{u}-\mathbf{v}|(\mathbf{u}-\mathbf{v}) + \frac{3}{4}C_L \frac{(\mathbf{u}-\mathbf{v}) \times \boldsymbol{\omega}}{\alpha} + \frac{3}{R}(\mathbf{u}-\mathbf{v})\frac{dR}{dt}. \tag{10}$$

Here $\alpha = |\omega| R / |\mathbf{u} - \mathbf{v}|$, C_L is a lift coefficient and C_D is a drag coefficient as given in [15].

While the equation of motion (10) provides the bubble's trajectory, the Rayleigh-Plesset equation—including a term accounting for the effect of the slip velocity between the bubble and the carrier phase—determines the time-varying radius of the bubble [7]:

$$R\ddot{R} + \frac{3}{2}\dot{R}^2 = \frac{1}{\rho_f}\left[p_v + p_g - p_\infty - \frac{2\sigma}{R} - \frac{4\mu_f}{R}\dot{R} \right] + \frac{(\mathbf{u}-\mathbf{v})^2}{4}. \tag{11}$$

In turbulent flows it might be important to take into account bubble break-up which occurs due to turbulent fluctuations and bubble/turbulent eddy collision. The probability based model of Martinez-Bazan et al. [11, 12] displayed very good agreement with experiments for a bubble break-up in a fully developed turbulent flow over a wide range of bubble sizes. This model has been implemented into the current framework.

To obtain fluid parameters needed for solution of Eqs. (10) and (11) a gradient-based interpolation from the fluid mesh cell center to the position of the bubble (defined by the position of it's center) is performed for each bubble [23]. The vapor-volume fraction is calculated from the discrete bubbles for each fluid mesh cell via Gaussian kernel-based interpolation as suggested by Shams et al. [20].

4.1 Hybrid MPI/OpenMP Parallelization

The parallelization of the Navier-Stokes algorithm (Eulerian part of the model) is done using a traditional domain-decomposition technique based on a Single Program Multiple Data (SPMD) message-passing model, i.e. each process runs the same program on its own subset of data. Inter-process communication employs the MPI communications protocol. Load balancing is achieved using the ParMETIS partitioning software.

Parallelization of the Lagrangian part is done based on the domain decomposition of the fluid field, i.e. a bubble is computed on the same process the local fluid field is assigned to. The latter is motivated by the fact that Eqs. (10) and (11) reveal strong links between the background fluid and the particles but no inter-particle information exchange. No additional data exchange between the fluid and the discrete phase is needed. When a bubble crosses a domain boundary it is transferred to another process. Such an approach can lead to an unbalanced parallelization because in general spatial distribution of bubbles is non-uniform and bubbles tend to get trapped in local areas of the flow (e.g. recirculation zones). An alternative option would be to uniformly distribute bubbles among processes. But in this case data exchange between phases based upon inter-process communications is needed which would lead to significant loses in efficiency.

To reduce unbalance, a hybrid MPI/OpenMP approach is implemented. At the moment, typical computational cluster has a certain amount of nodes, each of them has about 8–24 CPU cores which share the same memory. OpenMP routines work quite efficiently on these cores. For example, if we start 8 MPI processes on 8-cores node, each of these processes can start specific number of OpenMP threads, proportional to the number of bubbles, belonging to this process.

In order to assess the parallel performance, validation tests were done with a coupled Euler/Lagrange model for a cavitating propeller flow on a computational grid of about $4 \cdot 10^7$ cells and 100,000 bubbles. Figure 7 shows parallel performance for this case. Although the bubble distribution in the domain is highly non-uniform, the performance is still satisfactory. The overall speed-up on 128 processors is about 63 times. Such performance allowed to compute one revolution of the propeller for 3 wall-clock hours which is more than affordable for engineering applications.

4.2 Cavitating Hydrofoil Flow

Cavitating flow over a 2D hydrofoil at $\sigma = 3$ analysed in Sect. 3.1 was also computed with the Euler-Lagrange model. The only free parameter for this model is initial bubble distribution/bubble density which are released under no slip conditions

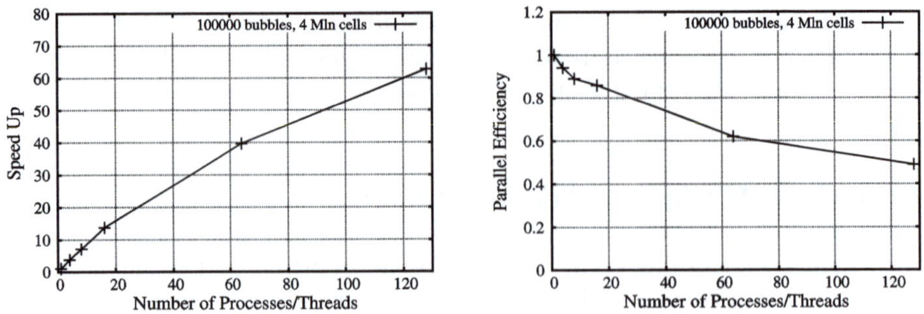

Fig. 7 Parallel performance of the coupled Euler-Lagrange simulations using 100,000 bubbles inside 4 Mio control volumes grid on an Intel cluster (XEON X5570 8-core CPUs)

Fig. 8 NACA66$_2$-415 hydrofoil. Measured bubble spectrum upstream the foil

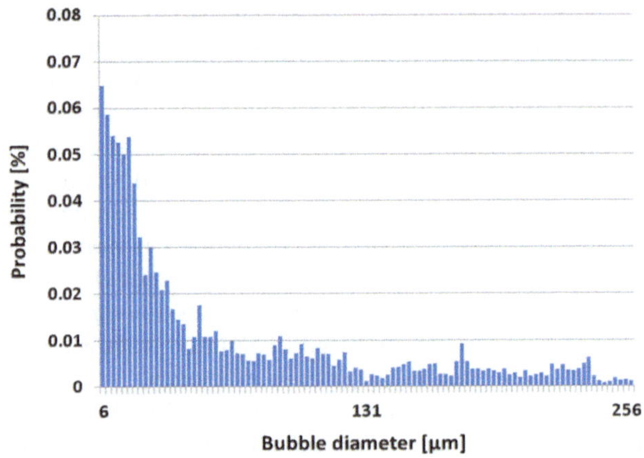

upstream the profile. The bubble distribution is defined by water quality effects and flow conditions. Simulations were performed with a fixed bubble diameter or with the use of an experimentally reported size distribution. Figure 8 shows typical bubble diameter distribution reported for the cavitation tunnel K22 of the HSVA. One can see that bubble diameter is within range of 6 to 256 microns.

Euler-Lagrange simulations were performed with use of this distribution, as well as with two fixed-diameter bubbles of 50 and 90 microns which all featured the same vapor-volume rate. As indicated by Fig. 9, results obtained from the experimentally reported bubble spectrum provides the best predictive agreement for cavitation length and thickness. This is most probably due to presence of bigger (diameter of 150 microns and more) bubbles in the initial distribution as compared to used fixed values. The smaller the bubble diameter is, the more pressure difference is needed to overcome surface tension forces and initiate the growth of vapor bubbles. Thus a small number of big bubbles produces substantially more cavitation then a large number of small bubbles. In general, the Euler-Lagrange model produces more realistic results as compared to Euler-Euler model.

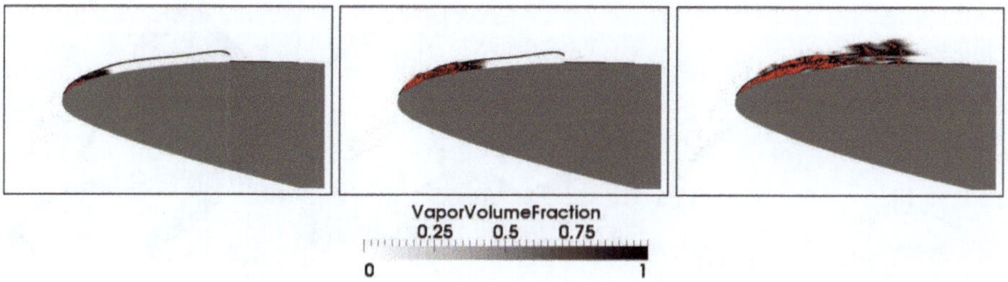

Fig. 9 Sensitivity of the predictions returned by the Euler-Lagrange cavitation model to the nuclei diameter spectrum imposed upstream the 2D flow over a NACA66$_2$-415 hydrofoil at $\sigma = 3$ and a frozen nuclei-volume rate. Fixed initial bubble diameter of 50 micron (*left*) and 90 micron (*centre*), measured initial bubble distribution from P.D.F. (*right*)

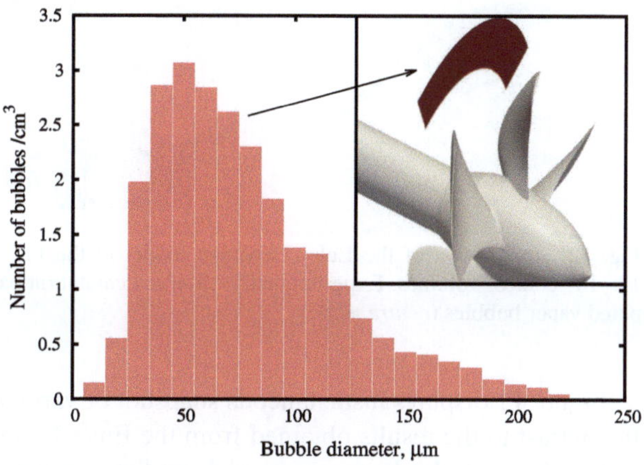

Fig. 10 Measured nuclei distribution and bubble-release area upstream the tip for the PPTC propeller case at $\sigma = 1.96$ and 80 % saturation

4.3 Cavitating Flow over the PPTC Propeller

The Euler-Lagrange model was used to compute cavitating propeller flow (Case 1, as described in Sect. 3.2). The nuclei distribution required for the Euler-Lagrange model has been measured experimentally with the Phase Doppler Anemometry (PDA) technique under the aegis of the BMWi-project *KonKav-I*. Experimental work was performed by the Institute for General Electrical Engineering of the University of Rostock. Figure 10 displays measured bubble spectra for the considered test case. Measured bubble diameters lay within range of 10 to 200 microns with most of bubbles within 20–120 microns range. At every Eulerian time step 140 bubbles were injected in the area upstream the propeller at radius of tip blade as shown in Fig. 10. Number of injected bubbles corresponds to the number of nuclei measured in the experiment and their radius is randomly chosen according to the measured probability density function. Injected bubbles travel downstream and enter the tip vortex region of one blades. Note that the bubble-release area is restricted to the upstream region of one blade tip and doesn't cover the whole domain in order to reduce the amount of injected bubbles.

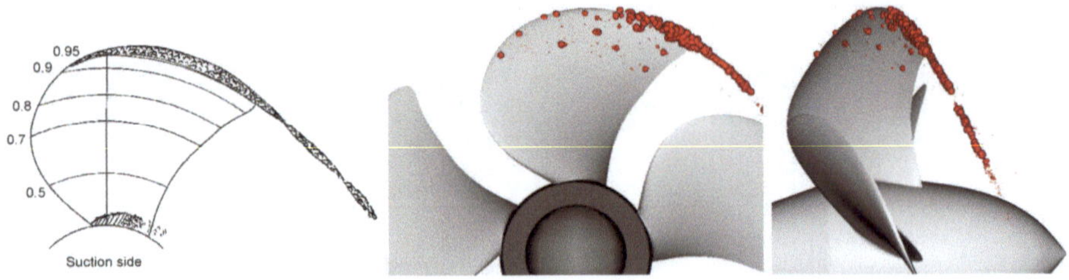

Fig. 11 Application of the Euler-Lagrange model to the PPTC propeller (Case 1) employing $4 \cdot 10^6$ control volumes. Experimentally observed cavitation pattern (*left*) and snapshot of computed vapor bubbles (*centre*, *right*)

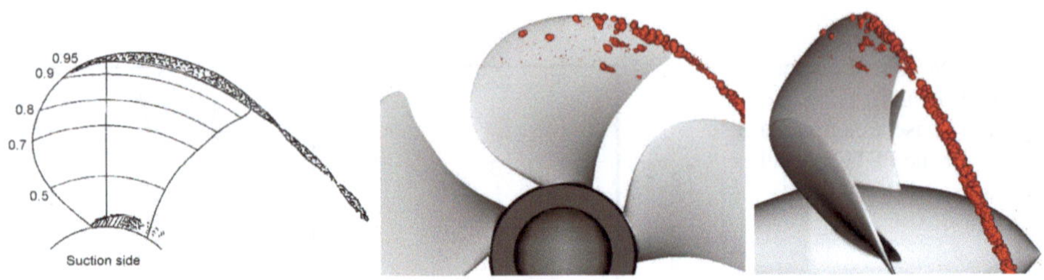

Fig. 12 Application of the Euler-Lagrange model to the PPTC propeller (Case 1) employing $12 \cdot 10^6$ control volumes. Experimentally observed cavitation pattern (*left*) and snapshot of computed vapor bubbles (*centre*, *right*)

Figure 11 displays instantaneous snapshot of vapor bubbles on $4 \cdot 10^6$ cells grid. In contrast to the results obtained from the Euler-Euler cavitation model displayed in Fig. 5, the Euler-Lagrange model predicts tip vortex cavitation even with this grid. However, the cavitating vortex weakens and disappears downstream the blade quite rapidly. Vapor bubbles grow in the low pressure region at the blade tip and are convected downstream. They are subsequently exposed to premature condensation due to the inability of the coarse grid to accurately resolve the confined low-pressure regime of the vortical flow aft of the propeller. A few single vapor bubbles are also found near the upper part of the blade. Such phenomena cannot be observed with the Euler-Euler model, because of the assumed unified kinematics shared by both phases and the relatively high pressure in this region. In conjunction with the Euler-Lagrange model, single bubble may penetrate into high pressure regions due to their individual momentum. Mind that no cavitation is predicted in the near-hub region, since no bubbles were injected that could be convected into that zone.

Similar to the Euler-Euler simulations, the low-pressure regime continues aft of the propeller when the tip vortex resolution is improved. As indicated by the bubbles displayed in Fig. 12, a substantial amount of vapor remains in the downstream region of the vortex. Generally, the cavitation pattern predicted by the Euler-Lagrange model is in a fair agreement with the experiment as well as with the Euler-Euler model results. It should be mentioned, that in spite of the observed similarity for the cavitation pattern, the Euler-Lagrange model offers potential benefits arising from the discrete treatment of the vapor phase, which better corresponds to cavitation physics. Among these benefits are erosion predictions, which require to mimic impact pressures from individual bubbles located near the wall, acoustic pressure calculations, which also rely on individual bubble information and water quality or scale effects [10], which are not considered by the Euler-Euler approach.

5 Combined Euler-Euler/Euler-Lagrange Cavitation Model

The main drawback of the Euler-Lagrange approach is the increased computational demand due to the large number of bubbles. Typically, momentum and Rayleigh-Plesset equations are advanced in time for several millions bubbles. Note that the Lagrangian time step is about thousand times smaller than the Eulerian time step due to the very small bubble sizes. Although the hybrid MPI/OpenMP parallelization has been implemented as described in Sect. 4.1, the model is still computationally expensive particularly as regards full-scale simulations. In order to reduce the computational cost, a combined Euler-Euler/Euler-Lagrange approach is suggested. Within this approach bubbles are only initialized upstream a selected area of particular interest (e.g. tip vortex) and Euler-Lagrange solution is used in those cells which host bubbles. In the remainder of the domain an Euler-Euler cavitation model is employed. The latter is tightly coupled to the mapped vapor-volume fraction of the Lagrangian approach. The algorithm of the combined model for one Eulerian time step (non-parallel version) can be described as follows:

1. Update Eulerian time $t_f^{n+1} = t_f^n + \Delta t_f$
2. Calculate new mixture pressure, velocities and turbulent quantities by solving equations (2)
3. Inject specified number of nuclei at initial position. The initial nuclei spectra (nuclei concentration and initial diameter distribution) depends on a water quality and is either known from experiments or estimated. The initial position is defined by a release area upstream of the region of interest. Bubbles are started randomly within this area
4. For each of the bubbles proceed while Lagrangian time $t_{v_i} \leq t_f^{n+1}$:
   - Update Lagrangian time for ith bubble $t_{v_i} = t_{v_i} + \Delta t_{v_i}$
   - Update bubble size and position solving iteratively equations (10) and (11)
5. Using above described probability based model compute bubbles break-up
6. Calculate vapor-volume fraction from bubbles
7. Solve vapor volume fraction equation (5) imposing vapor-volume fraction from bubbles (in cells where this is greater than zero)
8. Update mixture density and viscosity

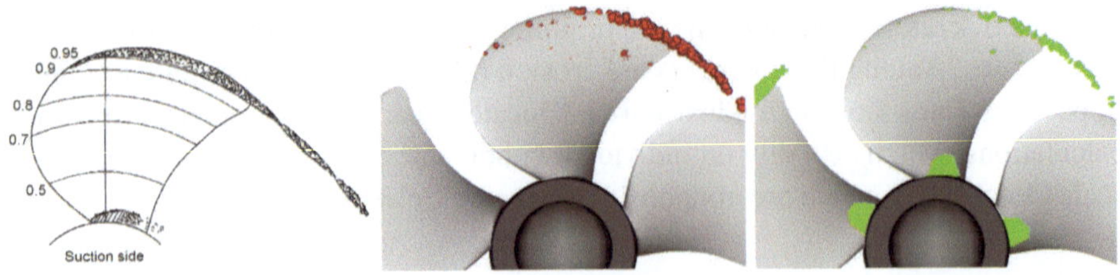

Fig. 13 Application of the combined Euler-Euler/Euler-Lagrange model to the PPTC propeller (Case 1) using $12 \cdot 10^6$ control volumes and 50,000 bubbles. Experimentally observed cavitation pattern (*left*), snapshot of computed vapor bubbles (*centre*), snapshot of computed vapor volume fraction ($\alpha = 0.2$, *right*)

Table 5 Computational effort associated to the investigated cavitation models for one revolution of the PPTC propeller case 1 on an 8-core Intel XEON Gainestown CPU (X5570, 2.93 GHz) machine

Cavitation model	CPU h
Euler-Euler	83
Euler-Lagrange (tip vortex only)	101
Euler-Lagrange (whole domain)	373
Combined Euler-Euler/Euler-Lagrange	112

5.1 Cavitating Flow over the PPTC Propeller

As discussed above, a combination of the Euler-Euler and Euler-Lagrange approaches inheres most of the predictive benefits of the Euler-Lagrange model in regions of interest. Since the simplified Euler-Euler approach usually covers the major portion of the computational domain, the combined model can be performed at significantly reduced computational costs. Figure 13 shows results for the Case 1 (see Table 3) obtained with the combined model with the grid of $12 \cdot 10^6$ cells described in Sect. 3.2. As expected, the predicted tip vortex cavitation is very similar to results returned by the Euler-Lagrange model (cf. Fig. 12). The fair prediction of root cavitation offered by the Euler-Euler approach (cf. Fig. 6) is also transfered to the combined modelling approach.

Table 5 analyses CPU-effort by means of the wall-clock time required to simulate one propeller revolution (360 time steps) with different models. All simulations were performed with the $12 \cdot 10^6$ cells grid on the North-German Supercomputing Alliance (HLRN-II) supercomputing system ICE2 (www.hlrn.de) using 128 Intel Xeon Gainestown CPU cores with 8 cores per CPU. The Euler-Euler simulation required 5.2 wall-clock hours (83.2 CPUh) to simulate one revolution. On the contrary, the simulation with the Euler-Lagrange model required 30 % more time although the bubble-release was restricted to a focal area upstream the tip vortex region. Approximately 50,000 bubbles were employed to investigate the tip vortex. Performing Euler-Lagrange simulations with bubbles injected to the whole 5-blade domain, requires fifteen times more bubbles, which drastically increases the computational time. In conjunction with the combined Euler-Euler/Euler-Lagrange model the simulation effort increased by 35 % as compared to the Euler-Euler model. Again bubbles were released in a restricted area upstream the tip vortex region.

6 Conclusions

Euler-Euler, Euler-Lagrange and a combined hybrid approach to cavitation modelling are investigated for cavitating marine engineering flows. The Euler-Euler approach is computationally inexpensive, but restricted to simplified dynamics and requires case-specific calibration of constants. The Euler-Lagrange approach allows to take into account inhomogeneous and transient water-quality effects, depending on air content and bubble spectra. At the same time, it requires significantly more computational resources and is prone to a successful implementation of a sophisticated parallelization strategy.

Results obtained from an Euler-Euler model for a cavitating flow over a 2D hydrofoil and a marine propeller in open water conditions reveal a strong dependence of the predictive success on appropriately chosen model constants. As opposed to this, the Euler-Lagrange model delivered a fair predictive agreement with experiments when using measured initial nuclei distribution. It should thus outperform traditional Euler-Euler techniques when calibrated model parameters are not available. Moreover, as illustrated by the investigated marine propeller flow, the Euler-Euler model usually fails to predict tip vortex cavitation unless using locally extremely fine grids. Having no convection term, the Euler-Lagrange model also seems to be slightly less demanding to the local grid arrangement/cell size but still requires a reliable prediction of the pressure and velocity fields.

The combined model returns tip-vortex predictions which are very close to the results of the Euler-Lagrange model at significantly lower computational expenses. It seems a viable approach to utilize the benefits of both approaches and can be applied to a wide range of multiphase engineering applications. Potential areas of application include acoustic pressure calculation and erosion risk estimation, study of water quality aspects as well as scale effect simulations.

Acknowledgements The current work is a part of the research project funded by German Ministry of Economics and Technology (BMWi; Grant Nr. 3SX286A) in the framework of the "Schifffahrt und Meerestechnik für das 21. Jahrhundert" research initiative. The experimental data were obtained by our colleagues from the University of Rostock. The simulations were performed on the HLRN-II supercomputer system at the North German Cooperation for High-Performance computing (HLRN). This support is gratefully acknowledged by the authors.

References

1. SMP'11 workshop on cavitation and propeller performance (2011) In: International symposium on marine propulsors, Hamburg, Germany
2. Abdel-Maksoud M, Hänel D, Lantermann U (2010) Modeling and computation of cavitation in vortical flow. Int J Heat Fluid Flow 31(6):1065–1074
3. Chahine GL (2004) Nuclei effects on cavitation inception and noise. In: 25th symposium on naval hydrodynamics, St John's, Newfoundland and Labrador, Canada

4. Elghobashi S (1994) On predicting particle-laden turbulent flows. J Appl Sci Res 52(4):309–329
5. Fuster D, Colonius T (2011) Modelling bubble clusters in compressible liquids. J Fluid Mech 688:481–512
6. Hsiao C-T, Chahine GL (2003) Scaling effect on prediction of cavitation inception in a line vortex flow. J Fluids Eng 125:53–60
7. Hsiao C-T, Chahine GL, Liu H-L (2000) Scaling effect on bubble dynamics in a tip vortex flow: prediction of cavitation inception and noise. Technical report, Dynaflow, Inc
8. Lauterborn W, Luther S, Mettin R (2000) Modeling acoustic cavitation by a Lagrangian approach. In: Nonlinear acoustic at the turn of the millenium: ISNA 15, 15th international symposium. AIP conference proceedings, vol 524, pp 351–354
9. Manzke M, Rung T (2010) Propeller flow predictions using turbulent vorticity confinement. In: Proceedings of the v European conference on computational fluid dynamics ECCOMAS CFD 2010, Lisbon, Portugal
10. Maquil T, Abdel-Maksoud M, Cankurt B, Rung T, Schiller P, Yakubov S (2011) Simulation of water-quality effects for cavitating engineering flows. In: CD-ROM proceedings of the WIMRC 3rd international cavitation forum, Warwick
11. Martinez-Bazan C, Montanes J, Lasheras JC (1999) On the breakup of an air bubble injected into a fully developed turbulent flow. Part 1. Breakup frequency. J Fluid Mech 401:157–182
12. Martinez-Bazan C, Montanes J, Lasheras JC (1999) On the breakup of an air bubble injected into a fully developed turbulent flow. Part 2. Size PDF of the resulting daughter bubbles. J Fluid Mech 401:183–207
13. Mattson M, Mahesh K (2010) Euler-Lagrangian simulation of bubble migration in a turbulent boundary layer. In: Proc 28th symposium on naval hydrodynamics, Pasadena, California, 12–17 September 2010
14. Maxey MR, Riley JR (1983) Equation of motion for a small rigid sphere in a nonuniform flow. J Phys Fluids 26(4):883–889
15. Oweis G (2005) Capture and inception of bubbles near line vortices. J Phys Fluids 17:022105
16. Epstein MPPS (1950) On the stability of gas bubbles in liquid-gas solutions. J Chem Phys 18:1505–1509
17. Rayleigh L (1917) On the pressure developed in a liquid druing collapse of a spherical cavity. Philos Mag Ser 6 34(200):94–98
18. Rung T, Wöckner K, Manzke M, Brunswig J, Ulrich C, Stück A (2009) Challenges and perspectives for maritime CFD applications. Jahrbuch der Schiffbautechnischen Gesellschaft, vol 103
19. Sauer J (2000) Instationär kavitierende Strömungen – Ein neues Model, basierend auf Front Capturing (VoF) und Blasendynamik. PhD thesis, Universität Karlsruhe
20. Shams E, Finn J, Apte S (2011) A numerical scheme for Euler-Lagrange simulation of bubble flows in complex systems. Int J Numer Methods Fluids 67(12):1865–1898
21. Singhal A, Athavale M, Li H, Jiang Y (2002) Mathematical basis and validation of the full cavitation model. J Fluids Eng 124(3):617–624
22. Sommerfeld M (2000) Theoretical and experimental modelling of particulate flow: overview and fundamentals. Lecture series, vol 2000-6. Von Karman Institute for Fluid Dynamics, pp 1–62
23. Yakubov S, Cankurt B, Abdel-Maksoud M, Rung T (2012) Hybrid MPI/OpenMP parallelization of an Euler-Lagrange approach to cavitation modelling. Comput Fluids. doi:10.1016/j.compfluid.2012.01.020
24. Yakubov S, Cankurt B, Maquil T, Schiller P, Abdel-Maksoud M, Rung T (2011) Cavitation simulations of the Potsdam propeller test case. In: Proceedings of the workshop on cavitation and propeller performance. Second international symposium on marine propulsors—SMP'11, Hamburg, Germany, pp 84–92
25. Zwart P, Gerber A, Belamri T (2004) A two phase flow model for predicting cavitation dynamics. In: ICMF 2004 international conference on multiphase flow, Yokohama, Japan, May 30–June 3

Part VII
Ship Hydrodynamics

CFD in Ship Hydrodynamics—Results of the Gothenburg 2010 Workshop

Lars Larsson, Frederick Stern, and Michel Visonneau

Abstract The Gothenburg 2010 Workshop on Numerical Hydrodynamics gathered 33 groups with computations for one or more of 18 test cases. All results were collected and discussed at a meeting in Gothenburg in December 2010. In the present paper some representative examples from the workshop are presented. The complete results are found in the workshop Proceedings.

Keywords Computational fluid dynamics · Hydrodynamics · Validation · Assessment · Workshop

1 Introduction

In 1980 an international workshop on the numerical prediction of ship viscous flow was held in Gothenburg [1]. The purpose was to assess the state-of-the-art and to find directions for the future developments in the field. Participants in the workshop had been invited long before and had delivered results for two well specified test cases to the organizers. Detailed information on the features of each participating

L. Larsson (✉)
Department of Shipping and Marine Technology, Chalmers University of Technology, 41296 Gothenburg, Sweden
e-mail: lars.larsson@chalmers.se
url: http://www.chalmers.se/smt/EN

F. Stern
IIHR—Hydroscience & Engineering, The University of Iowa, 100 C. Maxwell Stanley Hydraulics Laboratory, Iowa City, IA 52242-1585, USA
e-mail: frederick-stern@uiowa.edu
url: http://www.iihr.uiowa.edu

M. Visonneau
Fluid Mechanics Laboratory—UMR6598, ECN/CNRS, Centrale Nantes BP 92101, 44321 Nantes Cedex 3, France
e-mail: michel.visonneau@ec-nantes.fr
url: http://www.ec-nantes.fr

L. Eça et al. (eds.), *MARINE 2011, IV International Conference on Computational Methods in Marine Engineering*, Computational Methods in Applied Sciences 29,
DOI 10.1007/978-94-007-6143-8_14,

method had also been submitted and compiled in a table. By comparing the computed results on the one hand, and the details of the methods on the other, the most promising approaches could be sorted out.

Now, more than 30 years have passed since this first workshop and the event has been repeated a number of times. In 1990 the second workshop was held, again in Gothenburg [2]. While practically all methods participating in the 1980 workshop had been of the boundary layer type, now all but one were of the RANS type. A huge improvement in the prediction of the flow around the stern was noted. The workshop idea was picked up in Japan in 1994 and the third workshop was held in Tokyo in 1994 [3]. Notable from this workshop is that free-surface capabilities had become available in many of the methods. The fourth workshop in the series was held in Gothenburg in 2000 [4, 5]. Now, three modern hull forms were introduced as test cases, and these hulls have been kept ever since. At this time formal verification and validation (V&V) procedures were introduced. While in the previous workshops the emphasis had been on the wake and waves for a towed hull, self-propulsion was introduced in 2000. This was kept in the fifth workshop in Tokyo in 2005 [6], where some sea-keeping and manoeuvring cases were introduced as well. Even though the same three hulls were used this increased the number of test cases significantly. A further step in this direction was taken at the most recent workshop, held in Gothenburg in December of 2010. This workshop is the topic of the present paper.

2 Hulls

The three hulls used in the workshop were:

(1) The KVLCC2, a second variant of a Korean VLCC
(2) The KCS, a Korean container ship
(3) The DTMB 5415, a US combatant

The KVLCC2 was designed at the Korea Research Institute for Ships and Ocean Engineering (now MOERI) around 1997 to be used as a test case for CFD predictions. Extensive towing tank tests were carried out, providing data for resistance, sinkage and trim, wave pattern and nominal wake at several cross-planes near the stern [7–9]. Mean velocity and turbulence data were obtained by Postech [10] in a wind tunnel. At the CFD Tokyo Workshop in 2005 [6] there was a slight modification of the stern contour of this ship and it was therefore renamed as KVLCC2M. The modification is explained in Hino [6]. In the present workshop the original design was used.

Also the KCS was designed by MOERI for the same purpose as the KVLCC2, and similar tests were carried out for this hull [7–9]. Self-propulsion tests were carried out at the Ship Research Institute (now NMRI) in Tokyo and are reported in Hino [5]. Data for pitch, heave, and added resistance are also available from Force/DMI measurements [11].

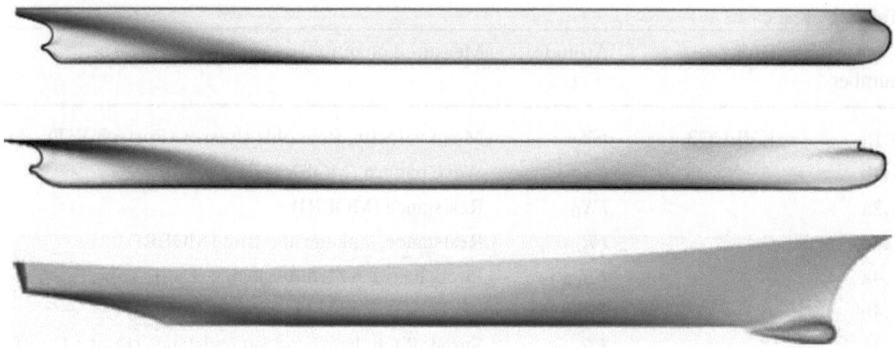

Fig. 1 The three ships used in the workshop (*upper*: KVLCC2; *middle*: KCS; *bottom*: 5415)

Table 1 Main particulars of the three ships

Main particulars (full scale)			KVLCC2	KCS	DTMB 5415
Length between perpendiculars		L_{PP} (m)	320.0	230.0	142.0
Maximum beam of waterline		B (m)	58.0	32.2	19.06
Draft		T (m)	20.8	10.8	6.15
Displacement		Δ (m^3)	312622	52030	8424.4
Wetted area w/o rudder		S_W (m^2)	27194	9424	2972.6
Wetted surface area of rudder		S_R (m^2)	273.3	115.0	30.8
Block coefficient (C_B)		$\Delta/(L_{PP} \cdot B \cdot T)$	0.8098	0.6505	0.507
Service speed	Speed	U (knots)	15.5	24.0	18.0, 30.0
	Froude number	Fr	0.142	0.26	0.248, 0.413

Model 5415 was conceived as a preliminary design for a Navy surface combatant around 1980. The hull geometry includes both a sonar dome and transom stern. Propulsion is provided through twin open-water propellers driven by shafts supported by struts.

The model test data for the 5415 includes:

- Local flow measurements (Mean velocity and cross flow vectors) [12].
- PIV-measured nominal wake in regular head waves (Mean velocity, turbulent kinetic energy, and Reynolds stresses) [13].
- Resistance, sinkage, trim, and wave profiles [12].
- Wave diffraction (Waves, 1st harmonic amplitude of mean velocities, turbulent kinetic energy, and Reynolds stresses) [13].
- Roll decay (Motion, free surface, mean velocities) [14].

Side views of the three hulls are seen in Fig. 1 and the main particulars are given in Table 1. No full scale ships exist.

Table 2 Test cases

Case number	Hull	Attitude	Measured quantity
1.1a	KVLCC2	FX_0	Mean velocity, Reynolds stresses (Postech WT)
1.1b		FX_0	Wave pattern (MOERI)
1.2a		FX_0	Resistance (MOERI)
1.2b		$FR_{z\theta}$	Resistance, sinkage and trim (MOERI)
1.4a		$FR_{z\theta}$	Pitch, heave, added resistance (INSEAN)
1.4b		$FR_{z\theta}$	Pitch, heave, added resistance (NTNU)
1.4c		$FR_{xz\theta}$	Surge, Pitch, heave, added resistance (Osaka Univ)
2.1	KCS	FX_0	Wave pattern, mean velocities (MOERI)
2.2a		FX_0	Resistance (MOERI)
2.2b		$FR_{z\theta}$	Resistance, sinkage and trim (MOERI)
2.3a		FX_0	Self propulsion at ship point (thrust, torque, force balance or RPM, mean velocity), local flow (NMRI)
2.3b		$FR_{z\theta}$	Self propulsion at model point (thrust, torque, force balance or RPM), sinkage and trim (FORCE)
2.4		$FR_{z\theta}$	Pitch, heave, added resistance (FORCE)
3.1a	DTMB 5415	$FX_{\sigma\tau}$	Mean velocity, resistance, wave pattern (INSEAN)
3.1b		$FX_{\sigma\tau}$	Mean velocity, resistance, wave pattern, Reynolds stresses (IIHR)
3.2		$FR_{z\theta}$	Resistance, sinkage and trim (INSEAN)
3.5		$FX_{\sigma\tau}$	Wave diffraction, Mean velocity (IIHR)
3.6		FR_{φ}	Roll decay (IIHR)

3 Test Cases

Several types of computations were requested, namely:

(1) Local flow at fixed condition, either zero sinkage and trim (denoted FX_0) or dynamic sinkage and trim ($FX_{\sigma\tau}$)
(2) Resistance, sinkage and trim either at FX_0 or at heave- and pitch-free condition ($FR_{z\theta}$)
(3) Self-propulsion at FX_0 or $FR_{z\theta}$
(4) Heave and pitch in waves either at $FR_{z\theta}$ or with free surge ($FR_{xz\theta}$)
(5) Forward speed diffraction at $FX_{\sigma\tau}$
(6) Free roll decay at $FX_{\sigma\tau}$ and free to roll (FR_{φ})

Note that maneuvering was not included, since it was the topic of the recently held SIMMAN workshop in the spring of 2008 [15].

All test cases for the three hulls are listed in Table 2. The measurements were taken at the organizations within brackets. See the references above. There are altogether 18 cases and the participants were free to select which cases to compute.

4 Participants and Methods

The workshop participants are listed in Table 3, together with the main features of their methods. In the first column the acronym of the participating group is given. This is used in combination with the code name of column three to identify each submission. The cases computed are given in column two. In the remaining columns the features of each method are given.

The majority of methods use two-equation turbulence models, $k - \varepsilon$ or $k - \omega$. There are also some one-equation models, either Spalart-Allmaras or Menter. The anisotropic models are either of the algebraic stress or Reynolds stress type. Note that there are also some LES/DES methods and even a DNS method.

Most of the participants use no-slip wall boundary conditions, but there are also several methods with wall functions, both with and without pressure gradient corrections.

The Volume of Fluid (VOF) technique is the most popular one for the free-surface modeling, but there are also several level set methods. There are only three entries with surface tracking.

The propeller is represented either as an actual rotating propeller or through a body force approximation.

Simulations were performed using both finite difference and finite volume codes, but there was no finite element method. 2nd or 3rd order accurate schemes were used and limited studies used 4th order schemes.

The grids used were either single- or multi-block structured ones (butt-joined or overlapping) or unstructured ones. Most methods are pressure based but there are also several solving the equations directly or with an artificial compressibility.

5 Selected Results

The Workshop Proceedings [16] contain a large number of figures and tables, as well as all papers contributed to the Workshop. A comprehensive analysis of all results will become available as a book [17], where additional calculations are also included to improve the evaluation of the results. With the space limitation of the present paper only a few examples from the Proceedings can be given. The examples will be presented by subject (cf. "Types of computations" in Sect. 3): resistance, self-propulsion, wave pattern, local flow, heave and pitch, and roll decay. Unfortunately, sinkage and trim, wave diffraction and the very extensive discussions on Verification and Verification (V&V) will have to be left out.

5.1 Resistance

In Table 4 a statistical analysis of all computed total resistance coefficients is presented. While the case and Froude number are presented in columns 1–3, column 4

Table 3 Workshop participants and methods

Organization	Cases submitted	Code	Turbulence (incl. non-RANS)	Wall model	Free surface	Propeller	Discretization		Grid type	Velocity pressure
							Type	Order		
CD-Adapco	2.2a, 2.2b	STAR-CCM+	Standard $k-\varepsilon$	N	VOF	–	FV	2	U	PR
CEHINAV TU Madrid	3.1a	STAR-CCM+	$k-\omega$ SST	N	VOF	–	FV	1	MU	PR SIMPLE
Chalmers	1.1a	SHIPFLOW4.3	$k-\omega$ SST EASM	N	–	–	FV	2	S	A
CSSRC	2.1, 2.2a, 2.2b, 2.3a, 3.1a, 3.2	FLUENT 6.3	$k-\omega$ SST RNG $k-\varepsilon$	N	VOF	Actual	FV	2	MS	PR SIMPLE
CTO	2.3a	STAR-CCM+	$k-\varepsilon$	N WO	VOF	Actual	FV	2	U	PR SIMPLE
ECN/CNRS	1.1b, 1.4a, 1.4b, 3.6	ISISCFD	$k-\omega$, EASM	WO	VOF	Body force	FV	2	U	PR SIMPLE
ECN/HOE	1.1a, 1.1b, 1.4a, 1.4b, 2.1, 3.1a, 3.1b, 3.5, 3.6	ICARE	Wilcox $k-\omega$ $k-\omega$ SST	N WO	Nonlin. track	Body force	FD	2	S	D
FLOWTECH	2.1	SHIPFLOW-VOF-4.3	$k-\omega$ SST	N	VOF	–	FV	2	MS	A
FOI	3.1a	OF	LES	WO	VOF	–	FV	2	U	PR
FORCE	2.4	CFDShip-Iowa	$k-\omega$ SST	N	Level set	–	FD	2	OS	PR PISO
GL&UDE Univ. Duisburg	1.4a, 1.4b, 1.4c, 2.2a, 2.2b, 2.4, 3.6	Comet OpenFOAM	$k-\varepsilon$	N WO	VOF	-	FV	Mixed	U	PR SIMPLE
HSVA	1.1a, 1.2a	FreSCo+	2E $k-\omega$	N	VOF	–	FV	3	U	PR

Table 3 (Continued)

Organization	Cases submitted	Code	Turbulence (incl. non-RANS)	Wall model	Free surface	Propeller	Discretization		Grid type	Velocity pressure
							Type	Order		
IHI/Univ. Tokyo	2.4	WISDAM-UTokyo	Baldwin-Lomax and DSGS	N	Density function ρ	–	FV	3	OS	PR MAC
IIHR	1.1a, 1.4a, 1.4c, 2.1, 2.3a, 2.3b, 2.4, 3.1a, 3.1b, 3.5, 3.6	CFDShip-Iowa V4, V4.5, V6	Hybrid $k-\varepsilon/k-\omega$ Based DES Hybrid ARS Based DES	WO N	Level set	Actual Body force	FD	2~4	S, OS	PR Fractional step
IIHR-SJTU	2.1, 2.3a	FLUENT12.0.16	Realizable $k-\varepsilon$	N, W	VOF	Body force	FV	3	MS	PR
IST	1.1a	PARNASSOS	$k-\omega$ SST	N	–	–	FD	2	S	D
Kyushu University	1.4b	RIAM-CMEN	DNS	N	THINC	–	FD	3	S	PR
MARIC	1.1a, 1.1b, 1.2a, 2.1, 2.2a, 2.3a, 3.1a, 3.1b, 3.2	FLUENT	Realizable $k-\varepsilon$	N	VOF	Actual	FV	2	MS	PR SIMPLE
MARIN	2.1, 2.3a, 3.1a, 3.2	PARNASSOS	1E Menter	N	Free-surface fitting	–	FD	2	MS	D
MOERI	1.1b, 1.2a, 1.2b, 2.1, 2.2a, 2.2b, 2.3a, 2.3b	WAVIS	Realizable $k-\varepsilon$	WO	Level set	Body force	FV	3	MS	PR SIMPLEC
NavyFOAM (NSWC/P S ARL)	1.1a, 3.1a, 3.1b, 3.2	NavyFOAM	Wilcox $k-\omega$	WO	VOF	–	FV	2, 3	MS MU	PR
NMRI	1.1a, 2.1, 2.3a, 3.1b, 3.5	SURF	1E Modified Spalart-Allmaras	N	Level set	Body force	FV	2	S, U	A
NSWC-PC ARL	3.2	CFDShip-Iowa V4.5	DES Hybrid $k-\varepsilon/k-\omega$	N	Level set	–	FD	4	S, OS	PR

Table 3 (Continued)

Organization	Cases submitted	Code	Turbulence (incl. non-RANS)	Wall model	Free surface	Propeller	Discretization		Grid type	Velocity pressure
							Type	Order		
NTNU	1.1a, 1.1b, 1.2a	FLUENT	$k-\omega$ SST	N	VOF	–	FV	2	MU	PR SIMPLE
SNUTT	2.1, 2.3a	FLUENT6.3	$k-\varepsilon$	W	VOF	Actual	FV	2	MU	D
Southampton Univ. QinetiQ	2.1, 2.2b, 2.3a	CFX 12	$k-\omega$ SST	W	VOF	Body force	FV	2	MS	D
SSPA	2.3a, 2.3b	SHIPFLOW4.3	EASM	N	–	Body force	FV	2	OS	A
SSRC Univ. Strathclyde	2.1, 2.3a, 3.1b, 3.5, 3.6	FLUENT12.1	$k-\omega$ SST	N	VOF	Actual	FV	2	MS	PR
SVA Potsdam	2.1, 2.2a	ANSYS-CFX12	$k-\omega$ SST	N	VOF	–	FV	2	MU	D
TUHH	1.1b, 2.4	FreSCo+	$k-\omega$ SST	N	VOF	–	FV	3	U	PR SIMPLE
TUHH ANSYS	2.1, 2.2a, 2.3a,	CFX12.1	$k-\omega$ SST	N WO	VOF	Actual	FV	2	MS	Fully coupled $w-p$, SIMPLER p-equation
Univ. Genova	3.1a, 3.2	StarCCM+	Realizable $k-\varepsilon$	N	VOF	–	FV	2	U	PR SIMPLE
VTT	1.1a, 1.1b, 1.2a, 2.1	FINFLO	$k-\omega$ SST	N	Nonlin. track	–	FVFD	3	MS OS	A, PR

A—Artificial compressibility; D—Direct method; FD—Finite difference; FV—Finite volume; MS—Multiblock structured; MU—Multiblock unstructured; N—No slip; OS—Overlapping structure; PR—Pressure correction; S—Single block structured; U—Unstructured; W/WO—Wall functions with/without pressure gradient correction

Table 4 Resistance statistics, all cases

Hull	Case No.	Fr	E_{mean} %D	σ %D	U_D %D	No. of entries
KVLCC2	1.2a (fixed)	0.1423	−1.7 (0.0)	1.3 (6.2)	1.0 (0.7)	5 (13)
	1.2b (free)	0.10∼0.15	−2.1	–	1.0	1
KCS	2.2a (fixed)	0.26	−1.3 (−1.1)	1.2 (4.2)	1.0	8 (11)
	2.2b (free)	0.11∼0.28	−0.2	1.2	1.0	27
	2.3a (fixed, prop.)	0.26	−0.3[a] (−0.9)	3.1 (1.0)	–	14[a] (4)
	2.3b (free, prop.)	0.26	7.2	3.3	–	3
5415	3.1a (fixed s&t)	0.28	2.5 (1.6)	3.8 (5.3)	0.6 (2.2)	5 (11)
	3.1b (fixed s&t)	0.28	−2.6	4.4	0.6	5
	3.2 (free)	0.138	−2.8	4.4	1.3	5
		0.28	0.1 (−1.9)	2.1 (–)	0.6 (2.2)	6 (1)
		0.41	4.3	1.4	0.6	5
Mean of all cases			$E_{mean} = -0.1$ %D	$\sigma_{mean} = 2.1$ %D		89 (40)

[a]Results from MARIC with a hub cap are excluded

gives the mean comparison error E_{mean} in per cent of the measured data value, D. According to the sign convention of the Workshop E_{mean} is defined as $D - S_{mean}$, where S_{mean} is the mean of all simulated values for the particular case. The standard deviation, σ, is given in column 5 in per cent of the data value, and in column 6 the estimated data uncertainty is presented. Finally, in the last column the number of entries for the case is seen. Values within brackets are from the 2005 Workshop [6].

Comparing the 2010 and 2005 results a substantial reduction in the standard deviation (%D) for the towed KVLCC2 and KCS cases is noted, from 6.2 to 1.3 and from 4.2 to 1.2 respectively in the fixed condition. Also, $|E_{mean}|$ for these conditions is well below 2 %D, which indicates that all predictions for this condition are quite accurate, although still not within the experimental accuracy. There is only one submission for the free KVLCC2 condition and $|E_{mean}|$ for all Froude numbers is of the same size 2.1 %D. The free KCS condition has several submissions and very small comparison errors (0.2 %D) and standard deviations, around 1 %D for both.

The self-propelled KCS has standard deviations around 3 %D and the comparison error is very small for the fixed case. However, for the free case $|E_{mean}|$ is quite high: 7.2 %D. All three submissions under predict the resistance significantly. It should be noted that the fixed KCS in self-propulsion is the only case for which the standard deviation has increased compared to 2005.

5415 with sinkage and trim fixed to the dynamic values have comparison errors below 3 %D and standard deviations around 4 %D. In view of the fact that the only difference between 3.1a and 3.1b is the Reynolds number (apart from a very small difference in sinkage and trim), the difference is large, but the statistical basis is too small for a comparison. For the free 5415 in 3.2 both the mean error and

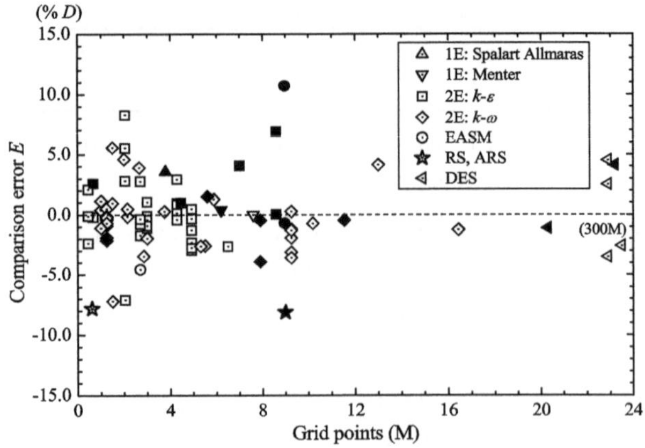

Fig. 2 Comparison error of all resistance submissions vs. grid size (turbulence model parameter)

the standard deviation seem to depend strongly on the Froude number. The best results are obtained at $Fr = 0.28$, where the water just clears the transom. For this condition the mean error is practically zero and the standard deviation among the 6 submissions is 2.1 $\%D$.

Table 4 shows the statistics for all cases, and indicates the accuracy obtainable for each case. Even more interesting is however the information found on the last line: the mean error and the mean standard deviation (weighted by number of entries) for all cases. The mean error for all computed cases is practically zero; only -0.1 $\%D$, while the mean standard deviation is 2.1 $\%D$; a surprisingly small value. In the 2005 Workshop the mean error of all 40 submissions was in fact equally small: 0.1 $\%D$, while the mean standard deviation was 4.7 $\%D$. While the distribution between "simple" and "difficult" cases is not the same in the two Workshops, it seems safe to conclude that the scatter has been reduced considerably. In fact, even the largest standard deviation in the present computations (cases 3.1b and 3.2) is smaller than the mean standard deviation in 2005.

Figure 2 shows the comparison errors for all submitted cases versus grid size. For each submission the turbulence model is denoted by a symbol. It is seen that about 90 % of all computations are made with grids smaller than 10M cells. The scatter within this range seems to be significantly larger than for the larger grids. However, this is mainly caused by the large scatter of the self-propulsion submissions (represented by filled symbols), so if these are excluded, and only towed resistance is considered, there is no error decrease above 3M grid points. All points seem to be within approximately $+/-4$ $\%D$. Not even the very large grid at 300M cells (moved into the figure and marked) shows any significant improvement; it is slightly below 3 $\%D$. However, below 3M cells the maximum errors increase to about 8 $\%D$.

There is a large number of entries for the 2-equation models and the results are generally good. For the others there are rather few entries. The relatively poor result for the more advanced methods is a surprise. However, EASM and RS, ARS suffer from one bad point for a very coarse grid and two bad points computed for the self-propulsion cases, which may be more difficult than the towed cases. The three very good results for the Menter model were obtained with the same code and user.

It would be interesting to investigate the performance of each model in different ranges of grid density, but that has not been done so far.

There is a large number of entries for the 2-equation models and the results are generally good. For the others there are rather few entries. The relatively poor result for the more advanced methods is a surprise. However, EASM and RS, ARS suffer from one bad point for a very coarse grid and two bad points computed for the self-propulsion cases, which may be more difficult than the towed cases. The three very good results for the Menter model were obtained with the same code and user. It would be interesting to investigate the performance of each model in different ranges of grid density, but that has not been done so far.

5.2 Self-propulsion

Self-propulsion results were requested only for the KCS hull and only at one Froude number: 0.26. In Case 2.3a the hull was kept fixed in the zero speed attitude, while in 2.3b the hull was free to sink and trim. Experimental data are available from NMRI for a 7.3 m hull in 2.3a and from FORCE for a 4.4 m hull in 2.3b. The NMRI hull was without a rudder, while a rudder was fitted to the FORCE hull.

In 2.3a computations were requested for the model at the ship point, i.e. the hull was towed to account for the larger skin friction at model scale compared to full scale. This force, the skin friction correction, SFC, was pre-computed and was the same as in the measurements. In the experiments the thrust T, was adjusted by varying the rpm, n, such that $T = R_T(SP) - SFC$, where $R_T(SP)$ is the resistance in self-propulsion. Most of the participants did the simulations in this way, i.e. the force balancing was automatically achieved by the flow code. An alternative was to avoid the balancing and use the measured rpm in the simulation. In the first case the achieved n was requested, while in the second case the resulting towing force $R_T(SP) - T$ was to be reported. In 2.3b computations were carried out for the model point, so no towing force was applied, but the balancing was carried out in the same way as in 2.3a.

The dependence on the grid density of the thrust coefficient, K_T, torque coefficient K_Q, achieved n for force balancing and towing force $R_T(SP) - T$ for given n are plotted in the Proceedings, but cannot be presented here due to the space limitation. There is a clear difference in scatter between the three predictions in the range 10–24M cells and those below 10M. For K_T, K_Q and n the maximum scatter in the upper range is around +/−7 %, 5 % and 2 %, respectively, while in the lower range it is within 19 %, 18 % and 6 %. For the towing force $R_T(SP) - T$ there are very few entries and the largest error is for an 11.5M grid. All quantities but n have considerably larger errors than resistance.

Of more interest is perhaps the difference between the actual and modeled propellers and between the force-balanced and fixed rpm cases. Difficulties of handling the free-to-sink-and-trim case may be revealed by comparing 2.3a and 2.3b, so the available set of results may be cut in different directions. To get a quantitative base

Table 5 Error statistics, Cases 2.3a and 2.3b

Items (No. Entries/Total)	K_T		K_Q		n		$R_T(SP) - T$	
	$E\ \%D$ mean	$\lvert E\ \%D\rvert$ mean	$E\ \%D$ mean	$\lvert E\ \%D\rvert$ mean	$E\ \%D$ mean	$\lvert E\ \%D\rvert$ mean	$E\ \%D$ mean	$\lvert E\ \%D\rvert$ mean
Actual prop. (9/17)	3.3	4.1	−1.4	2.9	−2.1	2.1	−7.8	7.8
Modeled prop. (8/17)	−2.4	6.5	−3.9	8.1	1.6	2.8		
Given SFC (12/17)	−0.2	6.0	−2.6	6.5	0.4	2.6		
Given n (5/17)	2.4	3.2	−2.6	2.6	–	–	−7.8	7.8
Case 2.3a (14/17)	−0.6	5.0	−4.6	5.1	0.6	2.3	−7.8	7.8
Case 2.3b (3/17)	6.2	6.3	6.7	6.7	−0.3	3.6	–	–
Mean (Case 2.3a&b)	0.6	5.2	−2.6	5.4	0.4	2.6	−7.8	7.8
Mean σ (Case 2.3a&b)	7.0		6.0		3.1		8.7	

for these comparisons Table 5 has been prepared. Here actual propeller results may be compared with those from modeled propellers, computations with a given SFC with those with a given n, and the fixed attitude results from 2.3a with the free attitude results from 2.3b. The comparisons are made in terms of the mean error E_{mean} and the mean absolute error $|E|_{mean}$, both in per cent of the experimental data. A standard deviation is not meaningful, except in the comparison between 2.3a and 2.3b, since the other comparisons include two cases.

There is a clear trend of smaller scatter for the actual propellers in K_T, K_Q and n (for $R_{T(SP)} - T$ there are only actual propellers represented). All three quantities have a smaller $|E|_{mean}$ for the actual propellers than for the modeled ones, and the difference is particularly large K_Q. For the mean error E_{mean} there is no clear trend. The actual propeller exhibits a considerably smaller error in K_Q, but for K_T and n the errors are slightly larger.

The scatter in the K_T and K_Q results is quite different between a given towing force and a given rpm. $|E|_{mean}$ for given n is only half of that for given towing force, while the mean signed error E_{mean} is larger for K_T. The most surprising result here is the large over prediction of the towing force for given rpm. If n is given, the towing force is significantly over predicted, while if SFC is given (and the forces balanced) the rpm is computed very well (see the relatively small values of E_{mean} and $|E|_{mean}$ for n). If the propeller is relatively lightly loaded a small (percentage) change in n may correspond to a relatively large (percentage) change in trust and a corresponding large change in towing force to acquire force balance.

It is seen in Table 5 that the small number of results for 2.3b (only 3) makes it very difficult to draw conclusions concerning the differences in accuracy between the fixed and free cases. E_{mean} and $|E|_{mean}$ for all quantities have been computed and presented in the table, but we will refrain from drawing any conclusions.

The last two lines of Table 5 are the most interesting ones. They present the mean values of all self-propulsion submissions, i.e. a weighted average of the results in 2.3a and 2.3b. These numbers may give a general indication of the accuracy obtainable in self-propulsion predictions. The last line gives the weighted mean of the standard deviations in the two cases. For K_T the mean error is 0.6 $\%D$ and the mean standard deviation 7 $\%D$ and the corresponding values for K_Q are -2.6 $\%D$ and 6 $\%D$, respectively. The predicted n for a given SFC has a mean error of 0.4 $\%D$ and a standard deviation of 3.1 $\%D$, while the numbers are larger for the towing force for given n: -7.8 $\%D$ and 8.7 $\%D$, respectively.

5.3 Wave Pattern

Wave pattern predictions were reported for cases 1.1b (KVLCC2), 2.1 (KCS) and 3.1a, b (5415). The hulls represent completely different ship types and Froude numbers, so the capability of the codes to predict the free surface was tested over a wide range of possibilities. Several different graphs were used to evaluate the performance of the codes. A general overview was provided in the wave contour plots, where the wave height was given in a region surrounding the hull. Wave cuts at three distances from the center plane were also presented for all hulls. These cuts enabled a very detailed comparison between computed and measured waves, since the measured data were presented in every plot.

Since the predicted waves are strongly dependent on the grid density near the surface every participant was requested to provide the following grid information: number of grid points per fundamental wave length along the waterline, number of grid points in the transverse direction on the surface at midship and step size in the vertical direction near the hull at midship. This information was plotted in a graph that is presented after the wave figures for each case in the Proceedings.

Due to the space limitation only one example will be given here, namely the most challenging one: case 1.1b. The Froude number for KVLCC2 is quite low, 0.142, which means that the fundamental wave length $2\pi Fr^2$ is only $1/8L_{PP}$, so a large number of cells are required to get a sufficient number of cells per wave length. Small cells are also required in the vertical direction, since the maximum wave height is less than 1 % of L_{PP}.

Wave contours for KVLCC2 are presented in Fig. 3. The measurements by MOERI reveal a complex wave pattern with very short waves essentially located at the edge of the Kelvin wedge. For a hull of this type with pronounced shoulders four wave systems should be expected: one from the high pressure regions at each end of the hull and one for each shoulder. However, the speed is so low in this case that no waves seem to be generated near the stern. The dominating wave system is that from the bow, but close inspection also reveals a more weak system originating at the forward shoulder and merging with the bow system after a short distance.

The best results were obtained by ECN/CNRS-ISISCFD, which is an unstructured grid solver with surface capturing and the VOF technique. The results, as seen in Fig. 3, reveal all the details of the wave pattern. In fact, the computations display the generated shoulder wave system more clearly than the experiments. It

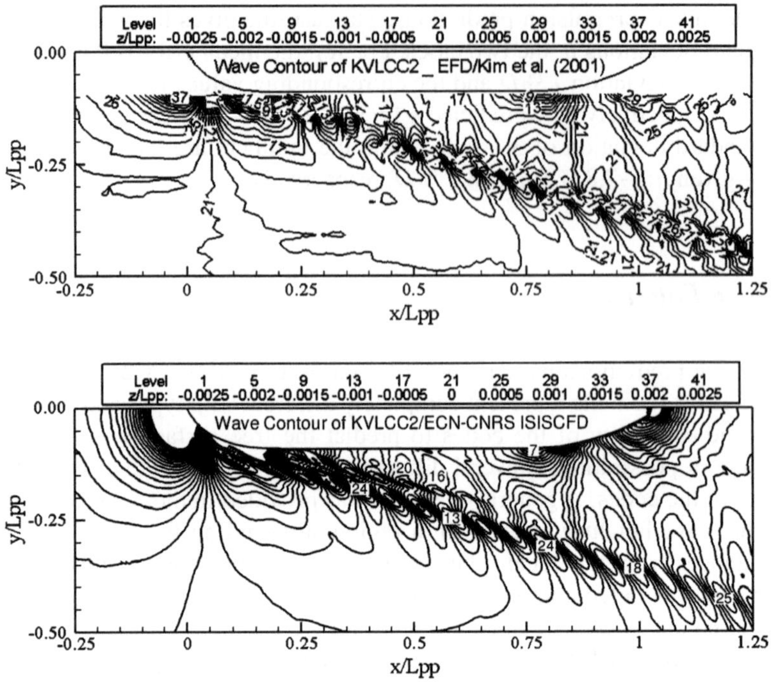

Fig. 3 Wave pattern around the KVLCC2 at $Fr = 0.142$. *Top*: measurements from MOERI, *bottom*: computations by ECN/CNRS-ISISCFD

is seen inside the main system from the bow and merges with the latter around $x/L_{PP} = 0.75$. ISISCFD has a newly implemented grid adaptation technique where the original grid is refined in several steps and concentrated in regions where a large grid density is required. The total number of grid points was 5.5M, but grid convergence was reported even with 2.7M grid points. The grid density plot (not shown here) shows that along the hull the step size has two peaks, like in most grid plots, one at the bow and one at the stern. In this case the maximum grid density at the bow is around 200 points per wave length (ppwl), while at the stern the density is somewhat lower, around 150 ppwl. Along the main part of the hull the density is around 70 ppwl. There is an interesting variation in the grid density in the transverse direction. Like in all methods the transverse grid density is very large close to the hull, in order to resolve the boundary layer, so there is first a rapid drop in ppwl, moving outwards. In most methods this drop is gradually reduced and a relatively smooth curve is obtained from the hull towards the outer edge of the free surface domain. However, in this case the adaptivity created a peak in the region $y/L_{PP} = 0.15–0.25$. This is where the bow wave system passes $x = 0$, where the grid density is reported. In the vertical direction the step size was reduced to about $3\times 10^{-4} L_{PP}$ close to the surface, which corresponds to about 20 cells per maximum wave height.

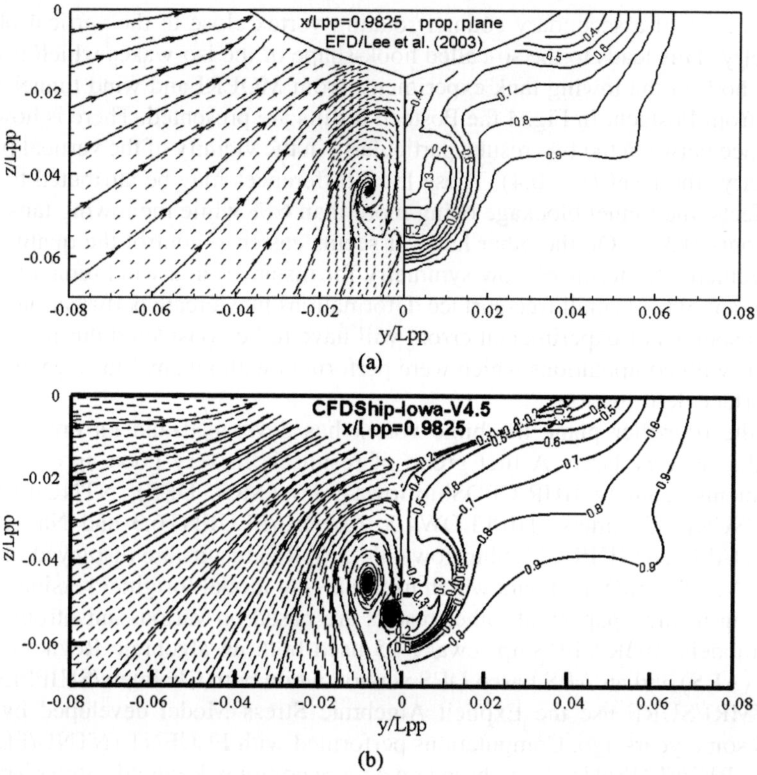

Fig. 4 Cross-flow vectors, 2D 'streamlines' and axial velocity contours at the propeller plane of the KVLCC2. (**a**) Measurements at Postech [10]; (**b**) Computations by IIHR-CFDShip Iowa-V.4.5 (ARS)

5.4 Local Flow (Steady Case)

A large number of local flow results were requested at the Workshop. Contours of axial velocity and cross-flow vectors were to be provided for all three hulls both under steady (all hulls) and unsteady (5415) conditions, even with an operating propeller (KCS). This information was requested for one or more cross-planes. Velocity profiles in the propeller plane were also asked for. Turbulence data at the propeller plane were requested for KVLCC2 and 5415. For KVLCC2, wake fraction contours and limiting streamlines were to be reported and surface pressure distributions should be given for KVLCC2 and KCS.

Here only one example can be given, and to be able to compare with previous workshops we select the flow at the propeller plane of the KVLCC2. The flow at model scale around this hull is characterized by the gradual development of an intense stern bilge vortex which creates a strong distortion of the axial velocity iso-contours at the propeller plane. See Fig. 4. This distortion is due to the transport of low momentum fluid from the vicinity of the hull to the center of the flow field under the action of the longitudinal vortex. Under the main vortex, one can guess the existence of a secondary counter-rotating vortex close to the vertical plane of symmetry. This leads to the so-called hook-shape of the iso-wakes which is clearly visible both in the towing tank experiments from MOERI and wind tunnel experiments from Postech. In Fig. 4 the Postech results are presented. There is however a difference between the two results particularly in the vicinity of the vertical plane of symmetry (the level $U = 0.4$). These local differences may be attributed to blockage effects, the tunnel blockage being more than 6 % while the towing tank blockage is only 0.3 %. On the other hand, it seems easier to control the quality of the measurements (in terms of flow symmetry for instance) in a wind tunnel than in a towing tank where small free-surface deformations may create perturbations. These various sources of experimental errors will have to be considered during the comparisons with computations which were performed without any blockage effect and free-surface deformation.

At the propeller plane the bilge vortex has developed and its impact on the iso-wakes is very large. A first group of results is in very good agreement with experiments, namely IIHR/CFDShip-Iowa-V4.5 (ARS, DES), NTNU/FLUENT, NMRI/SURF, Chalmers/SHIPFLOW4.3, MARIC/FLUENT6.3 and NavyFOAM/NavyFOAM. The IIHR/CFDShip-Iowa-V4.5 (ARS) results are displayed in Fig. 4. Except NavyFoam/NavyFoam which uses a $k - w$ model (original version of 1998 according to their paper), all other results are based on various anisotropic turbulence models. IIHR/CFDShip-Iowa-V4.5 used both an algebraic Reynolds Stress model (ARS) and an ARS based DES version while both Chalmers/SHIPFLOW4.3 and NMRI/SURF use the Explicit Algebraic Stress Model developed by ECN-CNRS some years ago. Computations performed with FLUENT (NTNU/FLUENT and MARIC/FLUENT6.3) are based on a more complex Reynolds Stress Transport Model which solves additional transport equations for the Reynolds Stress components. It is interesting to notice that all these results agree better with the towing tank than with the wind-tunnel experiments. The main stern bilge vortex is very accurately captured and the hook-shape of the iso axial velocity contours is very well reproduced. A second counter-rotating vortex, hardly visible in the experiments, is present in all these computations. In that region, the agreement between the best solutions and the experiments is less good. One can also notice that the ARS based DES solution contains more intense longitudinal vortices, a characteristic already observed in other test cases (case 3.1a for instance).

On the other hand, linear eddy viscosity models without ad-hoc rotation correction under estimate the intensity of the bilge vortex (ECN-BEC/Icare, HSVA/FreSCo+, IST-MARIN/PARNASSOS-SST, VTT/FINFLO). A noticeable exception is NavyFoam, based on the Wilcox's $k - w$ 1998 model, which gives a prediction similar to that obtained with algebraic Reynolds stress model. As this model is not as widely used as the SST model for example, and this peculiar performance needs to be further validated by other flow solvers. IST/MARIN presents some good predictions for the nominal velocity obtained with linear eddy-viscosity model with rotation correction or with a linear turbulence model. The improvement obtained by those ad-hoc modifications seems to be limited only to the mean velocity field at propeller plane. In particular, the recirculation region seems to be extended more upstream.

It is well known that turbulence anisotropy is an additional source of longitudinal vorticity production. In fact the turbulence anisotropy acts as a direct source term in the transport equation of the longitudinal vorticity. Having the normal Reynolds stresses available at the propeller plane makes possible a detailed verification of the amount of measured anisotropy in the plane. If one compares the relative values of the normal turbulent stresses, one can notice a strong anisotropy inside the characteristic hook shape found in the iso-axial velocity contours. For instance, $\max(uu) = 0.016$, $\max(vv) = 0.007$ and $\max(ww) = 0.008$ while $\max(k) = 0.016$. Most of the codes using explicit anisotropic turbulence models are able to predict with a reasonable agreement the turbulence structure at this cross-section. For instance, NMRI/SURF finds $\max(uu) = 0.014$, $\max(vv) = 0.008$ and $\max(ww) = 0.010$. The relative weights of the respective normal turbulent stresses are correctly predicted by DES, RSTM and EASM turbulence models while the linear isotropic models fail to reproduce the measured turbulence characteristics. Therefore, instead of using the right mechanism to enhance the longitudinal vorticity production, correction factors are used to limit the production of turbulence and consequently, to reduce locally the level of turbulent viscosity. This is illustrated by comparing the normal turbulent stresses computed by IST-MARIN/PARNASSOS(SST) to IST-MARIN/PARNASSOS(RCSST), for instance. One cannot see any significant difference on the normal turbulent stresses while the iso axial velocity contours differ. If one compares also the turbulent normal stresses predicted by the Reynolds Stress Transport model implemented in Fluent, one can notice a remarkable agreement between NTNU/FLUENT and MARIC/FLUENT6.3. Both organizations found $\max(uu)$ around 0.01, $\max(vv)$ around 0.004 and $\max(ww)$ around 0.005, results which are consistently smaller than the experiments. This is again an illustration of a consistent trend associated to a specific turbulence modeling, independently from the grid (which has to be fine enough) and from the user of the solver (who has to be experienced enough. . .).

This is also the first time that DES results are available for the KVLCC2. The spatial distribution of uu is organized around two peaks of uu with maximum values around 0.02, greater than the measurements supposed to be around 0.016. The same structure in two peaks is observed for vv and ww with maximum values around 0.003 and 0.004, respectively. The turbulence anisotropy is therefore more pronounced than what is observed with the anisotropic non-linear turbulence closures. This is probably related with the more pronounced longitudinal vorticity which is found in the DES computations. Although the boundary delimitating full RANSE and full LES formulations is not known, one can suspect that this effect is due to the LES formulations by comparing with the normal turbulent stress distribution obtained with the ARS model.

The turbulent shear stresses uv and uw were also measured and can be used to perform a detailed assessment of the computations. The agreement of all the computations is reasonable for uv except for the DES computations which again do not reproduce the measured spatial distribution. For uw, most of the contributors find a zone of $uw > 0.002$ which is consistently smaller than what is observed in the experiments, except the DES closure implemented in CFDShip-Iowa which is in good agreement with the experiments.

Compared to the situations in 2000 and 2005, one can observe that much progress has been made towards consistent and more reliable computations of afterbody flows for U-shaped hulls. The intense bilge vortex and its related action on the velocity field is accurately reproduced by a majority of contributors employing very similar turbulence models implemented in different solvers and on different grids. The debate on the relative importance of discretization vs. modeling errors opened in the mid-nineties should now be closed by the acknowledged prominent role played by turbulence anisotropy as long as a reasonable grid is used. From that point of view, around 3 million points are enough to assess the turbulence closures without any significant pollution from discretization errors. The turbulence data confirm that the turbulence anisotropy is large in the propeller disk and more specifically in the core of the bilge vortex. Explicit Algebraic Stress and Algebraic Reynolds Stress Models reproduce satisfactorily the measured structure of the turbulence and appear to be, up to now, the best answer in terms of robustness and computational cost for this specific flow field, compared to RSTM or DES strategies. Having recourse to Delayed DES will probably strongly improve the present DES results. A more detailed analysis of the turbulence characteristics remains to be done through the use of the Reynolds stress anisotropy tensor, anisotropy invariant maps and the analysis of the turbulent kinetic energy budget.

5.5 Heave and Pitch

Several seakeeping cases were included in the workshop. Test cases 1.4a,b and 2.4 are for pitch and heave in regular head waves for KVLCC2 and KCS, restrained in surge. In test case 1.4c for KVLCC2 the surge is also released. Test case 3.5 is forward speed diffraction for 5415 fixed at dynamic sinkage and trim, while case 3.6 is free roll decay for 5415 at the same attitude. These test cases have not been included in previous CFD Workshops, except test case 3.5, which was included at CFD Workshop Tokyo 2005.

New model tests were conducted for KVLCC2, without appendages, rudders, and propellers. The tests were carried out for all three 1.4 cases (a,b,c). The free surge tests were conducted in the Osaka University's towing tank for a 1/100 scaled model ($L_{PP} = 3.2$ m) using a spring to attach the model to the towing carriage. Fixed surge tests were conducted with the same model in the INSEAN 220 × 9 × 3.5 m towing tank and for a larger model ($L_{PP} = 5.5172$ m) at NTNU.

For cases 1.4a–c calm water, four experimental data sets are available which show different values for resistance, sinkage, and trim. Unfortunately, uncertainty of the data was not studied so the origin of the differences cannot be assessed. For the heaving and pitching motions the surge, neglected in 1.4a and b, may be important, since data for surge motion shows that its amplitude increases with wavelength, reaching up to 50 % of the wave amplitude for the range of wavelengths tested. For case 1.4a, there was no phase reference recorded for the data, so phases are not used for CFD error studies. Also, for test case 1.4b the surge motion was partially constrained by a spring system, and as a result the measured resistance amplitude is not used for comparison.

The CFD simulations of cases 1.4a–c included 12 submissions from 5 institutions with the number of grid points of 0.3–4.73M. Verification studies are performed by one submission for case 1.4a in calm water and waves. Assessment of CFD predictions for seakeeping separated capability for 1st order vs. higher order terms. The 1st order quantities are resistance calm water, 0th and 1st harmonic of resistance, and 1st harmonic of heave and pitch. The higher order quantities are calm water sinkage and trim and 0th harmonic of motions. CFD has achieved the prediction of 1st order quantities for cases 1.4a, b, c with the average error of 19 $\%D$ and 14 $\%D$ including and excluding the error for resistance amplitude, respectively. For higher order quantities, CFD overall prediction error is 33 $\%D$. For steady and 0th harmonic of resistance, the average error is around 8 $\%D$ and 18 %, respectively, while the 1st harmonic amplitude and phase are predicted by 31 $\%D$. For motions, the prediction error is 10 $\%D$ for steady and 33 $\%D$ for 0th harmonic while it is around 16 $\%D$ for 1st harmonic amplitude and phase. Therefore, for resistance, the largest error values are observed for the 1st harmonic amplitude and phase, followed by 0th harmonic amplitude and then steady. For both heave and pitch motions, the largest error values are observed for the 0th harmonic amplitudes followed by 1st harmonic amplitude and phase and then steady.

The number of grid points seems to have an obvious effect on both motions and resistance results. For most conditions, the smallest errors are for the submission with largest number of points. The other submissions are usually with higher errors based on how coarse their grids are. It may be that all codes would reach the small level of error, if using the finest grid. The smallest error averaged over amplitudes and phase for resistance is 11.19 $\%D$ for case 1.4c with $\lambda/L = 1.1$ for CFDShip-Iowa with 4.73M grid points. Also the smallest error averaged over amplitudes and phase for motions is 2.66 $\%D$ for case 1.4c with $\lambda/L_{PP} = 1.1$ for CFDShip-Iowa with 4.73M grid points.

Six submissions were available for test case 2.4, only one of which included verification studies. The mount used in the experiments allowed small surge motions which might have affected 1st harmonic resistance. Average prediction error was smaller for 1st order terms (17 $\%D$ and 8 $\%D$ including and excluding resistance amplitude) compared to higher order terms (83 $\%D$). For steady and 0th harmonic of resistance, the average error is around 6 $\%D$ and 21 %, respectively, while the 1st harmonic amplitude and phase are predicted by 40 $\%D$. For motions, the prediction error is 5 $\%D$ for steady and 83 $\%D$ for 0th harmonic while it is around 6 $\%D$ for 1st harmonic amplitude and phase. Increasing the number of grid points seemed to reduce only the resistance error.

5.6 Roll Decay at Forward Speed

The model-scale test for 1/46.6 scale 5415 bare hull with bilge keels free to roll-decay advancing in calm water was performed in the IIHR towing tank [13]. The

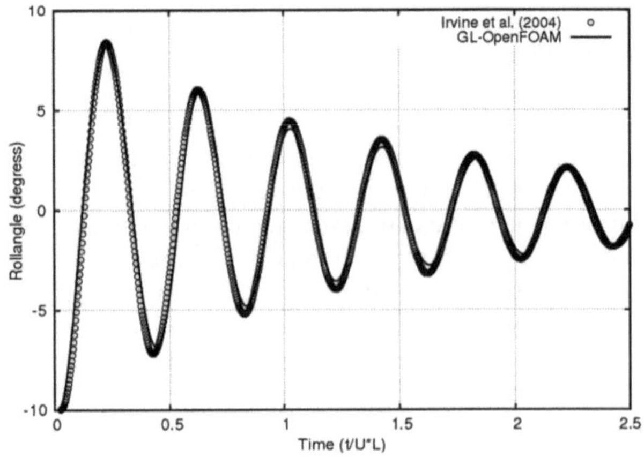

Fig. 5 Time history of roll angle (*Open circle*: Experiments; *Solid black line*: CFD (GL&UDE-OpenFOAM))

flow conditions were $Re = 2.56 \times 10^6$, $Fr = 0.138$, sinkage $= 2.93 \times 10^{-4} L_{PP}$, trim $= -3.47 \times 10^{-2}$ degrees and initial roll angle $\phi = 10$ degrees. Data were procured for the forces and moments using a strain gage load cell, the unsteady ship roll motion using a Krypton Motion Tracker, the unsteady wave elevation on the starboard side using four servo wave probes, and unsteady velocities at $x/L_{PP} = 0.675$ in a region near bilge keels using a 2D PIV system.

Four organizations (ECN-ICARE, ECN/CNRS-ISISCFD, GL&UDE-OpenFOAM, and SSRC/Univ of Strathclyde-Fluent 12.2) contributed for this test case, all using URANS methods. All the submissions show non-linear oscillations for the total resistance as observed in the experiments. The mean resistance is predicted within 10 $\%D$ of the experimental data. The amplitude and period of the roll motions are predicted very well within 0.85 $\%D$ as shown in Fig. 5.

The ECN-ICARE simulation on a 0.8M grid fails to predict the Kelvin wave pattern and the development of the wave troughs and crests due to the roll motion. Likewise, the SSRC/Univ of Strathclyde-Fluent 12.2 predictions on a 3M grid also show poor Kelvin wave predictions, and this was claimed to be due to grid resolution issues.

On the other hand, the ECN/CNRS-ISISCFD predictions on 4.9M grid shows overall good agreement for the Kelvin wave pattern and the development of wave troughs and crest at the shoulder, but the waves are closer to the hull compared to the data and dissipated away from the hull. Overall, the wave elevation predictions improve with grid resolution, but for such low *Fr* (which exhibits a small Kelvin wavelength) even larger grids are required to accurately predict the wave elevation pattern.

The coarse grid ECN-ICARE simulation over-predicts the boundary layer thickness and under-predicts the cross flow velocities by 30 $\%D$. However the general trends in the cross-flow pattern at the different roll phases are well predicted. This suggests that the generation of the bilge keel vortices is predicted well, but the vortex strengths are under-predicted and significantly diffused compared to the experimental data. ECN/CNRS-ISISCFD predictions on a finer grid compare very well

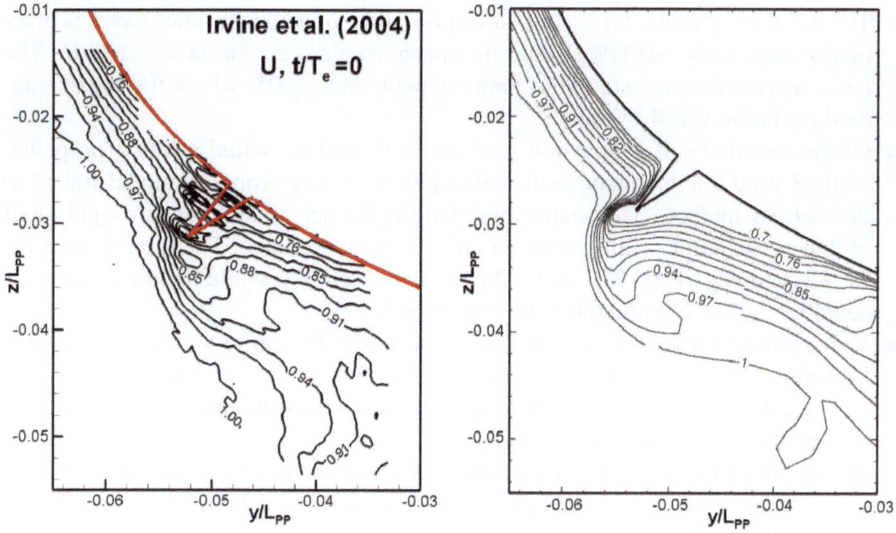

Fig. 6 Contours of U velocity at $x/L_{PP} = 0.675$ during second cycle of roll decay ($t/Te = 0$) (**a**) Experiment (IIHR, Irvine et al. 2004); (**b**) Computations by ECN/CNRS-ISISCFD

with the experiments, where the minimum and maximum velocities compare within 6 $\%D$, see Fig. 6. The cross-flow predictions are slightly better than the axial velocity, where latter shows over-prediction of the transport of low momentum fluid away from the bilge keel. This suggests that the vortex advection due to roll motion is predicted well, but the bilge keel vortex inception is not predicted accurately. Even finer grids near the bilge keel are required to capture the vortex inception accurately.

Overall, the results show that the force and roll motion predictions are not significantly dependent on grid resolution, but it is more important for the wave elevation and flow predictions. A further point to note is that the anisotropic turbulence models do not show significant improvement over the isotropic models for the global variable predictions. This is because the vortex generation is imposed by the geometry of the bilge keels, and thus the turbulence models do not influence the flow predictions. This is different from the KVLCC2 test case 1.1a or 5415 test cases 3.1 and 3.5, where the vortices are advected and the anisotropic turbulence models show improved predictions.

6 Conclusions

The Gothenburg 2010 Workshop on numerical Ship Hydrodynamics was huge effort by a large number of people. 33 groups participated and computed one or more of the 18 test cases for the three hulls. The results represent the state-of-the-art in computational hydrodynamics at present. For the areas covered in the present paper the main conclusions are:

- The mean error (data(D) – simulation) for all computed resistance cases is practically zero; only $-0.1\ \%D$, and the mean standard deviation is $2.1\ \%D$. The latter represents a considerable improvement since 2005, where the mean standard deviation was $4.7\ \%D$.
- Of the reported self-propulsion predictions 9 used an actual rotating propeller, while 8 used a hybrid approach with a potential flow propeller model linked to the viscous method. The scatter between the former predictions is roughly half of that of the latter, while the mean error is about the same. The total mean error for K_T and K_Q is $0.6\ \%D$ and $-2.6\ \%D$, respectively, while the mean standard deviation is $7.0\ \%D$ and $6.0\ \%D$, respectively.
- Very accurate predictions of the wave pattern over the entire discretized free surface were obtained in the most accurate solutions for all three hulls. Even for the most challenging case, at $Fr = 0.142$, all details of the wave pattern were captured by a method with adaptive gridding.
- The details of the nominal wake of the tanker hull can be predicted very well with methods using anisotropic turbulence models. This holds both for mean velocity and turbulence. The mean velocity distribution can also be captured with isotropic models with rotation correction, but with an erroneous distribution between the normal Reynolds stresses. DES seems to over-predict the anisotropy, thereby exaggerating the bilge vortex strength and the "hooks" in the wake contours.
- Three new sets of data were provided for the workshop in the seakeeping area. The tanker was tested in head waves free in pitch and heave at two organizations and free also in surge at a third laboratory.
- For seakeeping, the average error for steady and 0th harmonic of resistance is around $7\ \%D$ and 18 %, respectively, while the 1st harmonic amplitude and phase are predicted by $34\ \%D$. For motions, the prediction error is $9\ \%D$ for steady and $54\ \%D$ for 0th harmonic while it is around $13\ \%D$ for 1st harmonic amplitude and phase. Therefore, for resistance, the largest error values are observed for the 1st harmonic amplitude and phase, followed by 0th harmonic amplitude and then steady. For motions, the largest error values are observed for the 0th harmonic amplitudes followed by 1st harmonic amplitude and phase and then steady.
- For the roll decay the mean error in resistance was around 10 %, while the amplitude and period were predicted very well, 0.85 % on the average. Waves were predicted rather well by a method using adaptive grids and 4.9M cells, but other methods with coarser grids failed. This complicated case at low Froude number (0.138) calls for larger grids, at least if they are not adapted. This is the case also for the mean velocities in the boundary layer.

Acknowledgements The workshop was organized by a committee with six members: the authors and Dr. Emilio Campana, Dr. Suak Ho Van and Prof. Yasuyuki Toda, whose contributions are gratefully acknowledged. Very important contributions have also been made by Lu Zou, who compiled all results and prepared the Proceedings and by Dr. Alessandro Iafrati, who developed and maintained the web site. Prof. Rickard Bensow and Andreas Feymark prepared and compiled the questionnaire and Shanti Bhushan, Hamid Sadat-Hosseini and Maysam Mousaviraad contributed to the specification and analysis of the seakeeping cases. Finally, the great efforts by all workshop participants in the preparation and delivery of all computed results shall not be forgotten.

References

1. Larsson L (ed) (1981) SSPA-ITTC workshop on ship boundary layers. SSPA report 90, Gothenburg, Sweden
2. Larsson L, Patel VC, Dyne G (eds) (1991) SSPA-CTH-IIHR workshop on viscous flow. Flowtech research report 2, Flowtech Int. AB, Gothenburg, Sweden
3. Kodama Y, Takeshi H, Hinatsu M, Hino T, Uto S, Hirata N, Murashige S (1994) Proceedings, CFD workshop. Ship Research Institute, Tokyo
4. Larsson L, Stern F, Bertram V (eds) (2002) Gothenburg 2000—a workshop on numerical hydrodynamics, department of naval architecture and ocean engineering. Chalmers University of Technology, Gothenburg
5. Larsson L, Stern F, Bertram V (2003) Benchmarking of computational fluid dynamics for ship flow: the Gothenburg 2000 workshop. J Ship Res 47:63–81
6. Hino T (ed) (2005) Proceedings of CFD workshop Tokyo 2005. NMRI report
7. Van SH, Kim WJ, Yim DH, Kim GT, Lee CJ, Eom JY (1998) Flow measurement around a 300K VLCC model. In: Proceedings of the annual spring meeting, SNAK, Ulsan, pp 185–188
8. Van SH, Kim WJ, Yim GT, Kim DH, Lee CJ (1998) Experimental investigation of the flow characteristics around practical hull forms. In: Proceedings 3rd Osaka colloquium on advanced CFD applications to ship flow and hull form design, Osaka, Japan
9. Kim WJ, Van DH, Kim DH (2001) Measurement of flows around modern commercial ship models. Exp Fluids 31:567–578
10. Lee SJ, Kim HR, Kim WJ, Van SH (2003) Wind tunnel tests on flow characteristics of the KRISO 3600 TEU containership and 300K VLCC double-deck ship models. J Ship Res 47(1):24–38
11. Simonsen C, Otzen J, Stern F (2008) EFD and CFD for KCS heaving and pitching in regular head waves. In: Proc 27th symp naval hydrodynamics, Seoul, Korea
12. Olivieri A, Pistani F, Avanaini A, Stern F, Penna R (2001) Towing tank experiments of resistance, sinkage and trim, boundary layer, wake, and free surface flow around a naval combatant INSEAN 2340 model. Iowa Institute of Hydraulic Research, The University of Iowa, IIHR report No 421, pp 56
13. Longo J, Shao J, Irvine M, Stern F (2007) Phase-averaged PIV for the nominal wake of a surface ship in regular head waves. J Fluids Eng 129:524–540
14. Irvine M, Longo J, Stern F (2004) Towing tank tests for surface combatant for free roll decay and coupled pitch and heave motions. In: Proc 25th ONR symposium on naval hydrodynamics, St Johns, Canada
15. Stern F, Agdrup K, Kim SY, Hochbaum AC, Rhee KP, Quadvlieg F, Perdon P, Hino T, Broglia R, Gorski J (2011) Experience from SIMMAN 2008: the first workshop on verification and validation of ship maneuvering simulation methods. J Ship Res 55(2):135
16. Larsson L, Stern F, Visonneau M (2010) Gothenburg 2010—a workshop on numerical ship hydrodynamics. In: Proceedings, vol 2, Department of Naval Architecture and Ocean Engineering, Chalmers University of Technology, Gothenburg, Sweden
17. Larsson L, Stern F, Visonneau M (eds) (2013) Numerical ship hydrodynamics—an assessment of the Gothenburg 2010 workshop. Springer, Dordrecht, to be published

References

[illegible]

Prediction of the Transom Flow Regime with Viscous Free Surface Computations

Auke van der Ploeg and Bram Starke

Abstract A steady free-surface fitting RANS method is used to accurately and efficiently compute the flow near a ship's transom. The chosen block topology is very well suited for the prediction of such flows: viscous free-surface flow solutions have been obtained for all transom flow regimes (from completely dry, partly wetted to completely wetted) and the agreement with available measurements is good. We will show that scale effects on the flow aft of the transom can be significant: they can affect the wave length and wave amplitude aft of the transom, and can even change the transom flow regime from partly wetted to completely dry.

Keywords Computational fluid dynamics · Transom sterns · Scale effects

1 Introduction

Ship resistance and propulsion are principal aspects for the fuel efficiency of maritime transport. Modern CFD techniques in principle permit a further improvement of that efficiency by precisely predicting resistance and scale effects. Especially, an accurate prediction of viscous free-surface flows near a ship's transom is important, since this is a pre-requisite for the computation of the effect of stern wedges and stern flaps which can be used to improve the ship powering performance [4].

A (partly) wetted transom poses additional challenges on a computational method: the details of the free surface and the recirculating flow behind the wetted part of the transom, the transition between the wetted and dry regime along the transom, and the possible occurrence of a spilling breaker when the transom is close to becoming wetted. Scale effects affect the wave length and wave amplitude aft of the transom, but might become even more important when the transom flow regime changes between model and full scale.

In [10] a detailed study of the flow off a 2D transom stern was presented. For dry-transom flow, 'inviscid' computations using our RANS solver showed excellent

A. van der Ploeg (✉) · A.R. Starke
Maritime Research Institute Netherlands (MARIN), 6700 AA Wageningen, The Netherlands
e-mail: a.v.d.ploeg@marin.nl

L. Eça et al. (eds.), *MARINE 2011, IV International Conference on Computational Methods in Marine Engineering*, Computational Methods in Applied Sciences 29,
DOI 10.1007/978-94-007-6143-8_15, © Springer Science+Business Media Dordrecht 2013

agreement with nonlinear potential-flow solutions available in the literature. Viscous effects were shown to cause a substantial reduction of the trailing wave length, and a scale effect on that wave length. The transom immersion was systematically increased to investigate the limits of the dry-transom regime. A local vanishing of the longitudinal velocity at the wave surface near the first wave crest was used as an indication of wave breaking. The critical Froude number at which this happens is substantially increased by viscous effects, and much more so for model scale than for full scale. Therefore, in a range of transom Froude numbers a smooth flow will occur at full scale, but a spilling breaker just aft of the transom at model scale. It was shown that the width of this range may depend on the hull form (in 3D cases), as it results from two opposing scale effects.

For wetted transom flows, the predictions at model scale in [10] showed good agreement with experimental data for the water level at the transom and the trailing wave system. The transom immersion was systematically decreased to investigate the clearance of the transom. It was shown that at a given transom immersion the clearance is larger at full scale than at model scale. Thus it was shown that there is a range of conditions where the transom is just cleared at full scale, with a spilling breaker downstream, while the flow is still attached to the transom at model scale.

The question is whether such scale effects are similar for 3D ships. Strictly speaking, no direct comparability exists between 2D and 3D since in the 2D computations

(1) there is no effect of the waves coming from upstream;
(2) there is no decay of the waves as in 3D, so one can expect steeper waves at some distance aft in 2D computations;
(3) there is no effect of the wakes of any upstream appendages or the propulsion;

Therefore in this paper we will present viscous free-surface computations at both model and full scale for three different ships: the KRISO Container Ship (KCS), DTMB model 5415 (M5415) and the VIRTUE Container Ship (VCS) (Fig. 1). The first two test cases have extensively been used in many international workshops, most recently in the Gothenburg 2010 Workshop on Numerical Ship Hydrodynamics [11]. The last test case has been used in the EU-sponsored VIRTUE project. All three test cases are transom-stern vessels and for all three test cases experimental data at model scale is available. Some characteristics of the ships and the flow conditions are listed in Table 1. In Sect. 2 we will present our RANS method, with some emphasis on the free-surface treatment and the adopted block topology. In Sect. 3 we will present results of systematic grid-refinement studies, and comparison with available experimental data, followed by a discussion on the predicted scale effects near the transom in Sect. 4.

2 The Computational Method

2.1 RANS Solver

The method used is PARNASSOS, a code developed and used by MARIN and IST [3, 7]. It solves the discretised Reynolds-averaged Navier-Stokes (RANS) equations

Table 1 Characteristics of the three test cases

	Rn (model)	*Rn* (ship)	L_{PP} [m]	*Fn*	*T* [m]
Kriso Container Ship (KCS)	$1.40 \cdot 10^7$	$2.5 \cdot 10^9$	230	0.26	10.8
DTMB 5415 (M5414)	$1.19 \cdot 10^7$	$6.5 \cdot 10^8$	142	0.28	6.15
VIRTUE Container Ship (VCS)	$1.85 \cdot 10^7$	$2.0 \cdot 10^9$	246.4	0.272	12.5

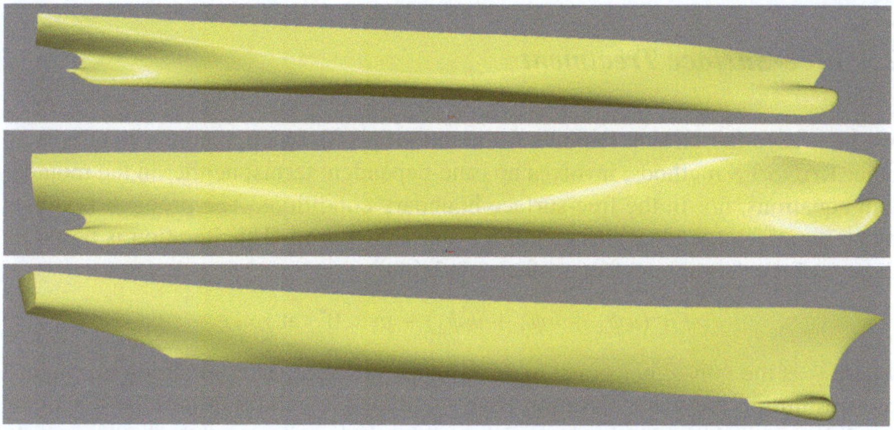

Fig. 1 Geometries for the three test cases: *top* KCS, *middle* VCS, *bottom* M5415

for a steady, 3D incompressible flow around a ship's hull. Various eddy-viscosity turbulence models are available. For the computations for the VCS and the M5415 presented in this paper, the one-equation turbulence model of Menter [6] was used. For the computation for the KCS we also used the two-equation $k - \omega$ SST model by Menter [5]. Both models were extended with a correction for the longitudinal vorticity [2].

The discretisation is of finite-difference type. All terms in the momentum and continuity equations are discretised by second or third-order accurate difference schemes. PARNASSOS can handle body-fitted, generally non-orthogonal HO-type grids, either single or multi-block structured.

For the cases studied here, the inflow boundary is located $0.5L_{PP}$ in front of the bow, and the outflow boundary was located $1.5L_{PP}$ behind the transom. Due to symmetry considerations, only the starboard side of the ships are taken into account. The lateral outer boundary is a quarter of a cylinder with axis $y = z = 0$ and radius $1.0L_{PP}$. At this boundary tangential velocities and pressure found from a potential-flow computation are imposed. Since that computation gives good results already for much of the wave pattern, these boundary conditions (although of Dirichlet type) hardly cause any wave reflection.

The momentum and continuity equations are solved in fully coupled form. Therefore, the continuity equation need not be recast in a pressure correction or pressure Poisson equation, but can simply be solved as it is. After discretisation and linearisation, the three momentum equations and the continuity equation give rise to a matrix equation containing 4×4 blocks, which is solved using preconditioned GMRES. This fully coupled solution has been found to be robust and quite insensitive to the mesh aspect ratio. This allows solving the discretized equations on extremely contracted grids close to the wall. As a result, wall functions are not necessary, not even at full scale. More details about the solution strategy can be found in [7].

2.2 Free-Surface Treatment

We have used the 'steady iterative formulation' [1, 9] which, contrary to almost all other RANS/FS methods, involves no time-dependent terms; neither in the momentum equations, nor in the free-surface boundary conditions. The problem is solved by an iterative procedure, instead of by time integration. This iteration is based on the use of the "combined free-surface condition":

$$Fn^2(u\psi_x + v\psi_y + w\psi_z) - w = 0 \quad \text{at } z = \zeta, \tag{1}$$

where ψ is the non-dimensional hydrodynamic pressure. This condition is obtained by substituting the wave elevation from the dynamic condition into the kinematic condition. Together with the dynamic condition it describes exactly the same problem as the original set of conditions; but it has the advantage of permitting a successful iterative procedure.

In the present applications, the 'balanced discretisation' derived in [9] has been used, which reduces the numerical damping of the waves to 5th order in the longitudinal step size Δx, and the numerical dispersion to 3rd order in the vertical spacing Δz. This contributes to a good accuracy of the wave pattern even at a distance from the hull.

2.3 Grids Suited for Both Dry and Wetted Transom

We use a special block topology that can handle both wetted and dry transoms, and even transoms that are partly dry and partly wetted. This topology, illustrated in Fig. 2, consists of four blocks. One block upstream of the transom contains an HO-type grid with a mild stretching towards the bow and the stern in longitudinal direction, and the usual strong contraction in wall-normal direction towards the hull in order to have y^+-values below 1 near the wall. This upstream block has a non-conformal matching with three blocks downstream of the transom, containing HH-type grids. The grid nodes in the block immediately behind the transom are both contracted towards the symmetry plane and the free surface to get sufficient resolution near the transom, also in vertical direction at a possibly wetted part of the transom.

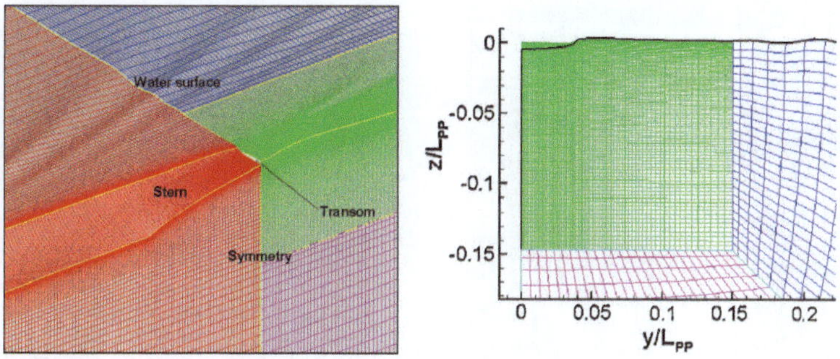

Fig. 2 *Left*: bird's-eye view of the adopted block topology. *Right*: a cross section of the computational mesh in the three blocks aft of the transom: the *black line* indicates the transom edge in the upstream block

2.4 Grid Density

The typical grid density we need for such computations is illustrated for the M5415 in Fig. 2. This grid consists of 3.7 million cells for all four blocks together. In the model-scale computations the block upstream of the transom (indicated in red in Fig. 2), consists of 290 × 120 × 52 cells in the streamwise, wall-normal and girthwise directions, respectively, adding up to a total of 1.8M cells for this block. For the full-scale computations, the number of cells in the wall-normal direction is increased (in order to keep y^+-values below 1), resulting in a total number of 4.2 million at full scale. The block immediately behind the transom (green in Fig. 2), consists of $144 \times 99 \times 99 = 1.4$M cells, and the remaining cells are distributed evenly over the two other blocks. For the other two test cases, similar grid densities are used.

3 Grid Dependence Studies and Comparison with Experiments

For all test cases, several grid dependence studies were performed and the grid dependence was shown to be limited, giving good agreement with measured wave heights, wake fields and resistance values [8, 11]. An example of such a study is shown in Fig. 3, in which both pictures show a comparison between the computed and measured wave patterns for the M5415. The computed wave elevation shown in the picture on the left was obtained using the 3.7M base grid described above, whereas the picture on the right compares the wave elevation computed on a grid that was refined by a factor of two in the main stream direction. Hence the total number of cells increased to 7.4 million. Even with the wave height multiplied by a factor of five in these figures hardly any difference can be seen between the predicted wave patterns, indicating that the basis grid resolution in the streamwise direction was sufficient. The predicted wave height corresponds well with the measured wave height.

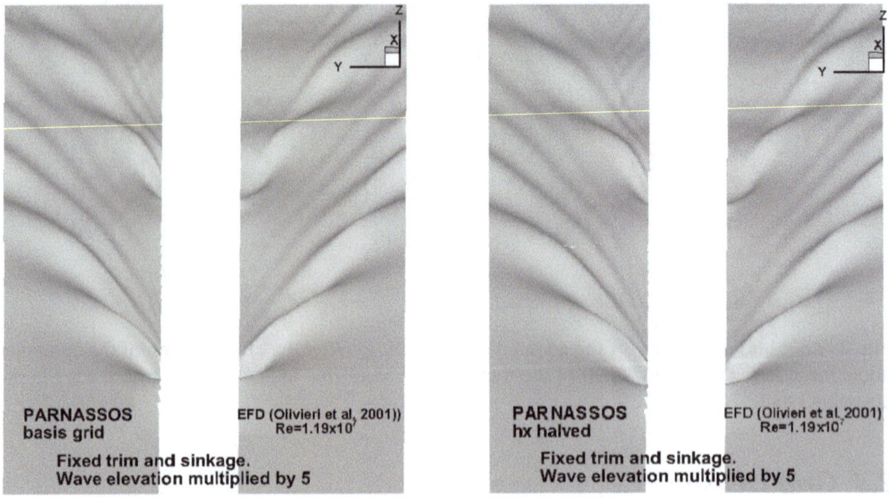

Fig. 3 Comparison between the computed and measured wave elevation for the M5415. *Left*: computed on the base grid. *Right*: computed on the base grid after halving the cell size in streamwise direction

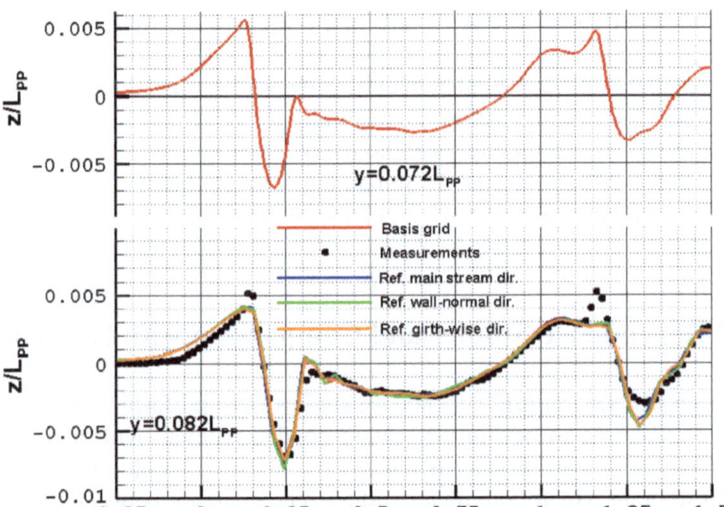

Fig. 4 *Top*: wave elevation along a longitudinal wave cut $y = 0.072L_{PP}$ for the M5415. *Bottom*: The influence of grid refinement in successive directions and comparison with experiments at a wave cut $y = 0.082L_{PP}$

Similar studies have been made for the grid resolution in the other directions, which is illustrated for a wave cut in the bottom picture in Fig. 4. At this longitudinal position the influence of grid refinement is negligible in all three directions, and

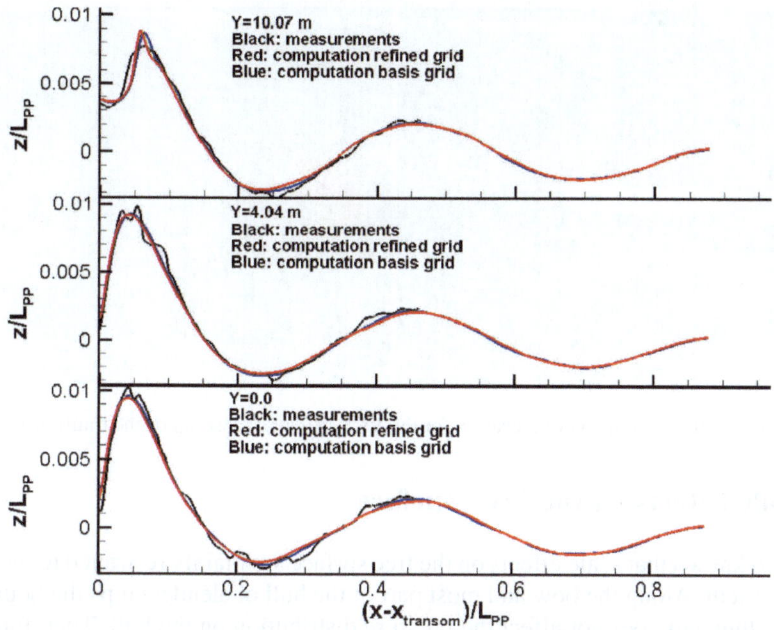

Fig. 5 Computed and measured wave elevations at three longitudinal positions aft of the transom of the VCS

very reasonable agreement is obtained with the experimental data along most part of the ship, with the exception of details near the bow and the stern. For example, the local wave top near the stern in the measurements near $x = 1.15L_{PP}$ seems not to be present in the computations. However, for a wave cut slightly closer to the symmetry plane this local peak is present in the computations as well, as is illustrated in the top picture in Fig. 4. Also the height of the computed bow wave at the wave cut $y = 0.072L_{PP}$ is closer to the measured height of this wave a the wave cut $y = 0.082L_{PP}$.

For the VCS, wave height measurements behind the transom of the model were obtained using a dedicated test rig supporting a set of line lasers in combination with a water spray curtain behind the transom that was installed on the towing carriage [8]. A comparison between measured and computed wave heights aft of the transom along three wavecuts is shown in Fig. 5. The agreement between computations and measurements is excellent, for the wave amplitude as well as for the wave length. Two sets of computational results are shown; the blue lines in Fig. 5 show results computed on the base grid, the red lines show results computed on a grid in which the mesh size in main stream direction was halved. As the blue lines and the red lines are practically indistinguishable, the grid dependence in the computed wave height is negligible.

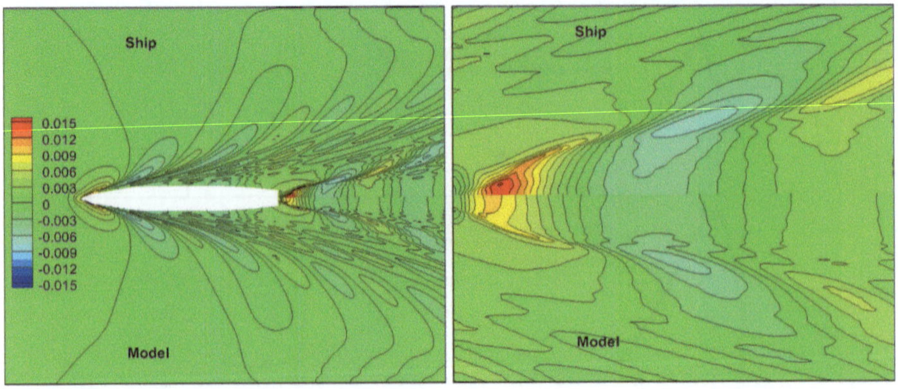

Fig. 6 Scale effects in the wave elevation for the VCS. *Right*: a close-up of the transom-region

4 Scale Effects on the Free Surface

It is well known that scale effects on the free surface are mainly restricted to the stern wave system. Along the bow and most part of the hull of slender ships the boundary layer is thin and does not affect the pressure distribution on the hull. Therefore, the wave patterns generated at model and full scale (and even for non-linear potential flow solutions) are similar, the very basis of Froude's hypothesis. Near the stern however the boundary can grow considerably (of course depending on hull form and flow conditions) thus reducing the pressure rise towards the stern and the velocity distribution in the wake. This in its turn affects the wave generation and propagation aft of the transom. All this is illustrated in Fig. 6 for the VCS. The entire bow-wave system and wave pattern along the hull are equivalent at model and at full scale. Only in the stern-wave system aft of the transom noticeable differences can be found. There viscous effects, which are larger at model scale than at full scale, result in a reduction of the wave elevation and, although not that clear from these graphs, a reduction of the transverse wave length aft of the transom. This is similar to what we have found in the 2D study discussed in Sect. 1. An illustration of the scale effect on the flow off the transom for all three test cases is given in Fig. 7. This figure shows the wave elevation aft of the transom with the predicted axial velocity in main stream direction projected on the free surface as colored contours. Black colors indicate regions with flow reversal. Figure 8 shows a more detailed study of the scale effect along a longitudinal wave cut along the symmetry plane, and comparison with experimental (model scale) data. Below we discuss the scale effects shown in Figs. 7 and 8 for each case.

KCS At model scale the flow is characterised by a region with low axial velocities in the top of the wave crest around the center line. This indicates that in reality a spilling breaker might occur, or that the flow is close to forming a spilling breaker. At full scale, however, the top of the first wave crest is located further downstream, while the velocity reduction towards the top of the wave crest is less strong than at model scale. Therefore, the flow at full

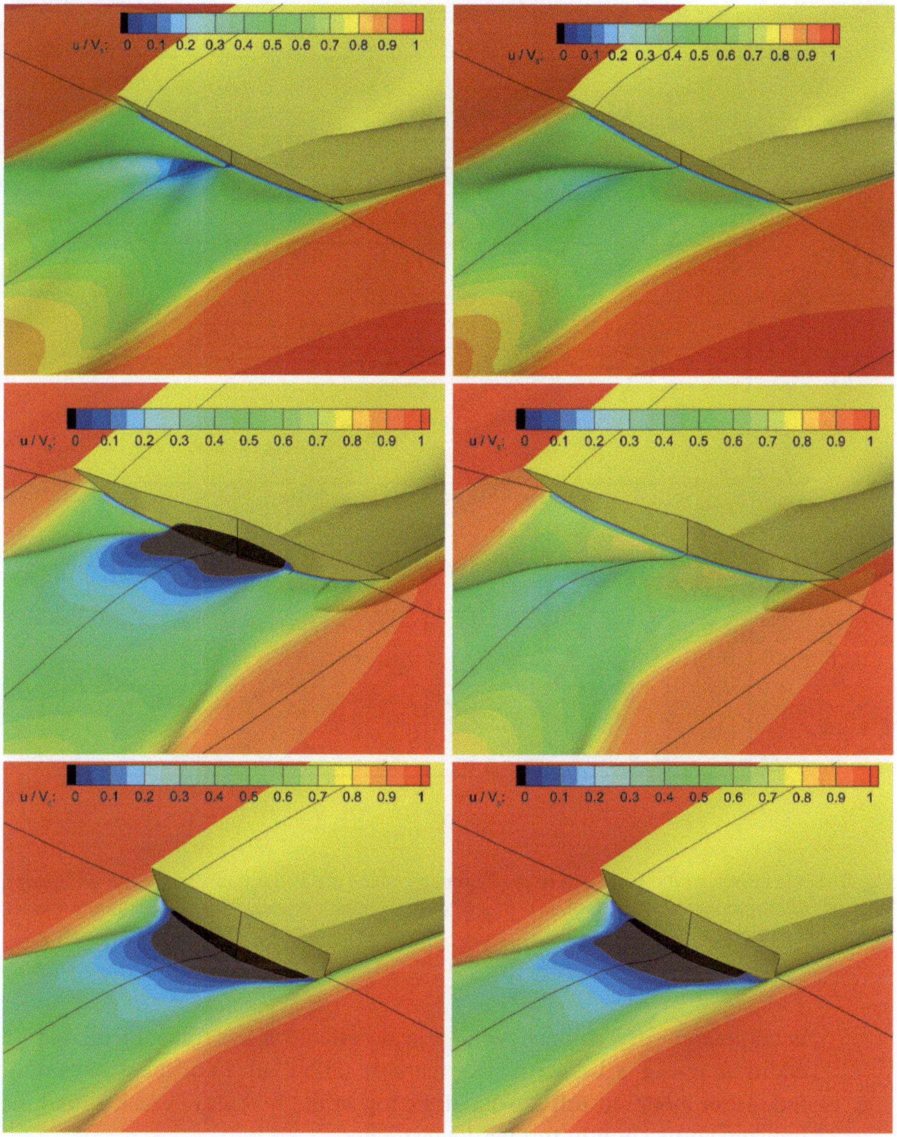

Fig. 7 Scale effects on the flow off the transom, model scale results on the *left*, full scale results on the *right*; KCS (*top*): dry at model scale, dry at full scale; VCS (*middle*): partly wetted at model scale, dry at full scale; M5415 (*bottom*): wetted at model scale, wetted at full scale

scale is further away from forming a spilling breaker than at model scale. In both conditions the transom is dry.

VCS At model scale the transom of the VCS is partly wetted, and in [11] it has been shown that the width of this region compares well with the experimental data. At full scale, however, our computations indicate that the transom

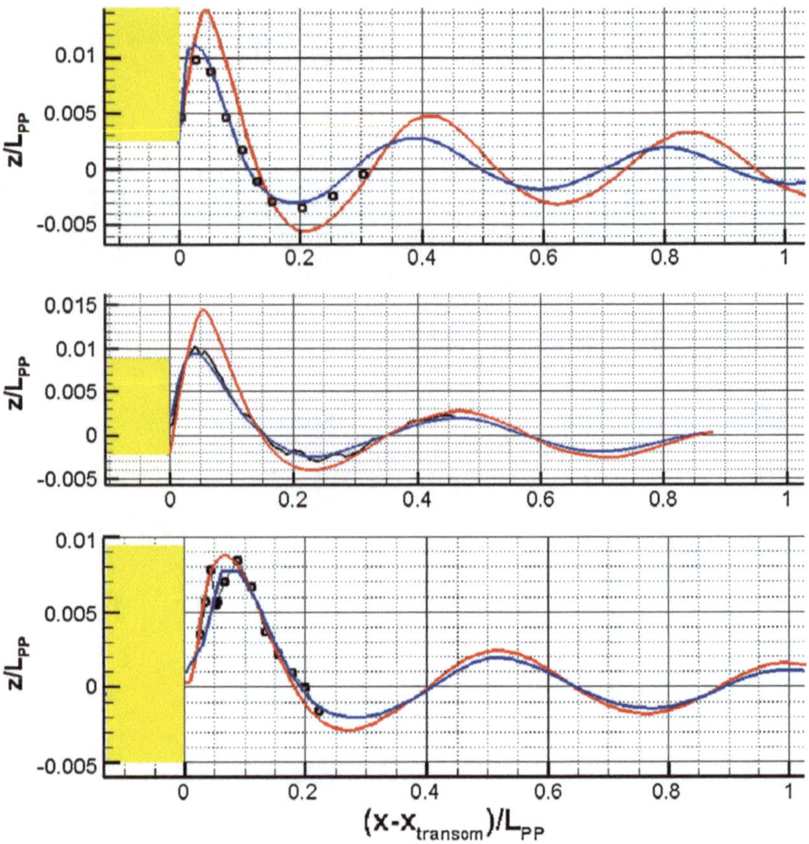

Fig. 8 Scale effects at the wave cut in the symmetry plane aft of the transom for the KCS (*top*), VCS (*middle*) and the M5415 (*bottom*). *Black*: measurements model. *Blue*: computation model. *Red*: computation ship

is completely dry, thus showing a scale effect on the transom regime. In addition, in Fig. 8 we see a clear increase (about 40 %) of the wave amplitude from model to full scale and the top of the first wave crest behind the transom shifts slightly further downstream.

M5415 The model of the M5415 has a completely wetted transom. At full scale the transom is predicted to be wetted as well, but the water level at the transom decreases from model to full scale. This is in agreement with the conclusions of our 2D study of the wetted regime, where the higher shear forces that act on the recirculation region at full scale result in a reduction of the pressure inside this region, and thus an increased clearance of the transom, compared to model scale. These effects can be seen more clearly in Fig. 9, which shows the scale effects in the pressure distribution in a cross section at the transom.

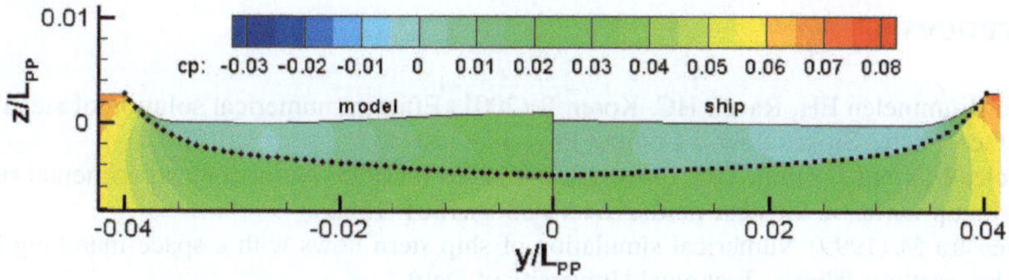

Fig. 9 Scale effects in the pressure coefficient in a cross section near the transom of the M5415. The *black dots* indicate grid points on the transom edge

5 Conclusions

With the free surface RANS method described in this paper, the viscous flow near the transom can be computed for several transom flow regimes, from completely dry to (partly) wetted. The steady iterative approach as used in PARNASSOS is very efficient to compute the free surface and the block structure as described in this paper is well suited to compute the viscous flow including the free surface accurately near the transom, since it enables local grid refinement near the wetted part of the transom.

Compared to many surface capturing methods, the surface fitting approach used in PARNASSOS has as advantage that the free surface is sharp: it is not 'smeared out' in the direction normal to the surface across several cells. The 'balanced discretisation' derived in [9] has been used, which reduces the numerical damping of the waves to 5th order which further enhances the accuracy of the computations. We have shown that for the three test cases the grid dependence in the results is limited and we get good agreement with model-scale measurements. Unfortunately, experimental data for full scale is not available. However, as mentioned earlier, we found excellent agreement between 'inviscid' computations using our RANS solver and nonlinear potential-flow solutions for the 2D test case mentioned in Sect. 1. Together with the similarity between model-scale computations and measurements this sustains the confidence in the computed scale effects:

- For all three test cases, scale effects on the free surface are mainly restricted to the stern wave system.
- For the test case for which the transom at model scale is dry, similar scale effects were observed as for the 2D test case described in [10]: the waves behind the transom become steeper, the wavelength decreases and the velocity in the top of the first wave behind the transom decreases when going from full scale to model scale. However, an additional scale effect for this 3D test case which was not observed for the 2D test case is the increase in wave amplitude from model to full scale.
- For the test case with a partly wetted transom at model scale, significant scale effects were observed: at full scale the transom is completely dry, and the wave amplitude increases significantly from model to full scale.
- The test case in which the transom is completely wetted at model scale showed a decrease of the water level at the transom from model to full scale.

References

1. Van Brummelen EH, Raven HC, Koren B (2001) Efficient numerical solution of steady free-surface Navier-Stokes flow. J Comput Phys 174:120–137
2. Dacles-Mariani J, Zilliac GG, Chow JS, Bradshaw P (1995) Numerical/experimental study of a wingtip vortex in the near field. AIAA J 33(9):1561–1568
3. Hoekstra M (1999) Numerical simulation of ship stern flows with a space-marching Navier Stokes method. Thesis, Technical University of Delft
4. Karafiath G, Cusanelli D, Cheng WL (1999) Stern wedges and stern flaps for improved powering—U.S. Navy experience. In: Sname annual meeting, Baltimore
5. Menter FR (1994) Two-equation eddy-viscosity turbulence models for engineering applications. AIAA J 32:1598–1605
6. Menter (1997) Eddy viscosity transport equations and their relation to the $k - \varepsilon$ model. J Fluid Eng 119:876–884
7. Van der Ploeg A, Hoekstra M, Eça L (2000) Combining accuracy and efficiency with robustness in ship stern flow computation. In: 23rd symp naval hydrodynamics, Val de Reuil, France
8. Van der Ploeg A, Chao K, Marzi J, Wackers J (2010) Computation of scale effects in free-surface flows near a ship's transom. In: 13th numerical towing tank symposium, Duisburg, Germany, October, 2010
9. Raven HC, Van der Ploeg A, Starke AR (2004) Computation of free-surface viscous flows at model and full scale by a steady iterative approach. In: 25th symposium on naval hydrodynamics. St. John's, Newfoundland, Canada, August 2004
10. Starke AR, Raven HC, Van der Ploeg A (2007) Computation of transom-stern flows using a steady free-surface fitting RANS method. In: 9th int conference on numerical ship hydrodynamics, Ann Arbor, Michigan, August 5–8, 2007
11. Starke AR, Van der Ploeg A, Raven HC (2010) Viscous free surface flow computations for self-propulsion conditions using PARNASSOS. In: CFD workshop Gothenburg

Anisotropic Mesh Refinement in Ship Flow Simulation with Free Surface

Jeroen Wackers, Ganbo Deng, Emmanuel Guilmineau, Alban Leroyer, Patrick Queutey, and Michel Visonneau

Abstract For the simulation of water flow with waves, grid refinement must be anisotropic to limit the total grid size. For these flows, the grid has to be refined at the water surface, to resolve the conservation law which indicates the surface position, and below the surface to resolve the water flow. A combined criterion is presented, based on the free-surface position and on the Hessian of the pressure. It is shown that this criterion creates suitable grids for two- and three-dimensional flows.

Keywords Combined criteria · Anisotropic refinement · Free-surface waves

1 Introduction

The simulation of water flows with a free water surface is inherently a multiphysics problem, as the surface deformation through waves interacts with the viscous, turbulent flow below the surface. In a hydrodynamics flow solver, a model for the surface position is added to the standard Navier-Stokes equations. Accurate simulation of water flow requires a good resolution of this additional model, as well as the flow equations, on the same grid. Thus, adaptive grid refinement applied to water flows must take into account both these models. And therefore, refinement criteria that are a combination of different sensors are essential.

An adaptive grid refinement method has been developed [1, 2] for ISIS-CFD, the unstructured Reynolds-averaged Navier-Stokes solver developed by LHEEA. This commercialised flow solver is aimed at the simulation of realistic flow problems in all branches of marine hydrodynamics. The method is therefore developed to be general and flexible, featuring anisotropic refinement on unstructured hexahedral grids, derefinement of previous refinements to enable unsteady flow computation, and full parallelisation including integrated dynamic load balancing. The anisotropic refinement is metric-based. Thus, the refinement criteria are 3×3 symmetric tensors in each cell, which indicate the local desired cell size in all directions. This formulation

J. Wackers (✉) · G. Deng · E. Guilmineau · A. Leroyer · P. Queutey · M. Visonneau
Laboratoire de recherche en Hydrodynamique, Energétique et Environnement Atmosphérique, Ecole Centrale de Nantes, CNRS UMR6598, 44321 Nantes Cedex 3, France
e-mail: Jeroen.Wackers@ec-nantes.fr

L. Eça et al. (eds.), *MARINE 2011, IV International Conference on Computational Methods in Marine Engineering*, Computational Methods in Applied Sciences 29,
DOI 10.1007/978-94-007-6143-8_16,

allows the straightforward implementation of highly differing refinement criteria. The refinement method has already been successfully applied to different test cases in marine flow simulation [2].

The focus of this paper is on the development of refinement criteria for hydrodynamic flows in this metric-based context. We discuss which features are relevant as refinement criteria and how multiple refinement criteria can be combined into one. Sections 2 and 3 introduce the flow solver and the mesh refinement method. Then Sect. 4 discusses the necessity of combined criteria for flows with a free water surface. Section 5 shows the construction of a criterion that combines directional refinement at the free water surface with a pressure Hessian criterion. Three test cases in Sect. 6 indicate that the criterion generates effective meshes for two- and three-dimensional free surface flows.

2 Finite-Volume Method

The grid refinement is applied to a finite-volume method on unstructured grids, with a surface-capturing discretisation of the water surface, as implemented in the flow solver ISIS-CFD developed by LHEEA. This section describes the finite-volume discretisation and the type of meshes used, concentrating on those aspects that are most important for grid refinement and the construction of refinement criteria. Full details of the discretisation can be found in [3].

The ISIS-CFD flow solver resolves the incompressible Unsteady Reynolds-Averaged Navier Stokes equations in a two-fluid formulation. Free-surface water flows are modelled with a convection equation for the volume fraction of water c_i. The flow equations are discretised in a finite-volume framework. Pressure-velocity coupling is obtained through a Rhie & Chow SIMPLE-type method: in each time step, the velocity updates come from the momentum equations and the pressure is given by the mass conservation law, transformed into a pressure equation. As the volume fraction c_i is discontinuous, its convection equation is discretised with compressive flux functions. The resolution of this equation in each time step is decoupled from the pressure and velocity updates.

The discretisation is face-based. While all unknown state variables are cell-centered, the systems of equations used in the implicit time stepping procedure are constructed face by face. Fluxes are computed in a loop over the faces and the contribution of each face is then added to the two cells next to the face. This technique poses no specific requirements on the topology of the cells. Therefore, the grids can be completely unstructured, cells with an arbitrary number of arbitrarily-shaped faces are accepted.

For this study, unstructured hexahedral meshes are used (see Fig. 1). These meshes are generated with the HEXPRESS grid generator from NUMECA International. They offer the flexibility of an unstructured grid, yet have large regions where the mesh is structured. Variations in mesh size are handled by having small cells laying next to larger cells, a situation called ‘hanging nodes’ by other authors.

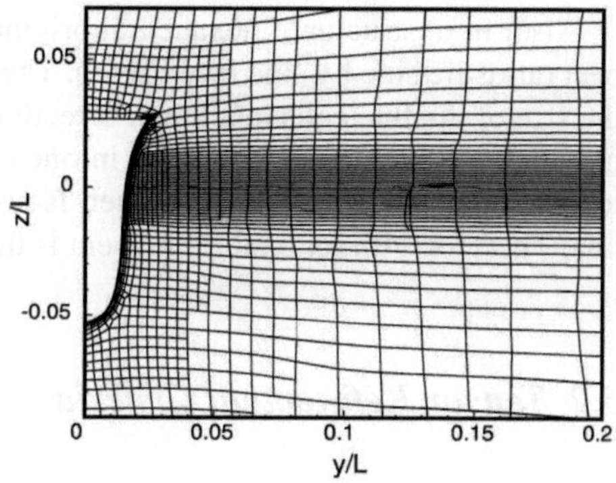

Fig. 1 Cut through an unstructured hexahedral mesh

In ISIS-CFD, due to the face-based algorithm, these cells are treated in exactly the same way as all the others: the larger cells are simply seen as cells with more than 6 faces.

Unstructured hexahedral grids are ideal for automatic grid refinement. Isotropic or anisotropic grid refinement can be applied to any of the hexahedral cells, the result will still be an unstructured hexahedral mesh. Therefore, locally refined meshes can be used directly in a flow solver that supports unstructured hexahedral meshes; no changes to the flow solver are needed to incorporate grid refinement.

3 Grid Refinement Procedure

The grid refinement procedure developed for ISIS-CFD [1, 2] is integrated completely in the flow solver. The method is entirely parallelised, including automatic redistribution of the grid over the processors. During a flow computation, the refinement procedure is called repeatedly. In such a call, first the refinement criterion is calculated, then in a separate step of the procedure the grid is refined based on this criterion. For steady flow, the refinement procedure converges: once the grid is correctly refined according to the criterion, further calls to the procedure no longer cause any changes.

3.1 Anisotropic Refinement

Anisotropic refinement is essential for our type of grid refinement. Isotropic refinement is very costly in three dimensions, since each refinement means a division in eight (for a hexahedron). Thus, creating very fine cells to accurately resolve a local flow phenomenon becomes almost impossible. However, by applying anisotropic refinement for flow features that need a fine grid in only one direction (notably, the water surface!), the total number of cells required can be greatly reduced or much finer flow details can be resolved.

Also, in unstructured hexahedral original grids, cells of completely different aspect ratios lie side by side (see Fig. 1). Therefore, when refining, we need to control the size of the fine cells in all their directions independently, otherwise refined grids may have smoothly varying sizes in one direction, but repeated changes from fine to coarse and back to fine in another. Isotropic refinement is not enough to prevent this. Therefore, directional refinement is the mandatory choice.

3.2 Tensor Refinement Criteria

For directional refinement, a way is needed to specify different cell sizes in different directions. The use of metric tensors as refinement criteria is such a way. This technique was first developed for the generation and refinement of unstructured tetrahedral meshes [4, 5]. It is also an extremely useful and flexible framework for the refinement of unstructured hexahedral meshes.

For tensor-based refinement, the refinement criterion in each cell is a 3×3 symmetric positive definite matrix $\mathcal{C}_i$. The refinement of the cells is decided as follows. Let the criterion tensors $\mathcal{C}_i$ in each cell be known (they are computed from the flow solution, see Sect. 5). In each hexahedral cell, the cell size vectors $\mathbf{d}_{j,i}$ ($j = 1, \ldots, 3$), which are the vectors between the opposing face centres in the three cell directions, are determined. Next, the modified sizes are computed as:

$$\tilde{\mathbf{d}}_{j,i} = \mathcal{C}_i \mathbf{d}_{j,i}. \tag{1}$$

Finally, a cell is refined in the direction j when the modified size exceeds a given, constant threshold value T_r:

$$\|\tilde{\mathbf{d}}_{j,i}\| \geq T_r. \tag{2}$$

The tensors $\mathcal{C}_i$ are direct specifications of the desired cell sizes: in the refined grid, the cell sizes are inversely proportional to the magnitude of the $\mathcal{C}_i$.

4 The Need for Combined Refinement Criteria

Water waves are usually generated by the pressure and velocity disturbances created when the water flow passes around a foreign body, either a stationary object or a floating body such as a ship. These disturbances are not only generated at the surface, but also well below it; even a fully submerged object may create waves (see for example Sect. 6.2). Once the waves are created, they propagate through orbital velocity fields in the water. Thus, to correctly resolve the generation and the propagation of a travelling surface wave, a good resolution is needed for the pressure and velocity fields below the surface.

Accurate resolution of the volume fraction equation is of prime importance as well. As the water surface is physically a discontinuity, the interface region for c_i

must be as sharp as possible. It can be shown [3], that a diffused interface increases numerical damping, so a too coarse grid at the surface will damp out waves. In our experience, the grid at the surface needs to be about twice as fine as the grid used in the vicinity of the surface.

Thus, to create suitable grids, a grid refinement criterion for water wave simulation must be based both on the pressure and velocity field and on the volume fraction. For these two, different indicators must be used. The reason for this, is that c_i is discontinuous at the surface and constant everywhere else, while the pressure and the velocity are smooth in the whole flow field except at the surface. Therefore, gradient- or second-derivative based error indicators can be used to identify the regions of importance for the flow field below the surface, but not for c_i. They go to infinity at the surface, thus causing infinite refinement unless the criterion is artificially limited. Also, grid misalignment must be avoided in the surface region, as it leads to large errors in the volume fraction. Therefore, the refined grid must be as uniform as possible near the surface. Numerical evaluations of the derivatives of c_i are never smooth, so they cannot guarantee uniform grids.

Therefore, a suitable refinement criterion for water flow with waves is an error indicator for the flow field and a simpler criterion for c_i, combined into one.

5 Pressure Hessian—Free Surface Criterion

A possible criterion for the simulation of flow with waves is based on the Hessian matrix of second derivatives of the pressure, combined with a criterion that refines in the normal direction of the surface for those cells where c_i is neither 0 nor 1.

5.1 *Free-Surface Criterion*

To resolve accurately the volume fraction, which is a discontinuity that is convected with the flow, it is sufficient to refine the grid in the direction normal to the water surface. When the surface is locally aligned with the cell directions, anisotropic refinement can be used to keep the total number of cells as low as possible.

The free-surface criterion is therefore based on a vector, normal to the surface, with length 1. Thus, from Eq. (2) it follows that the threshold value T_r directly indicates the desired cell size at the surface. The normal direction to the surface is computed from a c_i field that is smoothed out by averaging over a cell and its neighbours, a given number of times. The gradient of this field gives the normal directions. The criterion vectors $\mathbf{v}_i$ are then chosen as the unit vectors in this normal direction for those cells where the smoothed c_i field is non-zero, and as zero everywhere else. Switching based on the smoothed field guarantees that the mesh is refined also next to the surface, to create a margin of safety.

In tensor form, the free-surface criterion is implemented as matrices having only one non-zero eigenvalue, associated with the direction of the vector. In the directions normal to the vector, the eigenvalues are zero, so the desired grid size is infinity. Thus, the grid is only refined in the direction of the vectors. The tensors $\mathcal{C}_{S,i}$ are computed as follows (with $\otimes$ representing the tensor product):

$$\mathcal{C}_{S,i} = \mathbf{v}_i \otimes \mathbf{v}_i . \tag{3}$$

The free-surface criterion has been used on its own in our earlier work [1, 2].

5.2 Pressure Hessian

Hessian-based criteria are often used to control anisotropic grid refinement [4]. We base this criterion on the pressure as this variable is insensitive to boundary layers, where we consider that the original unrefined grid should be sufficiently fine to ensure the best grid quality.

$$\mathcal{H}_i = \begin{bmatrix} (p_i)_{xx} & (p_i)_{xy} & (p_i)_{xz} \\ (p_i)_{xy} & (p_i)_{yy} & (p_i)_{yz} \\ (p_i)_{xz} & (p_i)_{yz} & (p_i)_{zz} \end{bmatrix} . \tag{4}$$

We compute the Hessian in each cell using a least-squares fit of a third-order polynomial to the solution in the cell, its neighbour cells and its neighbours' neighbours. There is no better third-order polynomial fit to these points, so the fit is fourth-order accurate. Therefore, its second derivatives are second order accurate, independent of the configuration of the neighbour cells. This is very important on refined meshes.

To compute the refinement criterion, the Hessian is modified with a power law:

$$\mathcal{C}_{H,i} = (\mathcal{H}_i)^p , \tag{5}$$

where $(\mathcal{H}_i)^p$ has the same eigenvectors as $\mathcal{H}_i$ and eigenvalues that are those of $\mathcal{H}_i$ (in absolute value) to the power p. In general, we use $p = \frac{1}{2}$.

5.3 The Combined Criterion

For the final criterion, the two criteria above are combined. Even if the current implementation of the Hessian criterion has a behaviour similar to a free-surface criterion at the water surface, the real free-surface criterion is used as well because it guarantees that the grid at the surface is absolutely regular and that a safety zone of refined cells is generated around the surface.

The criteria are combined into one tensor criterion by taking a weighted maximum of the two tensors. We want T_r to indicate directly the desired cell size (as for the free-surface criterion), so we apply a weighting factor c only to the Hessian criterion:

$$\mathcal{C}_{C,i} = \max(\mathcal{C}_{S,i}, c\,\mathcal{C}_{H,i}) . \tag{6}$$

The (approximate) maximum of the two tensors is computed using a procedure similar to the one in [5]. First, the eigenvalues and eigenvectors of the two tensors are computed. Then new eigenvalues are set for each tensor, as the maximum of the original eigenvalue and the length of the corresponding eigenvector when it is multiplied by the other tensor. This gives two approximations to the maximum tensor; the final tensor in each cell is a weighted average of these two.

6 Test Cases

6.1 Prism Impacting on a Water Surface

As a first test, we present the impact of a freely falling 2D prism on a water surface. This test is based on experiments by Peterson et al. [6]. In the present context, the interest of the case is that the 2D nature allows us to see in detail the functioning of the directional refinement. The prism for this test has a bottom angle of 20°, a width W and an initial height H both of 61 cm, and a mass of 50 kg/m. The water has a density of 998.4 kg/m^3, the gravity is 9.81 m/s^2. Viscosity is neglected. The only movement allowed for the prism is vertical translation.

The numerical setup of the case is as follows. The free motion of the prism is computed using Newton's laws based on the integrated fluid forces on the prism, the motion is incorporated in the flow computation by block movement of the mesh. The same original mesh is used for all computations, this mesh has some local refinement around the prism. Refinement is called every 4 time steps. For this first test, the free surface criterion is used on its own; refinement is applied for c_i between 0.3 and 0.9 to prevent perturbations from the tiny droplets in front of the main jet.

Figures 2a–c show the mesh for three settings of the refinement threshold T_r (Eq. (2)). The figures show how the criterion adapts the refined mesh both to the shape of the water surface and to the original mesh. Directional refinement is applied where possible, notably around the horizontal water surface and in the wall-aligned grid on the prism. Cells in the fine part of the original grid are not refined as often as cells further from the prism, so that the grid around the surface has a uniform size (Fig. 2a is an extreme example, where the finest part of the mesh is not refined at all). Note that the mesh is not refined around the tip of the water jet, due to the limitation to $c_i > 0.3$.

The solution is characterised by a very strong pressure peak that moves outward over the prism during impact. As this peak is a highly localised feature, grid refinement is an effective way of resolving it well. Figure 3 shows the highest pressure on the prism as a function of time for the three simulations, compared with an asymptotic analytic solution by Scolan et al. [7]. As can be seen from this figure, the threshold has a very strong influence on the magnitude of the pressure peak; the agreement for the finest threshold is excellent. Also in Fig. 3, while the total number of cells increases with the reduction of the threshold, this increase is moderate, due mainly to the use of directional refinement.

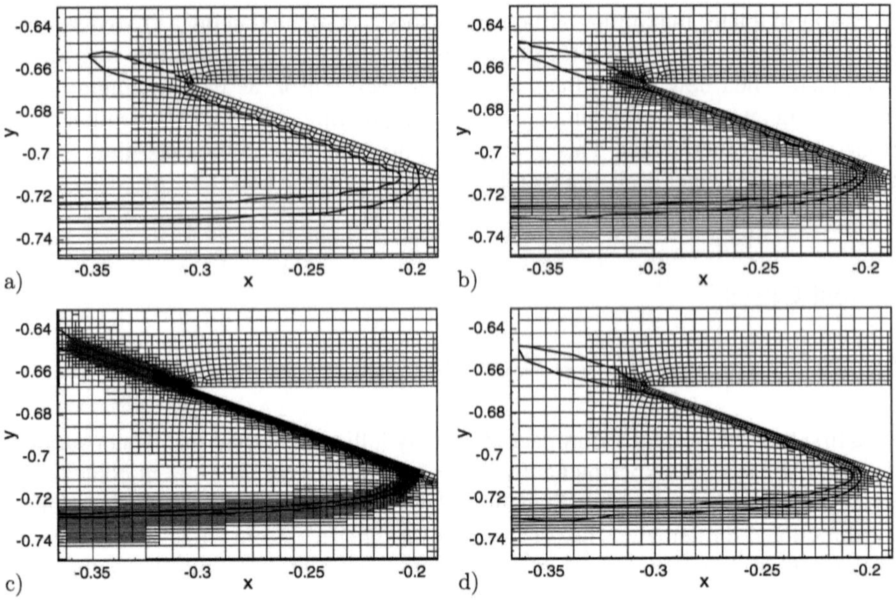

Fig. 2 Mesh and $c_i = 0.05$ and 0.95 isolines for the falling prism at $t = 0.0175$ s after initial impact. The figure is zoomed in on the left tip of the prism. The refinement threshold is $T_r = 0.0033$ (**a**), $T_r = 0.00165$ (**b**), and $T_r = 0.000825$ (**c**). Figure (**d**) represents the $T_r = 0.00165$ case with convection buffer layers

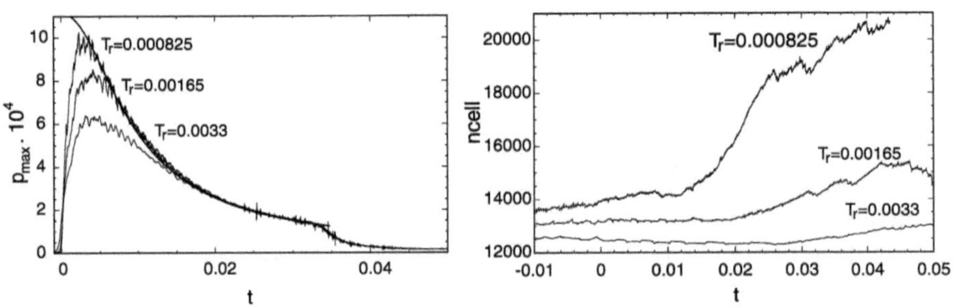

Fig. 3 Pressure peak (*left*) and total number of cells (*right*) as a function of the time after initial impact, for the computations at three settings of the refinement criterion T_r, without convection buffer layers. The pressure peak is compared with the asymptotic solution of Scolan et al. [7]

6.2 Immersed NACA0012 Wing

The second test case is meant to evaluate the behaviour of the combined refinement criterion for a two-dimensional wave field. We study the influence of the ratio parameter c from Eq. (6). The case is the geometry studied by Duncan [8], a wave train generated by an immersed NACA0012 profile of chord 0.203 m at 5 degrees angle

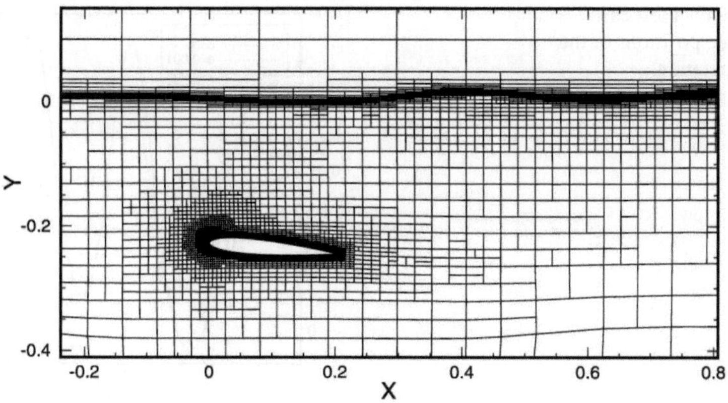

Fig. 4 Refined mesh around the immersed profile for the Duncan case with $c = 0.001$

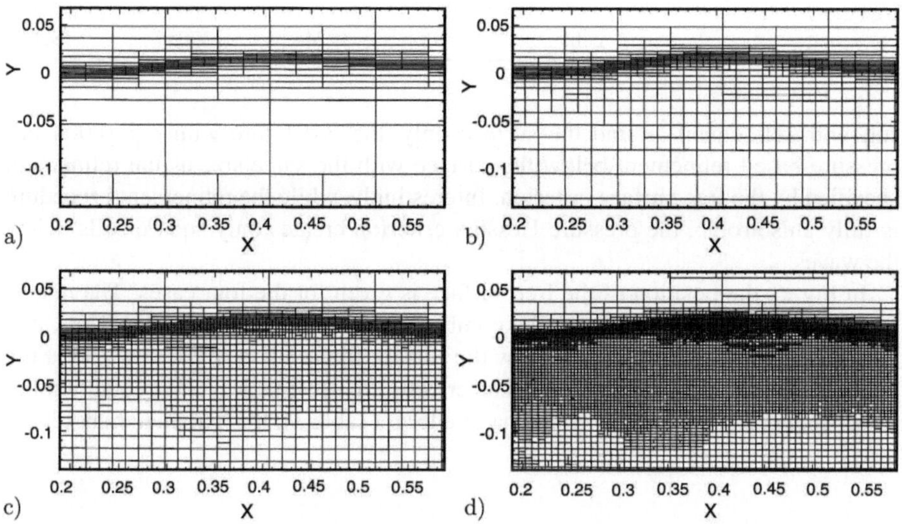

Fig. 5 Refined meshes at the first wave for the Duncan case with surface-only refinement $c = 0.0$ (**a**), with $c = 0.001$ (**b**), $c = 0.002$ (**c**), and $c = 0.004$ (**d**)

of attack, with its centre point at 0.236 m below the surface. The inflow velocity is 0.8 m/s and $Re = 1.42 \cdot 10^5$. The problem geometry can be found in Fig. 4.

Four simulations are performed, starting from an original mesh that has some refinement around the profile but none at the surface. For each simulation, the grid is refined around the free surface to a target size $T_r = 0.002$ m. Different values of c give different sizes of the grid below the surface. The grid for $c = 0.001$ is given in Fig. 4, it shows refinement around the profile (notably at the leading edge), in a specific region between the profile and the surface, and in the wave field. A zoom of the meshes around the first wave is given in Fig. 5; the first figure is the re-

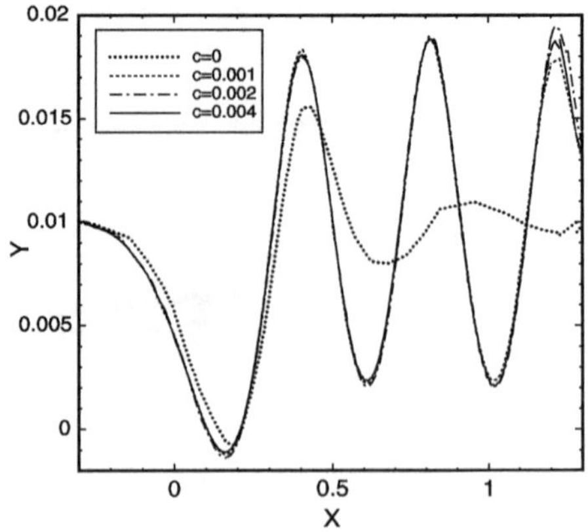

Fig. 6 Duncan test case, free-surface positions of the four computations

Table 1 Number of cells in the refined meshes, Duncan test case

c	0	0.001	0.002	0.004
cells	3539	10244	37035	189860

sult with refinement around the surface only, the last figure with $c = 0.004$ has pressure-based refinement below the surface with the same size as the refinement specified by the free-surface criterion. Interestingly, while the refinement procedure is fully anisotropic, the pressure Hessian criterion creates only square cells below the waves.

In Fig. 6, the position of the free surface is given for the four cases. The results on the three grids produced with the combined criterion are very similar. Thus, it is not necessary to refine the grid below the surface to the same size as the grid at the surface, twice and even four times coarser cells are acceptable. As the total number of cells increases strongly with the parameter c (Table 1), this parameter should be kept low.

6.3 Series 60 Wave Pattern

The final test concerns the flow around a Series 60 ship in still water at Froude number $Fr = 0.316$ and $Re = 5.3 \cdot 10^6$. The computation is started from a coarse mesh that has no initial refinement at all around the free surface, it is used to show that a sensible refined mesh for free-surface ship flow can be obtained entirely with automatic grid refinement. The grid is obtained with a target cell size $T_r = 0.001L$ and a ratio $c = 0.004$, the original grid has 253k cells and the final grid 2.81M cells.

Four X cross-sections of the refined mesh can be seen in Fig. 7. The free-surface criterion applies directional refinement around the undisturbed surface; without this refinement, the interface would be dispersed in front of the ship, so accurate computations would be impossible. Refinement in all directions appears in the strongest

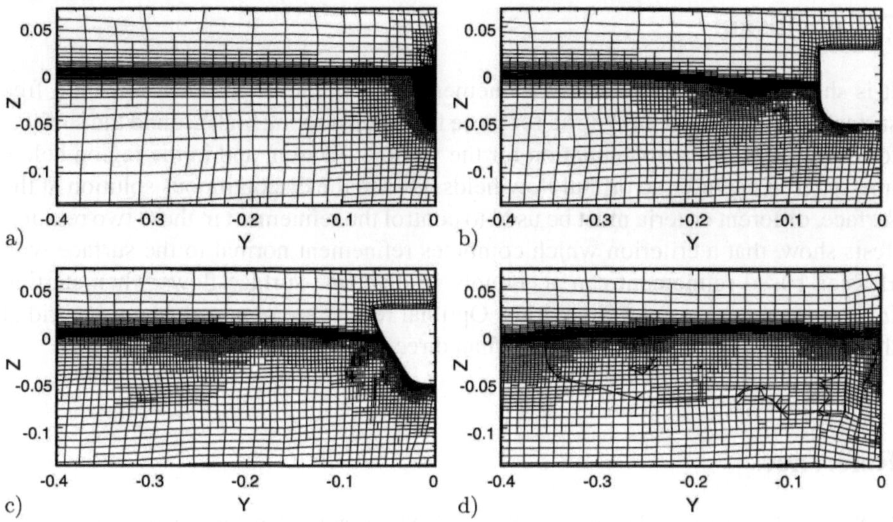

Fig. 7 Series 60, cross-sections of the grid at $X/L = 0.0$ (**a**), at $X/L = 0.4$ (**b**), at $X/L = 0.8$ (**c**) and behind the ship at $X/L = 1.2$ (**d**)

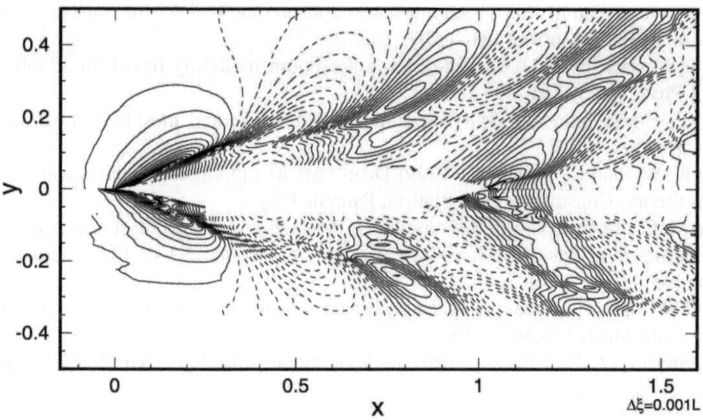

Fig. 8 Wave pattern of Series 60, compared with experimental results from IIHR [9]

waves. Pressure-based refinement is seen at the bow of the ship, below the hull at the stern, and below the waves. It is concentrated near the sharp peaks of the wave system. As for the Duncan test case, the refined cells below the waves are predominantly square.

The wave pattern, compared with experiments from IIHR [9], is given in Fig. 8. The correspondance is good, comparable with the results obtained in [3] on structured grids of 3.8M cells. Given the results of the Duncan test, a similar accuracy can probably be obtained with fewer cells if c is reduced. This is a subject for further study.

7 Conclusion

It is shown that anisotropic grid refinement is essential for water flow with a free surface. Also, refinement criteria for these flows must refine both around the surface, to resolve the convection equation for the volume fraction, and in the region below it in order to capture the orbital flow fields. Due to the discontinuous solution at the surface, different criteria must be used to control the refinement in these two regions. Tests show, that a criterion which combines refinement normal to the surface with Hessian-based refinement can accurately resolve free-surface flows when starting from uniformly coarse original grids. Optimal results are obtained when the grid at the surface is two to four times finer than directly below it.

References

1. Wackers J, Visonneau M (2009) Adaptive grid refinement for ship flow computation. In: ADMOS 2009, Brussels, Belgium
2. Wackers J, Ait Said K, Deng GB, Queutey P, Visonneau M, Mizine I (2010) Adaptive grid refinement applied to RANS ship flow computation. In: 28th symposium on naval hydrodynamics, Pasadena, California
3. Queutey P, Visonneau M (2007) An interface capturing method for free-surface hydrodynamic flows. Comput Fluids 36(9):1481–1510
4. Alauzet F, Loseille A (2010) High-order sonic boom modeling based on adaptive methods. J Comput Phys 229(3):561–593
5. George PL, Borouchaki H (1998) Delaunay triangulation and meshing: application to finite elements. Hermes, Paris
6. Peterson R, Wyman D, Franck C (1999) Drop tests to support water-impact and planing boat dynamics theory. Coastal Systems Station, Panama City
7. Scolan Y-M, Coche E, Coudray T, Fontaine E (1999) Etude analytique et numérique de l'impact hydrodynamique sur des carènes dissymétriques. In: 7es journées de l'hydrodynamique, Marseille, France
8. Duncan JH (1983) The breaking and non-breaking wave resistance of a two-dimensional hydrofoil. J Fluid Mech 126:507–520
9. Longo J, Stern F (2002) Effects of drift angle on model ship flow. Exp Fluids 32:558–569

MIX
Papier aus verantwortungsvollen Quellen
Paper from responsible sources
FSC® C105338

If you have any concerns about our products,
you can contact us on
ProductSafety@springernature.com

In case Publisher is established outside the EU,
the EU authorized representative is:
Springer Nature Customer Service Center GmbH
Europaplatz 3, 69115 Heidelberg, Germany

Printed by Libri Plureos GmbH
in Hamburg, Germany